SURFACE PHOTOVOLTAGE
ANALYSIS OF
PHOTOACTIVE MATERIALS

SURFACE PHOTOVOLTAGE ANALYSIS OF PHOTOACTIVE MATERIALS

Thomas Dittrich

Helmholtz-Zentrum Berlin für Materialien und Energie, Germany

Steffen Fengler

Helmholtz-Zentrum Geesthacht für Materialien und Küstenforschung, Germany

World Scientific

NEW JERSEY · LONDON · SINGAPORE · BEIJING · SHANGHAI · HONG KONG · TAIPEI · CHENNAI · TOKYO

Published by

World Scientific Publishing Europe Ltd.

57 Shelton Street, Covent Garden, London WC2H 9HE

Head office: 5 Toh Tuck Link, Singapore 596224

USA office: 27 Warren Street, Suite 401-402, Hackensack, NJ 07601

Library of Congress Cataloging-in-Publication Data

Names: Dittrich, Thomas, author.

Title: Surface photovoltage analysis of photoactive materials / Thomas Dittrich,
    Helmholtz-Zentrum Berlin für Materialien und Energie, Germany, Steffen Fengler,
    Helmholtz-Zentrum Geesthacht für Materialien und Küstenforschung, Germany.

Description: New Jersey : World Scientific, [2020] | Includes bibliographical references and index.

Identifiers: LCCN 2019033014 | ISBN 9781786347657 (hardcover)

Subjects: LCSH: Photoelectric cells. | Optoelectronic devices--Materials.

Classification: LCC TK8322 .D58 2020 | DDC 621.3815/42--dc23

LC record available at https://lccn.loc.gov/2019033014

British Library Cataloguing-in-Publication Data

A catalogue record for this book is available from the British Library.

For any available supplementary material, please visit
https://www.worldscientific.com/worldscibooks/10.1142/Q0227#t=suppl

Desk Editors: Anthony Alexander/Jennifer Brough/Shi Ying Koe

Typeset by Stallion Press
Email: enquiries@stallionpress.com

Preface

Surface photovoltage (SPV) signals arise as a change of the contact potential difference whenever charge carriers photogenerated in a photoactive material are separated in space. Photoactive materials consist of light absorbers and charge-selective surfaces and interfaces. SPV signals depend on photogeneration, recombination, charge transfer and charge transport and provide information about parameters limiting related processes. Therefore, SPV techniques are powerful tools for the characterization of photoactive materials in research and technology and analysis of dominating processes.

In the last decade(s), we had the great pleasure to apply several SPV methods to the characterization and analysis of numerous crystalline, disordered, nanostructured, porous, organic and hybrid photoactive materials in fields of mainly photovoltaics and photocatalysis. Continuous interest, especially of young researchers and developers, in the application of SPV methods inspired us to share our experience about the measurement and application principles of SPV techniques in this book. We felt that a combination of elements of a textbook with elements of scientific publications will be useful in order to encourage other researchers and developers to apply SPV techniques for the characterization and analysis of their specific photoactive materials. Furthermore, we set special focus on progress made in the fields of SPV measurements over wide time ranges and of random walk simulations of SPV signals since the comprehensive review of Kronik and Shapira in 1999.

In Chapter 1, some basics about photogeneration and recombination, relaxation of a disturbance of a space charge, contact potential difference and mechanisms leading to SPV signals are given. In Chapter 2, the principles of SPV measurements with macroscopic and microscopic Kelvin probes, fixed capacitor arrangements, electron beams and photoelectrons are explained. Depending on the specific method, SPV signals can be measured in the range from ps–ns to hours and longer times at sensitivities from 10–100 nV to about 10–30 mV. Chapter 3 is devoted to the practical aspects of continuous, modulated and transient SPV spectroscopy. Attention is paid to transient SPV measurements with high-impedance buffers at short times and to the variation of external parameters during SPV measurements. In Chapter 4, principles of random walk simulations in small systems with localized and delocalized states are introduced and applied to the simulation of transient and modulated SPV signals and multi-parameter fitting. Some applications of SPV techniques in research and development are illustrated in Chapter 5. The selected examples belong to the characterization of electronic states at silicon surfaces, in layer systems with gallium arsenide and in colloidal quantum dots as well as to the charge separation in metal oxides and local charge separation at bismuth vanadate crystallites.

We are grateful to all scientists and engineers contributing to applications and interpretation of SPV measurements during the last years. T. D. is especially grateful to Prof. Yoram Shapira (Tel Aviv University), Prof. Helmut Tributsch (former Hahn-Meitner-Institute), Prof. Frank Osterloh (University of California) and Prof. Can Li (Chinese Academy of Sciences, Dalian) for inspiration and to Prof. Fred Koch (Technical University Munich), Prof. Victor Timoshenko (Moscow State University) and Prof. Juan Bisquert (Universidad Jaume I) for numerous discussions about SPV. S. F. thanks Prof. Thomas Klassen and Dr. Mauricio Schieda for their strong support in creating a new SPV laboratory at the Helmholtz Zentrum Geesthacht. Dr. Michael Franke (Elektronik Manufaktur Mahlsdorf) and Prof. Sven Bönisch (Brandenburg University of Technology Cottbus–Senftenberg) are acknowledged for the development of high-impedance buffers. We always enjoy(ed) experiments and discussions about SPV with friends, for example, with Prof. Luis Otero (Universidad Nacional de Rio Cuarto), Prof. Fengtao Fan (Chinese Academy of Sciences, Dalian), Prof. Iván Mora-Seró (Universidad Jaume I), Prof. Andrey Rogach (City University of Hong Kong), Dr. Elisabeth Zillner, Dr. Thomas Bitzer, Dr. Jörg Rappich, Dr. Heike Angermann, Philipp Zabel, Prof. Sebastian

Fiechter and Dr. Marin Rusu (Helmholtz-Zentrum Berlin), Prof. Nopporn Rujisamphan (King Mongkut's University), Prof. Pongthep Prajongtat and Prof. Thidarat Supasai (Kasetsart University), Dr. Andrey Susha and Dr. Dieter Gross (Ludwig Maximilian University), Dr. Maria Nowotny and Dr. Leigh Sheppard (University of Western Sydney), Dr. Volodimyr Duzhko (University of Massachusetts Amherst), Dr. Igal Zidon (Tel Aviv University) and Dr. Igal Leviné (Weizmann Institute).

About the Authors

Thomas Dittrich received his Ph.D. from the Faculty of Physics of the Lomonosov Moscow State University in 1990 and got his habilitation degree from the Technical University of Munich in 2002. More than 20 years ago, he started to apply surface photovoltage techniques for the investigation of photoactive materials. At present, he researches about energy materials at the Helmholtz-Zentrum Berlin and teaches Materials Concepts for Solar Cells at different universities.

Steffen Fengler received his Ph.D. from the Faculty of Physics of the Freie Universität Berlin in 2016. Aside from applying surface photovoltage techniques, he develops algorithms for random walk simulations of transient and modulated surface photovoltage signals. At present, he mainly characterizes photocatalytic materials at the Helmholtz-Zentrum Geesthacht.

Contents

Symbols and Abbreviations

Those Beginning with Latin Letters

a/b	opening ratio of a monochromator
A	acceptor state, area, A-channel of an oscilloscope, amplitude
$A1(2,3)$	constant numbers
A^-	occupied acceptor state
A^0	unoccupied (neutral) acceptor state
A_0	set point
A^*	Richardson constant (for the free electrons: $120A/(cm^2K^2)$
A_{LCR}	amplitude of damped oscillations
AC, ac	alternating current
AFM	atomic force microscopy
Ag	silver
Al_2O_3	aluminum oxide
AM1.5	air mass 1.5
AM-KPFM	amplitude modulated Kelvin probe force microscopy
Au	gold
B	B-channel of an oscilloscope
$BiVO_4$	bismuth vanadate
$B_{rad,np}$	radiative recombination rate constant

C	capacitance
C_{in}	input capacitance
C_m	measurement capacitance
C_{par}	parasitic capacitance
CdSe	cadmium selenide
CdTe	cadmium telluride
$C_{A(e)}$	Auger recombination coefficients of electrons
$C_{A(h)}$	Auger recombination coefficient of holes
COOH	carboxyl group
C_3N_4	carbon nitride
$CH_3NH_3PbI_3$	methyl amine lead iodide
CPD	contact potential difference
c-Si	crystalline silicon
CoO_x	cobalt oxide
Cr	chromium
CTS	charge transfer state
$CuAlO_2$	copper aluminum oxide
$CuInSe_2$	copper indium diselenide
$Cu(In,Ga)S_2$	copper indium gallium disulfide
Cu_2O	copper oxide
d	thickness, width, distance, length, diameter
d_{avg}	average distance between localized states
d_{dr}	drift length
d_{eff}	effective charge separation length
d_{flight}	flight distance
d_H	layer thickness
d_{in}	width of the entrance slit of a monochromator
d_{MnO_x}	diameter of the MnO_x nanoparticles
d_{out}	width of the exit slit of a monochromator
d_s	diameter of a sphere
d_{rec-dl}	distance between a delocalized and a recombination state
d_{s-r}	minimum distance between a surface and a sphere
d_{SCR}	width of the space charge region
d_{slit}	width of the slit of a monochromator
d_{sep}	charge separation length
$d_x(+)$	projected position of a positive charge carrier
$d_x(-)$	projected position of a negative charge carrier

dx	width of a slice in x direction
D	donor state, demodulator
D^+	unoccupied donor state
D^0	occupied (neutral) donor state
D_e	diffusion constant of electrons
D_h	diffusion constant of holes
D_{it}	density of surface (interface) states
$D_{it,\mathrm{min}}$	density of surface (interface) states in the minimum
dQ	charge in a slice in x direction
dSPV	change of the SPV
DC, dc	direct current
dI/dU	differential conductivity
e^-	electron
e^{-*}	excited electron
E	energy
E_A	activation energy
E_C	energy of the conduction band edge
E_d	energy of defect states at the distribution maximum
E_{dl}	energy of a delocalized state
E_e	energy of an electron
E_{eh}	exciton transition energy
E_f	energy of a final (localized) state
E_F	Fermi energy
$E_{F,\mathrm{bulk}}$	Fermi energy in the bulk
$E_{F,n}$	quasi-Fermi energy of electrons
$E_{F,p}$	quasi-Fermi energy of holes
$E_{F,\mathrm{ref}}$	Fermi energy of reference electrode
$E_{F,s}$	Fermi energy of the sample electrode
$E_{F,s}$	Fermi energy at a semiconductor surface
E_g	energy of a band gap
E_i	energy of an initial (localized) state
E_{ig}	energy of an indirect band gap
E_k	kinetic energy of a photoelectron
$E_{k,0}$	kinetic energy of a photoelectron in equilibrium
$E_{k,\mathrm{dark}}$	kinetic energy of a photoelectron in the dark
$E_{k,\mathrm{ill}}$	kinetic energy of a photoelectron under illumination
$E_{\mathrm{on,x}}$	onset energy of in-phase signals

$E_{on,y}$	onset energy of phase-shifted by 90° signals
E_{rec}	energy of a recombination state in random walk simulations
E_{redox}	redox potential (energy)
$E_{S/S*}$	transition energy of a molecule
E_t	energy of a defect or trap state
E_t	energy of exponential tails
E_V	energy of the valence band edge
E_{vac}	vacuum energy
$E_{vac,ref}$	vacuum energy of the reference electrode
$E_{vac,s}$	vacuum energy of the sample electrode
E_{field}	electric field
eh	exciton

f	function
f_{meas}	measured function
f_{power}	power decay function
f_{sim}	simulated function
$f_{stretched}$	function of a stretched exponential
F_{es}	electrostatic force
$F_{es,ac}$	modulated electrostatic force
F_{on}	onset function
f_d	frequency of damped oscillations
f_{mod}	modulation frequency
f_0	bandwidth
FM-KPFM	frequency modulated Kelvin probe force microscopy
fcc	face centered cubic

g	distribution function
$g_e(E_i)$	distribution function of localized states in energy
g_{exp}	exponential distribution function
g_{Gauss}	Gaussian distribution function
g_{hom}	homogeneous distribution function
$g_s(x_i y_i z_i)$	distribution function of localized states in space
G	generation rate
G^{max}	maximum generation rate
GaAs	gallium arsenide
GaO_x	native gallium oxide

H	sample thickness
h^+	hole
HF	hydrofluoric acid
HOMO	highest occupied molecular orbitals
H_2O	water
H_2SO_4	hydrosulfuric acid
$h\nu$	photon energy
I	current
I_{ac}	alternating current
I_{bias}	bias current
$I_{diff,e}$	diffusion current of electrons
$I_{diff,h}$	diffusion current of holes
$I_{drift,e}$	drift current of electrons
$I_{drift,h}$	drift current of holes
I_e	electron current
I_{em}	current caused by thermionic emission
I_{SC}	short circuit current density
ILGAR	ion layer gas reaction
Int	intensity
In_2S_3	indium sulfide
ITO	indium-doped tin oxide
j	current flow
K	parameter for averaging procedure
k_B	Boltzmann constant $(1.38 \cdot 10^{-23} \, \text{J/K})$
KPFM	Kelvin probe force microscope
L	inductivity
L_{diff}	diffusion length of photogenerated charge carriers
LCR	inductivity-capacitance resistance
LED	light emitting device
$LiAlO_2$	lithium aluminum oxide
LUMO	lowest unoccupied molecular orbital
m_e	electron mass
m_e^*	effective electron mass
m_x	-1 to $+1$ switching function for in-phase signals
m_y	-1 to $+1$ switching function for phase-shifted by $90°$ signals

M	matrix of transfer rates
M_{res}	matrix of results from a multi-parameter fit
MES-FET	metal semiconductor field effect transistor
MnO_x	manganese oxide
Mo	molybdenum

n	density of free electrons
n^{++}	highly n-type doped
n_s	density of free electrons at the surface
n_0	density of free electrons in thermal equilibrium
N	density
N_A^-	density of ionized acceptors
N_C	effective density of states at the conduction band edge
N_D^+	density of ionized donors
N_Q	total areal density of photogenerated charge carriers
N_{st}	density of surface defects
N_t	density of trap states
N_V	effective density of states at the valence band edge
N	number
N	number of localized states
N_d	number of localized states in a small system
N_{avg}	number of read-outs
N_{meas}	total number of fitted transients
N_{tra}	maximum number of simulated transients
N_{ts}	number of localized states at the surface of a QD
N_{turns}	maximum number of sign variation in iterations
NNS_{board}	number of samples for averaging
NS_{board}	number of a sample on board
NS_{reg}	number of a certain sample to be registered
Nd:YAG	neodymium yttrium aluminum garnet
NH_4F	ammonium fluoride
NIR	near infrared
N_2	nitrogen

O_2	oxygen
OA	oleic acid
OPA	operational amplifier

p	density of free holes
p_0	density of free holes in thermal equilibrium

p_s	density of free holes at the surface
p_{meas}	measurement parameter
p_{sim}	simulation parameter
$p(O_2)$	partial pressure of oxygen
PbI_2	lead iodide
PbO_2	lead oxide
PbS	lead sulfide
Pd	palladium
PCBM	phenyl-C_{61} butyric acid methyl ester
PDDA	poly(diallyldimethylammonium chloride)
PES	photoelectron emission spectroscopy
PPV	poly(phenylene-vinylene)
PSD	position sensitive detector
Pt	platinum
q	elementary charge ($1.602 \cdot 10^{-19}$ As)
Q	charge
Q_{fix}	fixed charge
Q_{SC}	density of charge in the space charge region
Q_{SS}	density of charge in surface states
Q_i	density of influenced charge
QD	quantum dot
QE	quantum efficiency
r	radius, distance
r_i	distance between a lattice site and a localized state
r_{ij}	distance between localized states
r_{if}	distance between the initial and finals states
r_{min}	minimum distance
R	reflectivity
R	resistance
R_{crit}	critical resistance
R_d	series resistance for damping
R_{load}	load resistance
R_L	load resistance
R_m	measurement resistance
R_{par}	shunt resistance
R_{ser}	series resistance
R_{source}	source resistance

R	amplitude of modulated signals
R	recombination rate
$R_{A(e)}$	Auger recombination rate of electrons
$R_{A(h)}$	Auger recombination rate of holes
$R_{rad,np}$	rate of radiative band-band recombination
R_{SRH}	SRH recombination rate
R_{surf}	surface recombination rate
R_{rdm}	random number
R	processes of transfer steps in random walk simulations
R_{rec-d}	transfer from a localized to a recombination state
R_{rec-dl}	transfer from a delocalized to a recombination state
R_{therm}	transfer from a localized to a delocalized state by thermal activation
$R_{therm-rec}$	recombination from a localized state via thermal activation into a delocalized state
R_{tunn}	transfer between localized states by tunneling
$R_{tunn-rec}$	tunneling recombination by transfer from a localized state
ra	random rank number
r(ra)	distribution of random rank numbers
Ru	ruthenium
RWS	random walk simulation
s	surface recombination velocity
S	ground state of a molecule
S^*	excited state of a molecule
S_{max}	maximum number of simulations
$S_{1,2,3,4}$	integration parameters for random walk simulation of modulated SPV signals
SC	semiconductor
SCR	space charge region
SH	thiol group
SiO_2	quartz
Si(111):H	hydrogenated Si(111) surface
SiN_x:H	hydrogenated silicon nitride
SI GaAs	semi-insulating gallium arsenide
SnO_2:F	fluorine-doped tin oxide
SPV	surface photovoltage
SPV_{max}	amplitude of SPV transients
SPV_{SCR}	SPV at a surface space charge region

SRH	Shockley–Read–Hall recombination
SRSPS	spatially resolved surface photovoltage spectroscopy
STM	scanning tunneling microscope
t	time
t_{bias}	time of applying a bias potential
t_ℓ	total time at the number ℓ of simulated transients
t_{off}	time of switching off (a laser pulse or a potential)
t_{on}	time of switching on (a laser pulse or a potential)
t_{tr}	transit time
$t_{trigger}$	trigger time
T	temperature
T_{cold}	temperature of cold reservoir
T_{reg}	regulation temperature
T_s	temperature at a sample
T	period
T_0	period of the natural frequency
T_{mod}	modulation period
T_{tr}	transmission probability
TCO	transparent conductive oxide
TGA	thioglycolic acid
Ti	titanium
TiO_2	titania, titanium oxide
TOF	time-of-flight
TOP	trioctylphosphine
U	potential
U_{bias}	bias potential
U_{es}	electrostatic potential
U_{drift}	drift potential
U_{ij}	potential between two localized states
Uin	input potential
U_{LCR}	potential of excited damped oscillations
U_{meas}	measured potential
U_{offset}	offset potential
U_{osc}	potential of (damped) oscillations
U_{out}	output potential
U_{source}	undisturbed source voltage
U_D	Dember photovoltage
UPS	ultraviolet photoelectron spectroscopy
UV	ultraviolet

V, $V_{A,B}$	volume
V_{ac}	modulated potential
V_b	bias potential
$V_{b,dark}$	bias potential in the dark
$V_{b,ill}$	bias potential under illumination of the sample
V_{exc}	excitation potential for a piezo drive
V_{OC}	open circuit voltage
v_s	surface recombination velocity
V_{tip}	potential at the tip
v_{th}	thermal velocity
$V_{x,y,z}$	potentials of a x,y and z-piezo drives, respectively
W	work function
W_{ref}	work function of the reference electrode
$W_{ref,dark}$	work function of the reference electrode in the dark
$W_{ref,ill}$	work function of the reference electrode under illumination
W_s	work function of the sample electrode
$W_{s,dark}$	work function of the sample electrode in the dark
$W_{s,ill}$	work function of the sample electrode under illumination
x	position in the direction perpendicular to the sample electrode
$\langle x_n \rangle$	center of negative charge
$\langle x_p \rangle$	center of positive charge
X	in-phase signal
X_{max}	maximum in-phase signal
XPS	X-ray photoelectron spectroscopy
y_0	dimensionless surface band bending in thermal equilibrium
y	dimensionless surface band bending under illumination
Y	phase-shifted by 90° signal
Y_{max}	maximum phase-shifted by 90° signal
ZnO:Al	aluminum-doped zinc oxide
ZnPc	zinc phthalocyanine

Those Beginning with Greek Letters

α^{-1}	absorption length
α_p	power parameter
α_{tunn}	inverse tunneling length
β	stretching parameter
γ	photon
ΔCPD	light-induced change of the contact potential difference
ΔE_k	resolution of the kinetic energy
ΔE	spectral resolution of the energy
Δn	density of negative photogenerated charge carriers
Δp	density of positive photogenerated charge carriers
ΔQ_{ds}	charge injected into deep localized states in a reaction layer
ΔQ_{i}	change of the density of influenced charge
ΔQ_{SC}	change of the density of charge in a space charge region
ΔQ_{SS}	change of the density of charge at surface states
$\Delta\text{SPV}_{\text{max}}$	change the amplitude of a SPV transient
Δt	time interval
Δt_{if}	time interval for transfer from an initial state a final state (waiting time)
ΔU_{bias}	variation of a bias potential
ΔU_i	potential drops across the insulating layer
ΔU_L	voltage drop at a load
$\Delta\varphi_0$	change of a band bending
$\Delta\varphi_{\text{rl}}$	potential drop across a reaction layer
$\Delta\varphi_{\text{Hl}}$	potential drop across a Helmholtz layer
$\Delta\lambda$	spectral resolution of the wavelength
δ	diameter (extension) of a localized state
δ	decay constant
δ	small deviation
δ^-	negative polarization charge
δ^+	positive polarization charge
δ_{SPV}	noise in modulated SPV measurements
ε_0	dielectric constant of vacuum ($8.84\ 10^{-14}$ As/Vcm)
ε_{r}	relative dielectric constant
$\varepsilon_{\text{r,eff}}$	effective relative dielectric constant
θ_{i}	horizontal angle describing the position of a localized state
λ	wavelength

λ_{mfp}	mean free path of electrons
μ_e	mobility of electrons
μ_h	mobility of holes
ν_{if}	tunneling rate between the initial and final states
ν_{ij}	tunneling rate between two localized states
ν_{ij}	transfer rate between two localized states
$\nu_{i-\text{th}}$	de-trapping rate from a state i
$\nu_{0,\text{th}}$	maximum de-trapping rate
$\nu_{0,\text{tunn}}$	bouncing frequency
ρ	density of charge
σ	electrical conductivity
σ	standard deviation of a Gaussian Distribution
σ_d	standard deviation of defects
σ_b	electrical conductivity of the bulk
σ_e	capture cross section for electrons
σ_h	capture cross section for holes
σ_{se}	capture cross sections of electrons at the surface
σ_{sh}	capture cross sections of holes at the surface
τ	relaxation time, time constant
τ_{int}	integration time constant
τ_{M}	dielectric relaxation time constant
τ_{on}	onset time constant
τ_{RC}	RC time constant
$\tau_{\text{RC}-\text{th}}$	RC time constant for thermionic emission
$\tau_{\text{RC}-\text{tunn}}$	RC time constant for tunneling
τ_{res}	resolution time
τ_0	elementary time
$\tau_{0.5}$	decay time to half of the initial value
Φ_{B}	barrier height
φ	electric potential
φ_0	built-in potential
φ_{ill}	surface band bending under illumination
φ	phase angle
φ_{LCR}	phase angle of damped oscillations
φ_{i}	azimuth angle describing the position of a localized state
χ_{s}	electron affinity of the semiconductor
χ^2	error square

$\chi^2_{\min}$	minimum value of the error square
Ψ	wave function
ω	angular frequency
ω_0	natural frequency
ω_{ac}	angular modulation frequency
ω_{mod}	modulation angular frequency
ω_{res}	resonance angular frequency
$\bar{h}$	Planck constant $(1.05 \cdot 10^{-34} \, \mathrm{Js})$

Chapter 1

Light-Induced Charge Separation and Surface Photovoltage

1.1. Introduction

Surface photovoltage (SPV) signals arise whenever photogenerated charge carriers are separated in space. SPV signals depend on the fundamental processes of light absorption causing photogeneration of mobile charge carriers, recombination of charge carriers, charge transport and relaxation of a disturbance of a space charge. Phenomena related to those fundamental processes can be generally investigated by SPV techniques in photoactive materials.

This chapter begins with a brief overview of photogeneration and recombination mechanisms. It follows a description of the relaxation of a disturbance of a space charge by dielectric relaxation and by transport over a barrier. Orders of magnitude are given for relaxation times. The contact potential difference (CPD) between a sample and reference electrodes and the SPV as its negative light-induced change $(-\Delta\text{CPD})$ are defined. The center of charge approach, the charge separation length and the superposition principle are introduced. The SPV is described in terms of a parallel plate capacitor and the sensitivity of SPV measurements is estimated.

Different mechanisms leading to separation of photogenerated charge in space are described. SPV signals can be caused by separation of photogenerated charge carriers in space charge regions (SCRs), by diffusion of photogenerated charge carriers and the Dember effect, by asymmetric

1

transfer of photogenerated charge carriers and their injection across surfaces and interfaces and by dissociation of photogenerated excitons in combination with charge transfer. Furthermore, light-induced reversible and non-reversible changes of the CPD are possible due to polarization of molecules, surface chemical reactions and adsorption and desorption of molecules.

1.2. Light Absorption and Recombination Processes in Photoactive Materials

1.2.1. *Light absorption and photogeneration of charge carriers*

Any material can absorb light. During light absorption, the energy of a photon $(h\nu)$ is converted into an increase of the energy of an electron (E_e) whereas the photon (γ) disappears during the absorption event. Light absorption in photoactive materials for photovoltaics and photocatalysis is related to absorption of photons of the sun spectrum, i.e. to photons with energies from the near infrared (0.4–0.5 eV) to the ultraviolet (UV) range (4–5 eV).

During an absorption event, an electron is excited from an occupied into an unoccupied state at a higher energy. Electronic states are occupied or unoccupied if their energy is below or above the Fermi energy (E_F), respectively. Photoactive materials are distinguished by their distributions of occupied and unoccupied states. Due to the overlap of wave functions, electronic states form extended states or bands at high densities. Charge carriers are mobile (or free) in extended states and not mobile (or trapped) in localized states.

Absorption resulting in the excitation of charge carriers from states where they are not mobile into states where they are mobile is called photogeneration of free or mobile charge carriers. In solar cells, photogenerated electrons (e^-) are separated from an absorber at a charge-selective contact and their energy is converted into electric power at an external load. After that, the electrons flow back into the absorber at reduced energy. In photocatalysts, photogenerated electrons, for example, can be separated toward reactive sites where they participate in photocatalytic reactions.

Metals have a very high density of free electrons and can absorb photons of any energy (Figure 1.1(a)). Features can appear in absorption spectra of metals, for example, due to plasmonic absorption. During an absorption event in a metal, only the kinetic energy of an electron increases, i.e.

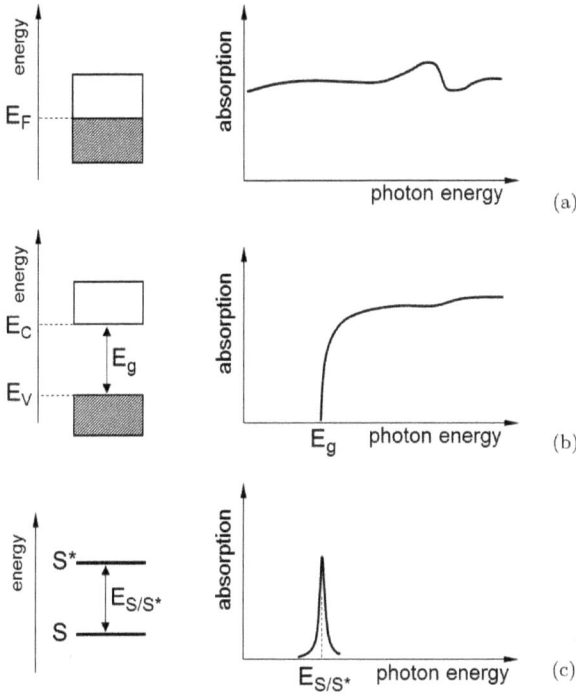

Figure 1.1. Schematic energy levels (left) and absorption spectra (right) of (a) a metal, (b) a semiconductor or insulator and (c) a molecule. E_F, E_V, E_C, E_g, S, S^* and $E_{S/S*}$ denote the Fermi energy of a metal, the conduction and valence band edges and the band gap of a semiconductor or an insulator and the ground and excited states and the transition energy of a molecule, respectively.

photogeneration does not take place (Equation (1.1)):

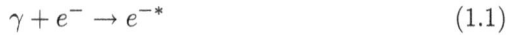

$$\gamma + e^- \to e^{-*} \qquad (1.1)$$

The density of free charge carriers can be very low in semiconductors and insulators so that absorption by free charge carriers can usually be neglected. In semiconductors and insulators, absorption is dominated by excitation of not mobile electrons from states in the valence band into states in the conduction band where the electrons are mobile (fundamental absorption). An electron missing in the valence band is a positively charged mobile quasiparticle called a hole (h^+):

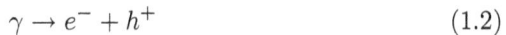

$$\gamma \to e^- + h^+ \qquad (1.2)$$

A certain minimum energy is needed for exciting an electron from an occupied valence state into an unoccupied conduction state. This

Table 1.1. Band gaps of some inorganic and organic semiconductors and insulators.

Material	E_g (eV)	Reference
PbS	0.37	Scanlon (1958)
CuInSe$_2$	1.06	Shay *et al.* (1973)
c-Si	1.12	Alex *et al.* (1996)
CdTe	1.43	Yamada (1960)
GaAs	1.42	Blakemore (1982)
CH$_3$NH$_3$PbI$_3$	1.57	De Wolf *et al.* (2014)
CdSe	1.74	Nimoiya *et al.* (1995)
Cu$_2$O	2.1	Meyer *et al.* (2012)
PbI$_2$	2.34	Ahuja *et al.* (2002)
BiVO$_4$	2.5	Stoughton *et al.* (2013)
TiO$_2$ (rutile)	3.0	Tang *et al.* (1994)
TiO$_2$ (anatase)	3.45	Tang *et al.* (1994)
ZnPc	1.53	Hamam *et al.* (2017)
PPV	2.4	Eckhardt *et al.* (1989)
muscovite	5.09	Kalita *et al.* (2016)
SiO$_2$	8–9	Nekrashevich *et al.* (2014)

minimum energy is the difference between the conduction and valence band edges (E_C and E_V, respectively) and is called the band gap (E_g). Semiconductors have a lower band gap than insulators. The band gap is the most important property of photoactive materials since it enables the realization of differences in chemical potentials of electrons and holes that drive, for example, solar cells and photocatalytic reactions. Table 1.1 gives the values of E_g at room temperature for some inorganic semiconductors (crystalline silicon, c-Si; gallium arsenide, GaAs; methyl amine lead iodide, CH$_3$NH$_3$PbI$_3$; copper oxide, Cu$_2$O; bismuth vanadate, BiVO$_4$; titania, TiO$_2$; copper indium selenide CuInSe$_2$; lead sulfide, PbS; lead iodide, PbI$_2$; cadmium telluride, CdTe; cadmium selenide, CdSe) and insulators (quartz, SiO$_2$; mica: muscovite) and organic semiconductors (poly(phenylene-vinylene), PPV; zinc phthalocyanine, ZnPc). Incidentally, the band gap of organic semiconductors is also called the HOMO–LUMO (highest occupied molecular orbitals and lowest unoccupied molecular orbitals, respectively) gap.

Ideal semiconductors and insulators are transparent for photons with energies below E_g (Figure 1.1(b)). Incidentally, direct and indirect band gaps are distinguished depending on whether an absorption event can occur without or only with the participation of a phonon (a phonon is a lattice vibration), respectively, in order to fulfill the condition of momentum conservation (see, for example, Pankove, 1971).

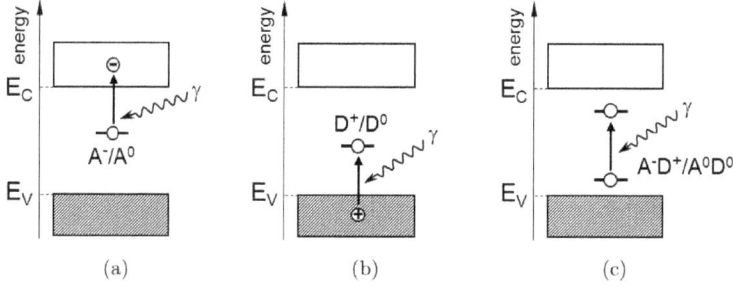

Figure 1.2. Schematic diagram of absorption events for photogeneration (a) of an electron from an occupied acceptor state, (b) of a hole from an unoccupied donor state and (c) for excitation of an electron from an occupied acceptor into an unoccupied donor state.

In molecules, electrons can be excited from a discrete energy level called the ground state (S) to an excited state (S^*). Molecules absorb light in a narrow range of photon energies (Figure 1.1(c)). Absorption spectra of electron transitions in molecules can be broadened due to interactions between atoms in molecules:

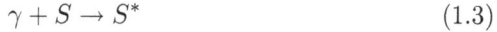

$$\gamma + S \to S^* \tag{1.3}$$

Defect states can be involved in absorption processes. Donor (D) and acceptor (A) states are distinguished. A donor is a defect or chemical species which is neutral when occupied (D^0) and positively charged when unoccupied (D^+). In analogy, an acceptor is a defect or chemical species which is negatively charged when occupied (A^-) and neutral when unoccupied (A^0). Donors and acceptors are also called traps since an electron at A^- and a hole at D^+ are localized, i.e. not mobile.

In bulk semiconductors and insulators, light can be absorbed at photon energies below E_g if donor and/or acceptor states are present in the band gap (Figure 1.2).

A free electron can be photogenerated by excitation of an electron from A^- into the conduction band (Figure 1.2(a)). A free hole can be photogenerated from an unoccupied donor state into the valence band (Figure 1.2(b)). The excitation of an electron from A^- into D^+ (Figure 1.2(c)) does not result in photogeneration of free charge carriers.

$$\gamma + A^- \to A^0 + e^- \tag{1.4}$$

$$\gamma + D^+ \to D^0 + h^+ \tag{1.5}$$

$$\gamma + A^- + D^+ \to A^0 + D^0 \tag{1.6}$$

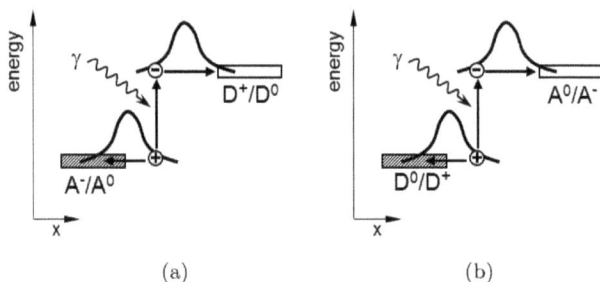

Figure 1.3. Schematic diagram of direct photon absorption due to a partial overlap (a) of wave functions of an occupied acceptor and an unoccupied donor states and (b) of an occupied donor and an unoccupied acceptor states.

Donor and acceptor states are not located at the same place in a photoactive material as supposed in the idealized Figure 1.2(c). Wave functions of donors and acceptors can partially overlap in space. Donors and acceptors with a partial overlap of wave functions are also called donor–acceptor complexes. The overlap of wave functions between donors and acceptors can lead to the formation of so-called charge transfer states (CTS), (see, for example, Vandewal, 2016).

Due to a partial overlap of wave functions of donor and acceptor states, photons can be directly absorbed. Depending on whether the energy of an unoccupied donor state is higher than the energy of an occupied acceptor state or whether the energy of an unoccupied acceptor state is higher than the energy of an occupied donor state, photon absorption leads to the formation of A^0 and D^0 (Figure 1.3(a)) or A^- and D^+ (Figure 1.3(b)).

$$\gamma + A^0 + D^0 \rightarrow A^- + D^+ \tag{1.7}$$

Photons with energies below the band gap can also be absorbed in semiconductors or insulators in the presence of strong electric fields or potential fluctuations (Figure 1.4(a)). Due to the extension of wave functions, there is a partial overlap of occupied and unoccupied states within the band gap depending on the strengths of the electric field. Electrons and holes excited at energies below the band gap are transferred by tunneling into the conduction and valence bands, respectively (tunneling-assisted absorption and photogeneration). The corresponding red-shift of the onset of fundamental absorption of semiconductors in electric fields is called the Franz–Keldysh effect (Franz, 1958; Keldysh, 1958).

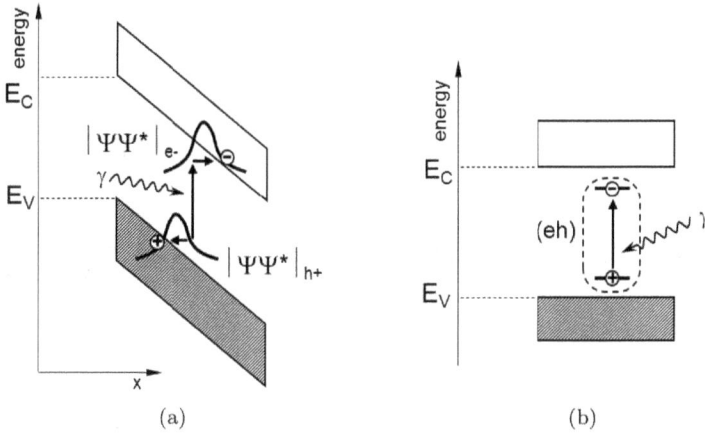

Figure 1.4. (a) Tunneling-assisted absorption of photons with energy below the band gap for semiconductors in electric fields (Franz–Keldysh effect) and (b) absorption leading to the formation of bond (eh) states in a bulk semiconductor (exciton absorption). $|\Psi\Psi^*|_{e-}$ and $|\Psi\Psi^*|_{h+}$ denote the probability density of electron and hole wave functions, respectively.

Electrons and holes can form bound states (eh) due to Coulomb interaction. An electron hole pair in such a bond state is an electrically neutral quasiparticle called an exciton (see, for example, Scholes and Rumbles, 2006, and references therein). An exciton is formed by an absorption event (exciton absorption, Figure 1.4(b)). The binding energy of excitons depends on the dielectric properties of the semiconductor and the local surrounding. In the case that the exciton binding energy is larger than the thermal energy, exciton absorption leads to peaks in the absorption spectrum below the band gap of bulk semiconductors. In semiconductor nanoparticles, the wave functions of electrons and holes are strongly confined in space so that the exciton transition energies can increase to values above the band gap (quantum confinement, Brus, 1984). The size of so-called quantum dots (QDs) is of the order of several nanometers. The onset of optical absorption (absorption gap) of QDs is blue-shifted in comparison to the bulk semiconductor. The absorption gap of excitons can be tuned with the size of QDs (quantum size effect).

The presence of surfaces and interfaces results in additional absorption processes which can lead, depending on the overlap of wave functions, to photogeneration. Figure 1.5 shows, as an example, photogeneration at a metal/semiconductor contact. Due to the overlap of wave functions

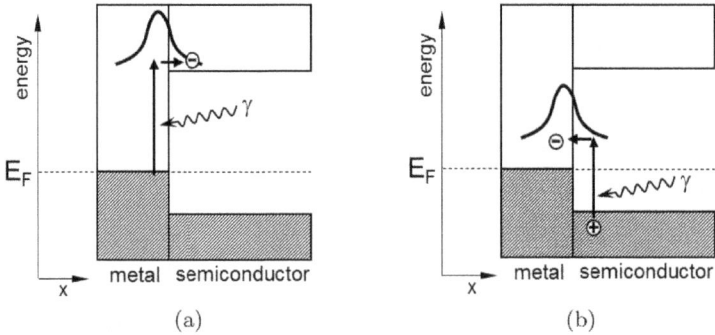

Figure 1.5. Schematic diagram of direct photon absorption caused by a partial overlap of wave functions of occupied and unoccupied electronic states in the metal with the conduction and valence bands in the semiconductor, respectively, leading to photogeneration of (a) electrons and (b) holes.

of unoccupied states in the metal with the conduction band in the semiconductor, electrons can be excited directly from states in the metal into the conduction band (Figure 1.5(a)). Vice versa, due to the overlap of wave functions of unoccupied states in the metal with the valence band in the semiconductor, electrons can be excited directly from the valence band into states in the metal resulting in photogeneration of holes (Figure 1.5(b)). Similar transitions are possible between conduction and valence bands of different semiconductors at heterojunctions. Incidentally, the examples shown in Figure 1.5 are also equivalent to photogeneration from occupied surface states into the conduction band and from unoccupied surface states into the valence band of a semiconductor. The excitation of photogenerated charge carriers across interfaces is also called internal photoemission.

1.2.2. *Recombination processes of charge carriers*

The density of free charge carriers in photoactive materials is limited by recombination, i.e. the annihilation of free charge carriers. The energy of a recombining charge carrier can be transferred to photons (radiative recombination), to another charge carrier (Auger recombination) or to lattice vibrations (Shockley–Read–Hall (SRH) and surface recombination) (Shockley and Read, 1952; Hall, 1952).

Radiative band–band recombination is defined as the annihilation of a free electron with a free hole under the emission of a photon (Figure 1.6(a)):

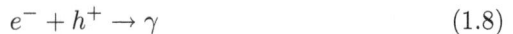

$$e^- + h^+ \to \gamma \tag{1.8}$$

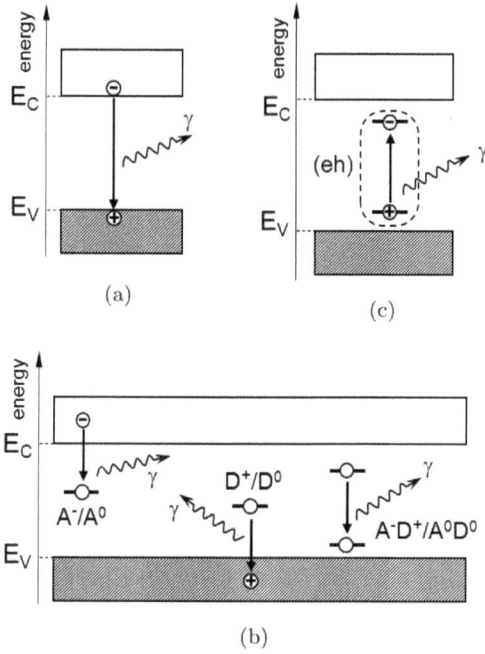

Figure 1.6. (a) Radiative band-to-band recombination, (b) trap-assisted radiative recombination mechanisms and (c) radiative exciton recombination.

The rate of radiative band–band recombination ($R_{rad,np}$) is proportional to the densities of free electrons (n) and holes (p):

$$R_{rad,np} = B_{rad} \cdot n \cdot p \qquad (1.9)$$

The proportionality factor is called the radiative recombination rate constant (B_{rad}). Typical values of $B_{rad,np}$ are of the order of $2 \cdot 10^{-10}$ cm^3/s and $3 \cdot 10^{-15}$ cm^3/s for direct (GaAs, Strauss *et al.*, 1993) and indirect (c-Si, Varshni, 1967) semiconductors, respectively.

Radiative recombination can also take place between free charge carriers and trap states (trap-assisted radiative recombination, Figure 1.6(b)). Free electrons or holes can recombine at unoccupied or occupied, donor or acceptor states, respectively:

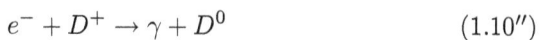

$$e^- + A^0 \rightarrow \gamma + A^- \qquad (1.10')$$

$$e^- + D^+ \rightarrow \gamma + D^0 \qquad (1.10'')$$

$$h^+ + A^- \rightarrow \gamma + A^0 \tag{1.10'''}$$

$$h^+ + D^0 \rightarrow \gamma + D^+ \tag{1.10''''}$$

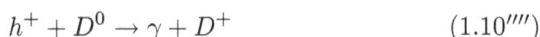

Depending on the energies of unoccupied and occupied donor and acceptor traps, radiative recombination can also be observed from donor–acceptor complexes:

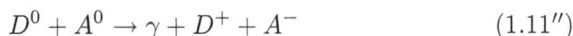

$$D^+ + A^- \rightarrow \gamma + D^0 + A^0 \tag{1.11'}$$

$$D^0 + A^0 \rightarrow \gamma + D^+ + A^- \tag{1.11''}$$

Radiative recombination of excitons (Figure 1.6(c)) plays an important role in QDs and organic semiconductors.

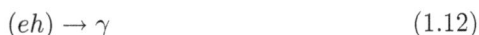

$$(eh) \rightarrow \gamma \tag{1.12}$$

Auger recombination implements the increase of the kinetic energy of a third charge carrier, i.e. an electron or a hole (Figures 1.7(a) and 1.7(b), respectively). The high kinetic energy of the excited third charge carrier is transformed into a cascade of lattice vibrations (phonons, Γ):

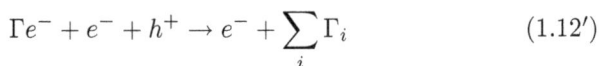

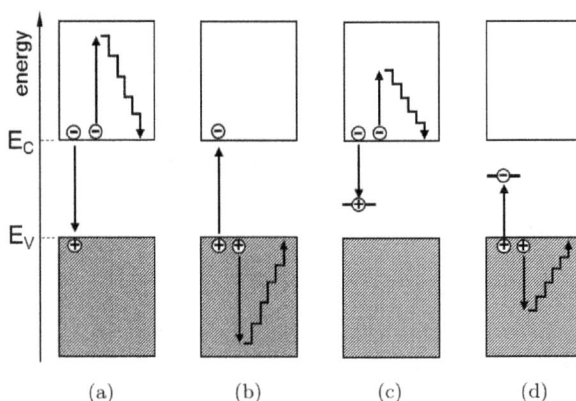

$$\Gamma e^- + e^- + h^+ \rightarrow e^- + \sum_i \Gamma_i \tag{1.12'}$$

$$e^- + h^+ + h^+ \rightarrow h^+ + \sum_j \Gamma_j \tag{1.12''}$$

Figure 1.7. Auger recombination of (a) two electrons and one hole, (b) two holes and one electron, (c) trap-assisted Auger recombination of two electrons and (d) trap-assisted recombination of two holes.

The Auger recombination rates are proportional to the density of holes and squared density of electrons ($R_{A(e)}$) or to the density of electrons and squared density of holes ($R_{A(h)}$):

$$R_{A(e)} = C_{A(e)} \cdot n^2 \cdot p \tag{1.13'}$$

$$R_{A(h)} = C_{A(h)} \cdot n \cdot p^2 \tag{1.13''}$$

Deep in the conduction and valence bands, the densities of states are very high for all inorganic semiconductors. Therefore, the Auger recombination coefficients for the energy transfer to electrons ($C_{A(e)}$) or holes ($C_{A(h)}$) are very similar and of the order of $10^{-30}\,\mathrm{cm}^6/\mathrm{s}$ (see, for example, Yablonovich *et al.*, 1986).

Energy can also be transferred to a free electron or hole if the second free electron or hole recombines with a trapped hole or electron, respectively (trap-assisted Auger recombination, Figures 1.7(c) and 1.7(d)):

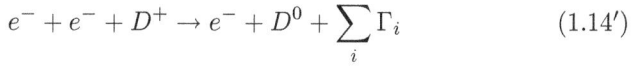

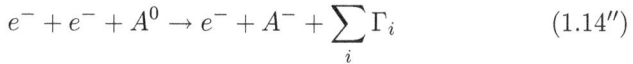

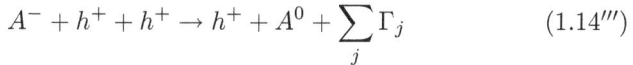

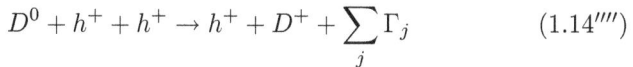

$$e^- + e^- + D^+ \rightarrow e^- + D^0 + \sum_i \Gamma_i \tag{1.14'}$$

$$e^- + e^- + A^0 \rightarrow e^- + A^- + \sum_i \Gamma_i \tag{1.14''}$$

$$A^- + h^+ + h^+ \rightarrow h^+ + A^0 + \sum_j \Gamma_j \tag{1.14'''}$$

$$D^0 + h^+ + h^+ \rightarrow h^+ + D^+ + \sum_j \Gamma_j \tag{1.14''''}$$

Free electrons and holes can recombine via defect states in the band gap (E_t — energy of the defect or trap state) due to trapping of electrons and holes SRH recombination (Shockley and Read, 1952; Hall, 1952). An electron can be captured from the conduction band at an unoccupied trap state (Figure 1.8(a)). In analogy, a hole can be captured from the valence band at an occupied trap state (Figure 1.8(b)). Furthermore, an electron can be emitted from an occupied trap state into the conduction band (Figure 1.8(c)) and a hole can be emitted from an unoccupied trap state into the valence band (Figure 1.8(d)). Both capture processes correspond to a non-radiative recombination event of a free electron with a free hole via a trap state.

For a single trap with E_t and a density N_t, the SRH recombination rate (R_{SRH}) depends on E_t, N_t, the densities of free electrons (n) and holes (p),

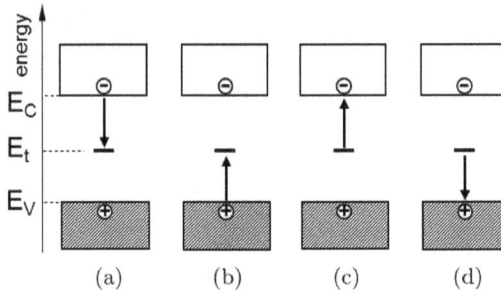

Figure 1.8. Elementary processes at a trap state in the band gap of a semiconductor: (a) capture of an electron from the conduction band at an unoccupied trap state, (b) capture of a hole from the valence band at an occupied trap state, (c) emission of an electron from an occupied trap state into the conduction band and (d) emission of a hole from an unoccupied trap state into the valence band.

E_g, the effective densities of states at the conduction and valence band edges (N_C and N_V, respectively), the capture cross-sections for electrons and holes (σ_e and σ_h, respectively) and the thermal velocity (v_{th}, about 10^7 cm/s):

$$R_{\text{SRH}} = \frac{n \cdot p - N_C \cdot N_V \cdot \exp\left(-\frac{E_g}{k_B \cdot T}\right)}{\frac{1}{\sigma_h \cdot v_{\text{th}} \cdot N_t} \cdot \left(n + N_C \cdot \exp\left(-\frac{E_C - E_t}{k_B \cdot T}\right)\right) + \frac{1}{\sigma_e \cdot v_{\text{th}} \cdot N_t} \cdot \left(p + N_V \cdot \exp\left(-\frac{E_t - E_V}{k_B \cdot T}\right)\right)} \tag{1.15}$$

Equation (1.15) is rather complex. A maximum R_{SRH} is reached if E_F and E_t are in the middle of the band gap and σ_e is equal to σ_h:

$$R_{\text{SRH}}^{\max} = \sigma_{h(e)} \cdot v_{\text{th}} \cdot N_t \cdot n(= p) \tag{1.16}$$

The value of a capture cross-section depends on the nature of the trap state and can be of the order of 10^{-18}–10^{-14} cm^2. In general, R_{SRH} increases with increasing N_t, decreases when E_t shifts closer to E_C or E_V and increases with increasing n and/or p. Trap states with E_t near the middle of the band gap are recombination active (also called recombination centers), trap states close to E_C or E_V are traps for electrons or holes, respectively.

In analogy to SRH, defect states at semiconductor surfaces and interfaces can act, depending on their energy, as surface electron or hole

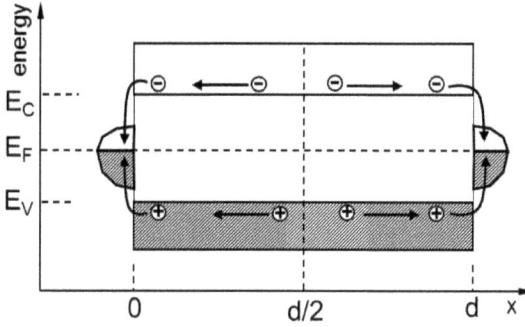

Figure 1.9. Schematic band diagram for surface recombination at a symmetric crystal with thickness d.

traps or surface recombination centers. In difference to SRH, the transport of free charge carriers to the surface can limit surface recombination (Figure 1.9).

The densities and capture cross-sections of electrons and holes at the surface are denoted as n_s, p_s, σ_{se} and σ_{sh}, respectively. The density of surface defects is denoted as N_{st}. In the (simplest) case that diffusion does not limit the access of free charge carriers to the surface, the surface recombination rate (R_{surf}) can be expressed for a symmetric crystal with the thickness d as follows:

$$R_{surf} = \sigma_{sh(se)} \cdot v_{th} \cdot N_{st} \cdot \frac{n_s(p_s)}{d/2} \tag{1.17}$$

The values of n_s, p_s and N_{st} are given in units of $1/cm^2$. The product of $\sigma_{se,sh}$, v_{th} and N_{st} has the unit of cm/s and is therefore also called the surface recombination velocity (v_s).

The total recombination rate is the sum of the recombination rates of all recombination processes. The recombination lifetime is defined as the ratio between the density of free charge carriers and the corresponding recombination rate. The total lifetime of free charge carriers (τ_{total}) can be approximated, for example, for a p-type doped semiconductor, by

$$\frac{1}{\tau_{total}} = B_{rad,np} \cdot p_0 + C_A \cdot p_0^2 + \frac{1}{\tau_{SRH}} + \frac{v_s}{d/2} \tag{1.18}$$

1.3. Relaxation of a Disturbance of a Space Charge

1.3.1. *Relaxation of photogenerated charge carriers separated in space*

Mobile photogenerated charge carriers are created in photoactive materials by light absorption. The total number of photogenerated charge carriers is limited by recombination. Photogenerated charge carriers can be separated in space and can therefore disturb the local distribution of space charge. A space charge can be disturbed within the bulk of a photoactive material (Figure 1.10(a)), for example, due to diffusion.

A space charge can also be disturbed due to charge separation over a barrier (Figure 1.10(b)). After switching off photogeneration, the total number of photogenerated charge carriers decreases in time and the disturbance of the space charge relaxes to equilibrium. The time-dependent densities of photogenerated positive and negative charge carriers depend on the rates of the different recombination processes and on the access of

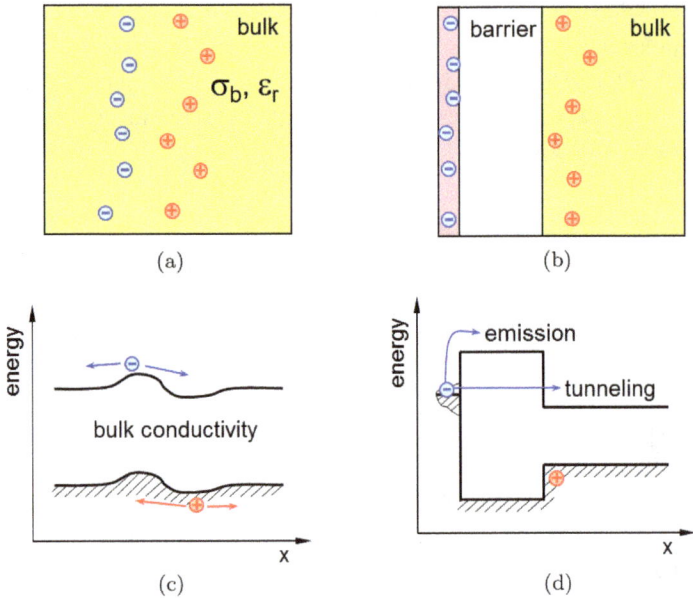

Figure 1.10. Schematic diagram of a disturbance of a charge distribution from equilibrium due to charge separation in (a) a homogeneous bulk or (b) across a barrier and (c) schematic band diagrams depicting relaxation due to bulk conductivity or (d) due to charge transfer over a barrier by emission or through a barrier by tunneling.

mobile charge carriers to regions where they can recombine, i.e. on charge transport.

In the bulk of a photoactive material, charge transport is limited by the electrical conductivity of the bulk (σ_b, Figure 1.10(c)). Charge carriers can overcome large barriers by thermally activated emission or, if the barrier layer is extremely thin, by tunneling (Figure 1.10(d)).

The redistribution of photogenerated charge carriers separated in space toward equilibrium is called relaxation of photogenerated charge carriers separated in space. Processes limiting the relaxation of photogenerated charge carriers separated in space can be investigated by SPV measurements. Incidentally, SPV and photoluminescence techniques are complementary in the sense that they register independently charge separated in space and charge, which is no longer separated in space and which disappears by radiative recombination.

1.3.2. *Conductivity and dielectric relaxation*

The current flow (j) through a sample is proportional to the electric field (E_field). The proportionality factor is called the conductivity (σ). The Ohm's law describes the dependence of j on E_field:

$$j = \sigma \cdot E_\text{field} \tag{1.19}$$

The conductivity depends on the densities of mobile charge carriers and their mobility. Charge transport by ions can be usually neglected inside photoactive materials. The elementary charge and the mobilities of electrons and holes are denoted by q ($1.602 \cdot 10^{-19}$ As) and μ_e and μ_h, respectively:

$$\sigma = q \cdot (n \cdot \mu_e + p \cdot \mu_h) \tag{1.20}$$

The density of free electrons can vary over many orders of magnitude from more than 10^{22} cm^{-3} for metals to about 10^{10} cm^{-3} for intrinsic crystalline silicon and to values smaller than 10^2 cm^{-3} for insulators.

The mobility of charge carriers depends on the transport mechanism (Figure 1.11). In crystalline semiconductors such as crystalline silicon or gallium arsenide, mobile charge carriers are delocalized over distances much longer than the lattice constant and electrons and holes can move freely in the conduction or valence bands, respectively (Figure 1.11(a)). The mobility of free electrons or holes is limited by scattering at lattice vibrations, ionic charge and other free charge carriers. The mobility, for example, can

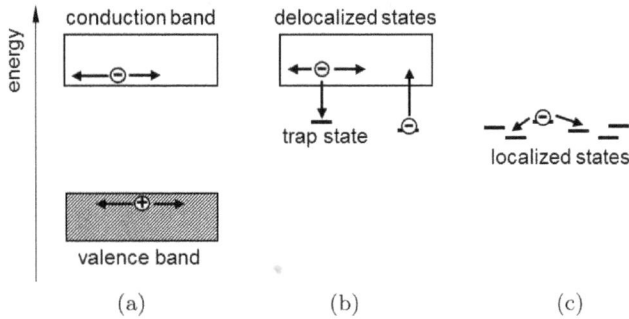

Figure 1.11. (a) Schemes of charge transport in bands, (b) in delocalized states limited by trapping (left) and de-trapping (right) and (c) between localized states by hopping.

be as high as $2 \cdot 10^5 \, \text{cm}^2/\text{Vs}$ in graphene layers (Bolotin *et al.*, 2008) or $1000 \, \text{cm}^2/\text{Vs}$ in crystalline silicon (Jacoboni *et al.*, 1977).

Localized charge carriers are not mobile. Localization of charge carriers is also called trapping. Charge carriers can be localized, for example, by trapping at defects (trap limited transport in crystalline semiconductors, Figure 1.11(b)) or local potential fluctuations (charge transport in disordered or amorphous semiconductors) or by trapping caused by polarization of the environment of a charge carrier (polaronic transport). Trapped charge carriers can be excited from trap states into delocalized states (detrapping) by thermal activation or by optical absorption. Furthermore, if the distance between localized states is very short, charge carriers can tunnel between localized states (hopping transport, Figure 1.11(c)). The mobility is low in disordered semiconductors, for example, about $2 \, \text{cm}^2/\text{Vs}$ in undoped amorphous silicon (Marshall *et al.*, 1986) and can be as low as 10^{-3} or $10^{-6} \, \text{cm}^2/\text{Vs}$ in layers of PCBM (phenyl-C_{61} butyric acid methyl ester, von Hauff *et al.* (2005) or PPV, Lebedev *et al.* (1997)), respectively.

The distribution of an electric field depends on the distribution of uncompensated or space charge (ρ) in a sample. The relative dielectric constant and the dielectric constant of vacuum are denoted by ε_r and ε_0 ($8.84 \, 10^{-14}$ As/Vcm), respectively. The space charge and the electric field are connected by the Poisson equation as follows:

$$\text{div}(E_{\text{field}}) = -\frac{\rho}{\varepsilon_r \cdot \varepsilon_0} \tag{1.21}$$

A disturbance of a space charge in time causes a current flow which is described by the continuity equation,

$$\text{div}(j) = -\frac{\partial \rho}{\partial t} \tag{1.22}$$

By taking into account Ohm's law, a disturbance of a space charge results in an exponential decay of ρ in time as follows:

$$\rho(t) = \rho_0 \cdot \exp\left(-\frac{t}{\tau_M}\right) \tag{1.23}$$

with

$$\tau_M = \frac{\varepsilon_r \cdot \varepsilon_0}{\sigma} \tag{1.24}$$

The time constant (τ_M) is called the dielectric or Maxwell relaxation time. Figure 1.12 gives an overview of the orders of magnitude of Maxwell relaxation times as a function of the density of free charge carriers for different values of the mobility. In Figure 1.11, ε_r was set to 12, which is the typical order for numerous inorganic semiconductors such as crystalline silicon or gallium arsenide. One has to keep in mind that ε_r can depend differently on the material (for example, $\varepsilon_r \sim$ 89–173 for TiO_2 (rutile, Schmidt (1902)) or about 2–6 for conjugated polymers (Torabi *et al.*, 2015) and on the frequency.

Depending on the density of free charge carriers and on the mobility, τ_M can change over a huge range of times. The dielectric relaxation time is extremely short (below ps) in metals and can be very long (up to years and longer) in insulators. In doped crystalline inorganic semiconductors, τ_M is of the order of ps. In undoped conventional crystalline inorganic semiconductors, τ_M can be in the μs–ms range.

1.3.3. *Charge transfer over barriers*

Barriers (barrier height denoted by Φ_B) can strongly limit charge transport or charge transfer (see, for example, Sze (1981)). In photoactive materials, barriers can range between tens of meV and 1–2 eV. For example, the offset between the conduction band edges at the SnO_2:F/TiO_2 contact corresponds to Φ_B for electron transfer from SnO_2:F into TiO_2 and is of the order of 0.1–0.2 eV (Hou *et al.*, 2014). Furthermore, electronic states can pin E_F near the middle of the band gap at the surface of covalently bonded semiconductors such as c-Si (Φ_B about 0.55 eV) or GaAs (Φ_B about 0.7 eV). Large barriers can occur for charge carriers trapped at

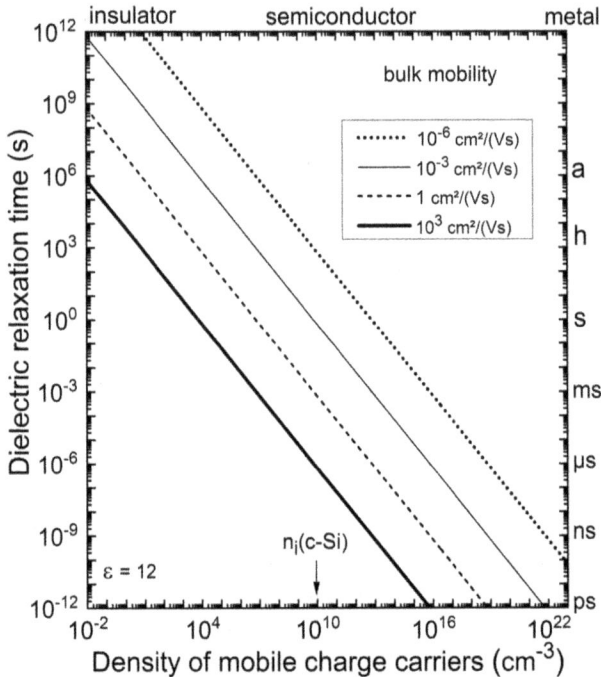

Figure 1.12. Dependence of the Maxwell or dielectric relaxation time for charge transport in a homogeneous bulk on the density of charge carriers for different values of the mobility of charge carriers of 10^{-6}, 10^{-3}, 1 and 10^3 cm^2/Vs corresponding to the mobility in a disordered polymer PPV (poly phenylene vinylene, see, for example, Lebedev *et al.* (1997)), in a layer of PCBM (phenyl-C$_{61}$-butyric acid methyl ester, von Hauff *et al.* (2005)), in a layer of undoped amorphous silicon (Marshall *et al.*, 1986) and in a silicon crystal (Jacoboni *et al.*, 1977) — thick solid, thin solid, dashed and dotted lines, respectively. The regions of densities of mobile charge carriers for insulators, semiconductors and metals are roughly marked and the time regions are denoted by ps, ns, μs, ms, s, h (hour) and a (year). The relative dielectric constant was set to the constant value of 12.

the surface of semiconductors or metal oxides with wide band gaps (Φ_B up to several eV).

Charge carriers can be thermally excited to high kinetic energies so that a transfer into delocalized states of the material of the barrier becomes possible (thermionic emission). The density of electrons with high kinetic energy increases exponentially with increasing temperature (T). Therefore, thermionic emission is thermally activated. Furthermore, if a barrier is very thin, charge carriers can tunnel through the barrier. Charge transfer limited by tunneling does not depend on temperature. Depending on the shape of the barrier, barrier height and measurement conditions, charge transfer by tunneling-assisted thermionic emission is also possible.

A barrier can be described by a capacitance (C) and a shunt resistance (R). The thickness and the area of the barrier region are denoted by d and A, respectively. A barrier region can be considered as a parallel plate capacitor:

$$C = \frac{\varepsilon_r \cdot \varepsilon_0}{d} \cdot A \qquad (1.25)$$

$$R = \frac{1}{\sigma} \cdot \frac{d}{A} \qquad (1.26)$$

The corresponding RC time constant (τ_{RC}) can be considered as a reciprocal charge transfer rate across the barrier. Incidentally, τ_{RC} is equivalent to τ_M:

$$\tau_{RC} = \frac{\varepsilon_r \cdot \varepsilon_0}{\sigma} \qquad (1.27)$$

The dependence of the current caused by thermionic emission (I_{em}) on Φ_B and T was derived from the Maxwell distribution of electron velocities and is given by the following equation (see, for example, Sze (1981), and references therein):

$$I_{em} = A^* \cdot T^2 \cdot \exp\left(-\frac{\Phi_B}{k_B \cdot T}\right) \qquad (1.28)$$

The parameter A^* is called the Richardson constant $(120\,\text{A}/(\text{cm}^2\text{K}^2)$ for the free electron (Crowell, 1965)) and k_B denotes the Boltzmann constant $(1.38 \cdot 10^{-23}\,\text{J/K})$.

A differential conductivity (dI/dU) can be estimated by, first, replacing Φ_B in Equation (1.28) by the sum of Φ_B and the product of q and the potential (U), second, taking the first derivative afterwards and, third, setting U to 0:

$$\frac{dI}{dU}\bigg|_{U \to 0} = \frac{q \cdot A^*}{k_B} \cdot T \cdot \exp\left(-\frac{\Phi_B}{k_B \cdot T}\right) \qquad (1.29)$$

The differential conductivity corresponds to the reciprocal resistance. Therefore, the RC time constant for thermionic emission $(\tau_{RC\text{-th}})$

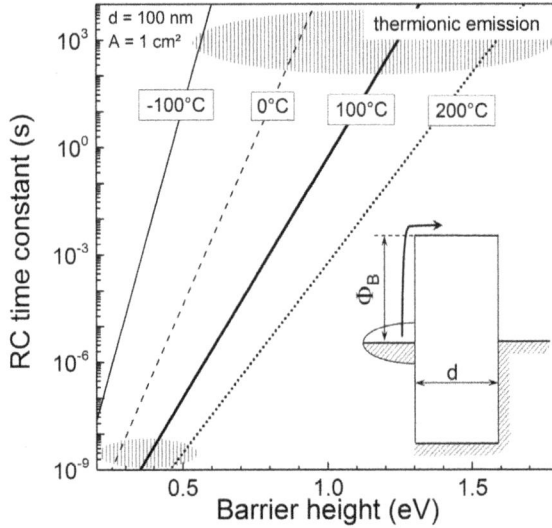

Figure 1.13. Dependence of the RC time constant on the barrier height for the relaxation of charge separated in space by thermionic emission over a barrier layer with a thickness of 100 nm for temperatures of −100°C, 0°C, 100°C and 200°C (thin solid, dashed, thick solid and dotted lines, respectively). The inset shows the corresponding barrier configuration. The areas marked give the approximate time limits at short and long times for time-resolved SPV measurements.

can be calculated if the values of ε_r and d are known by using Equations (1.25)–(1.27) and (1.29):

$$\tau_{RC-th} = \frac{\varepsilon_r \cdot \varepsilon_0 \cdot k_B}{q \cdot A^* \cdot T} \cdot \exp\left(\frac{\Phi_B}{k_B \cdot T}\right) \tag{1.30}$$

As an example, τ_{RC-th} is plotted as a function of Φ_B (varied between 0.2 and 1.8 eV) for d equal to 100 nm and for different temperatures in Figure 1.13. The RC time constants are of the order of 1 ns or less for Φ_B lower than about 0.4–0.5 eV at temperatures higher than −100°C. For larger barrier heights, τ_{RC-th} can reach values of 100 s and more, for example, τ_{RC-th} is about 100 s for Φ_B and T equal to 0.5 eV and −100°C or to 1.5 eV and 200°C, respectively. On the other hand, τ_{RC-th} increases with decreasing temperature for a given Φ_B. For example, τ_{RC-th} increases for Φ_B equal to 0.5 eV from 3 ns to 100 s for temperatures increasing from −100°C to 200°C.

Electrons can tunnel through very thin barriers with a thickness of the order of a nanometer. The transmission probability (T_{tr}) of a barrier increases exponentially with decreasing width and height of the barrier. Tunneling through a triangular barrier between a metal and a highly doped semiconductor (Dittrich, 2018) can serve as a model system for estimating orders of relaxation times. For a highly n-type doped semiconductor, T_{tr} is given by

$$T_{\text{tr}} = \exp\left(-\frac{8}{3} \cdot \sqrt{\frac{m_e^* \cdot \varepsilon_r \cdot \varepsilon_0}{N_D^+}} \cdot \frac{\Phi_B}{\hbar}\right) \tag{1.31}$$

The effective electron mass is denoted by m_e^* and the Planck constant ($\hbar$) is equal to $1.05 \cdot 10^{-34}$ Js. The tunneling current is proportional to T_{tr} and the differential conductivity can be obtained following the same procedure as for thermionic emission. For m_e^* and ε_r equal to $9 \cdot 10^{-31}$ kg and 12, respectively, the tunneling resistance can be obtained and the RC time constant for tunneling through a triangular barrier across a surface SCR of a highly n-type doped semiconductor ($\tau_{\text{RC-tunn}}$) can be approximated by the following equation (Dittrich, 2018):

$$\tau_{\text{RC-tunn}} \approx \frac{\varepsilon_r \cdot \varepsilon_0 \cdot A}{d} \cdot \frac{\Omega \text{cm}^2}{1.26 \cdot 10^9} \cdot \sqrt{\frac{10^{19} \text{cm}^{-3}}{N_D}}$$

$$\cdot \exp\left(79 \cdot \sqrt{\frac{10^{19} \text{cm}^{-3}}{N_D}} \cdot \frac{\Phi_B}{\text{eV}}\right) \tag{1.32}$$

Figure 1.14 shows the dependence of $\tau_{\text{RC-tunn}}$ on d of a triangular barrier for Φ_B equal to 0.6 and 1.0 eV. Furthermore, the N_D related to d of a SCR of highly n-type doped crystalline silicon is also shown. The values of $\tau_{\text{RC-tunn}}$ are of the order of nanoseconds for values of d between 2 ($\Phi_B = 1.0$ eV) and 2.5 nm ($\Phi_B = 0.6$ eV). Incidentally, the values of N_D are of the order of 10^{20} cm^{-3} for such low values of the width of the SCR. The RC time constant for relaxation by tunneling increases to values larger than 10^3 s for values of d of about 6 ($\Phi_B = 1.0$ eV) and 8 nm ($\Phi_B = 0.6$ eV). Therefore, the variation of the width of the barrier has a tremendous influence on charge transfer through barriers by tunneling.

Tunneling through a triangular barrier can be combined with thermal activation so that the width and the height of the barrier reduce for tunneling. This process is called phonon-assisted tunneling. Incidentally, the height of a triangular barrier is additionally reduced at metal-semiconductor contacts by image force.

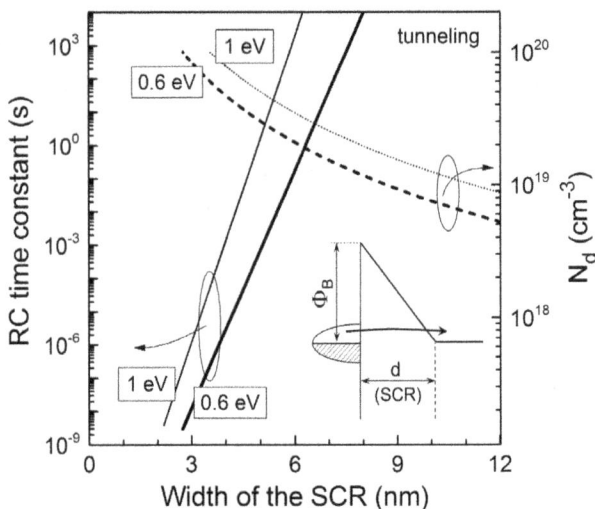

Figure 1.14. Dependence of the RC time constant for relaxation by tunneling on the width of a triangular barrier (inset) for barrier heights of 0.6, and 1 eV (thick and thin lines, respectively). The width of the barrier is given by the SCR of highly doped semiconductors, i.e. by the density of donors for barrier heights of 0.6, and 1 eV (thick dashed and thin dotted lines, respectively).

1.4. Contact Potential Difference and SPV

1.4.1. *Work function of sample and reference electrodes*

Electric potentials are measured between two electrodes. In SPV measurements, a sample electrode is coated or contacted with a photoactive material and a reference electrode is placed in front of the photoactive material. The sample and reference electrodes are metallic, i.e. they can be made, for example, from metals or transparent conductive metal oxides. A major property of metals is their work function (W), which is defined as the lowest energy required for removing an electron from the metal to vacuum. The bare sample and the reference electrodes are characterized by their work functions W_s and W_{ref}, respectively (Figures 1.15(a) and 1.15(b), respectively).

The sample and reference electrodes are electrically connected (Figure 1.15(c)). In the moment of connecting both electrodes, electrons flow from the electrode with the lower to the electrode with the higher work function. During the current flow, the Fermi energies of the sample ($E_{F,s}$) and reference ($E_{F,ref}$) electrodes change until one common Fermi energy (E_F) is established for both electrodes. At the same time, the work

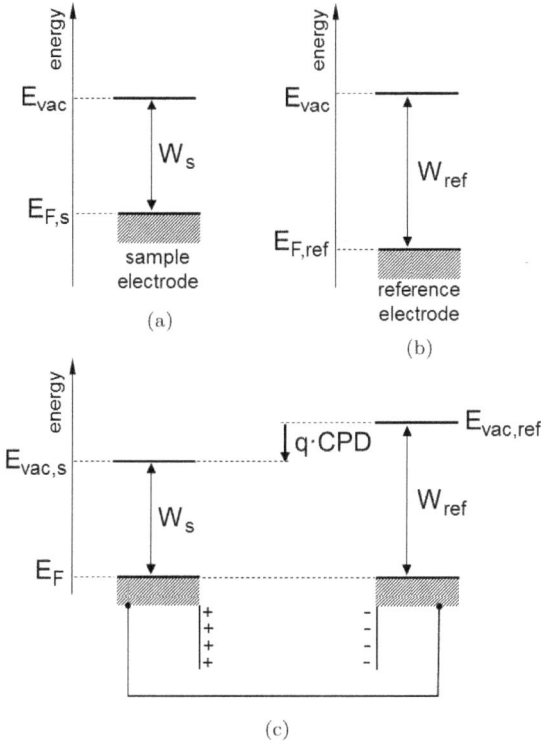

Figure 1.15. (a) Energies of isolated bare sample and (b) reference electrodes and (c) of electrically connected bare sample and reference electrodes. E_{vac}, $E_{F,s}$, $E_{F,\mathrm{ref}}$, E_F, W_s, W_{ref} and CPD denote the vacuum energy, the Fermi energies of the isolated sample and reference electrodes and of the connected electrodes, the work functions of the sample and reference electrodes and the CPD, respectively.

functions remain constant. The difference between the two work functions is related to the electrostatic potential and is called the CPD between both electrodes.

$$q \cdot \mathrm{CPD} \equiv W_{\mathrm{ref}} - W_s \qquad (1.33)$$

1.4.2. *Contact potential difference under illumination and SPV*

The work function of a sample electrode changes due to adsorption of molecules or deposition of a photoactive material. In the following, it is

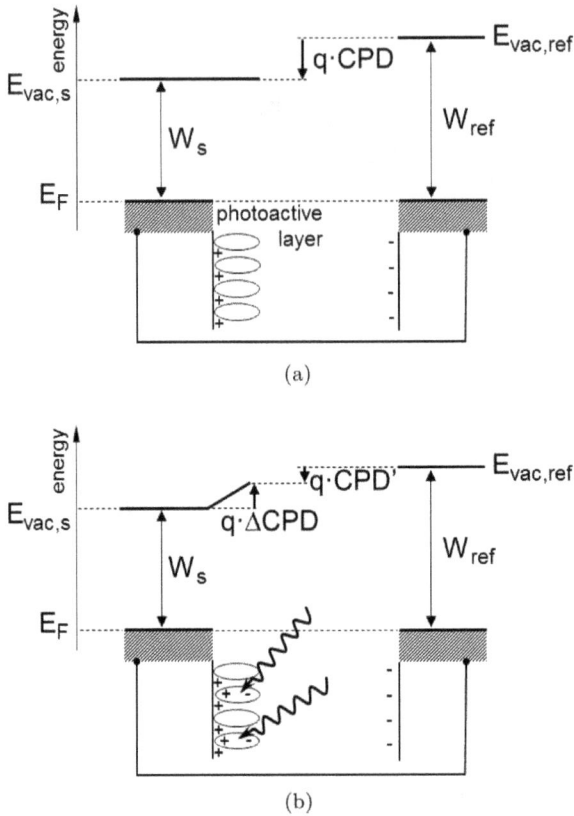

Figure 1.16. Schematic diagram of energies for a sample electrode coated with a photoactive layer and connected with a reference electrode in (a) the dark and (b) under illumination. ΔCPD denotes the change of the CPD due to the separation of photogenerated charge carriers in space.

assumed that W_s is independent of the deposited photoactive material in the dark (Figure 1.16(a)).

Under illumination, photogenerated charge carriers are separated in space, for example, photogenerated electrons and holes move toward the interface with the sample electrode or to the external surface, respectively (Figure 1.16(b)). An electrostatic potential arises in the region of the photoactive layer where photogenerated charge carriers are separated in space.

The work function of the sample electrode is changed under illumination due to the electrostatic potential between separated negative and positive charge carriers in the photoactive layer. The SPV is defined by the difference between the work functions of the sample electrode under illumination ($W_{s,\text{ill}}$) and in the dark ($W_{s,\text{dark}}$):

$$q \cdot \text{SPV} \equiv W_{s,\text{ill}} - W_{s,\text{dark}} \tag{1.34}$$

Following Equation (1.33), the change of the CPD caused by illumination (ΔCPD) can be written as

$$q \cdot \Delta\text{CPD} = W_{\text{ref,ill}} - W_{s,\text{ill}} - (W_{\text{ref,dark}} - W_{s,\text{dark}}) \tag{1.35}$$

Therefore, the SPV corresponds to the negative light-induced change of the CPD if assuming that the work functions of the reference electrode in the dark ($W_{\text{ref,dark}}$) and under illumination ($W_{\text{ref,ill}}$) are equal:

$$\text{SPV} = -\Delta\text{CPD} \tag{1.36}$$

Incidentally, charge carriers separated in space can also be described by a change of the surface dipole.

In Figure 1.17, the ΔCPD and the signs of the SPV are depicted for the separation of photogenerated electrons and holes toward the external surface and interface with the sample electrode, respectively, or toward the interface with the sample electrode and the external surface, respectively.

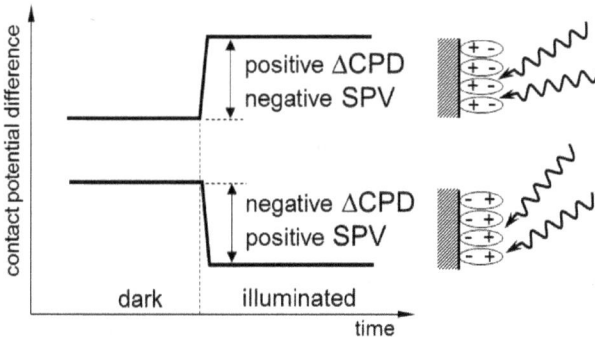

Figure 1.17. Schematic change of the contact potential of sample electrodes coated with a photoactive layer resulting in separation of photogenerated electrons and holes toward the external surface and interface with the sample electrode, respectively (positive ΔCPD and negative SPV) or toward interface with the sample electrode and the external surface, respectively (negative ΔCPD and positive SPV).

1.4.3. *The center of charge approach and the superposition principle*

For the description and for the purpose of simulation of SPV signals, the introduction of the so-called center of charge approach is very useful. Photogeneration leads to the evolution of distributions of mobile positive and negative charge carriers. Due to the nature of photogeneration, the total densities of positive and negative charge carriers are identical.

The densities of negative and positive photogenerated charge carriers are denoted by Δn and Δp, respectively, which are functions of the position in the sample (x, in the direction perpendicular to the sample electrode) and of time (t). Incidentally, Δn and Δp contain all photogenerated charge carriers independently whether they are mobile or trapped.

The dependence of the distribution of the electric potential can be obtained from the Poisson equation:

$$\frac{\partial^2 \varphi}{\partial x^2} = \frac{q}{\varepsilon_0} \cdot \left[\frac{\Delta n(x,t)}{\varepsilon_r(x)} - \frac{\Delta p(x,t)}{\varepsilon_r(x)} \right] \tag{1.37}$$

The relative dielectric constant and the dielectric constant in vacuum are denoted by ε_r and ε_0 ($8.8 \cdot 10^{-14}$ As/(Vcm), respectively. After integrating the Poisson equation twice and assuming that the electric field is 0 at the point H (thickness of the sample), the following expression can be obtained for the SPV:

$$\text{SPV} = \frac{q}{\varepsilon_0} \cdot \int_0^H dx \int_0^x \left[\frac{\Delta n(y,t)}{\varepsilon_r(y)} - \frac{\Delta p(y,t)}{\varepsilon_r(y)} \right] dy \tag{1.38}$$

In Equation (1.38), it has been considered that charge carriers can move across interfaces between different materials which have a different dielectric constant. The precise dependence of the dielectric constant on the position near interfaces is usually not well known. Therefore, the description of SPV signals caused by charge transfer across interfaces over relatively short distances is still challenging. Furthermore, dielectric screening near interfaces in thin layers and nanoparticles with very different dielectric constants (Keldysh, 1979) and image charge near interfaces with metals or metal nanoparticles shall be considered for the description of SPV signals especially in photocatalytic materials. In the following, it is assumed that ε_r is constant.

Figure 1.18 shows photogenerated positive and negative charge carriers in a homogeneous sample at a moment t and the corresponding distributions of Δn and Δp.

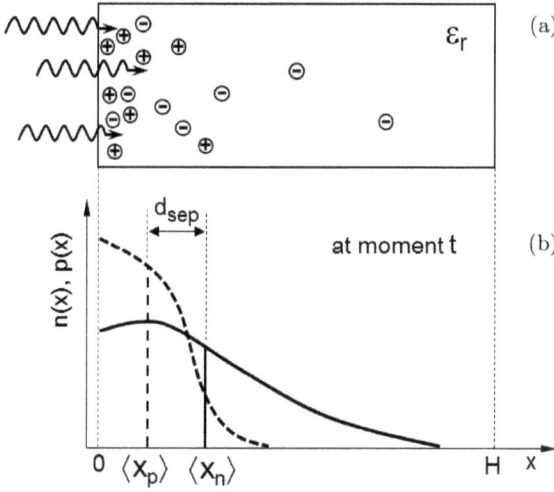

Figure 1.18. (a) Schematic diagram of positive and negative charge carriers in a homogeneous sample and corresponding distributions ($\Delta n(x)$ and $\Delta p(x)$, solid and dashed lines, respectively). (b) $\langle x_n \rangle$, $\langle x_p \rangle$ and d_{sep} denote the centers of negative and positive charge and the charge separation length, respectively.

The charge in a slice dx at the distance x can be calculated as follows:

$$dQ(x) = q \cdot A \cdot dx \cdot [\Delta n(x,t) - \Delta p(x,t)] \tag{1.39}$$

The charge dQ causes a change of the SPV at the position x:

$$d\text{SPV}(t) = \frac{x}{\varepsilon_r \cdot \varepsilon_0} \cdot \frac{dQ(x)}{A} \tag{1.40}$$

The combination of Equations (1.39) and (1.40) and integration results in the SPV:

$$\text{SPV}(t) = \frac{q}{\varepsilon_r \cdot \varepsilon_0} \cdot \int_0^H x \cdot [\Delta n(x,t) - \Delta p(x,t)] dx \tag{1.41}$$

The total areal densities of positive and negative photogenerated charge carriers are

$$N_Q(t) = \int_0^H \Delta n(x,t) dx = \int_0^H \Delta p(x,t) dx \tag{1.42}$$

The centers of negative and positive charge are defined as follows:

$$\langle x_n \rangle(t) = \frac{1}{N(t)} \cdot \int_0^H x \cdot \Delta n(x,t) dx \qquad (1.43')$$

$$\langle x_p \rangle(t) = \frac{1}{N(t)} \cdot \int_0^H x \cdot \Delta p(x,t) dx \qquad (1.43'')$$

The charge separation length (d_{sep}) is defined as the difference of the positions of the centers of positive and negative photogenerated charge carriers:

$$d_{\text{sep}}(t) \equiv \langle x_n \rangle(t) - \langle x_p \rangle(t) \qquad (1.44)$$

The SPV can be described by the following expression (see also Figure 1.18(b)):

$$\text{SPV}(t) = \frac{q}{\varepsilon_r \cdot \varepsilon_0} \cdot N_Q(t) \cdot d_{\text{sep}}(t) \qquad (1.45)$$

Equation (1.45) is the so-called center of charge approach. This approach is a general one since it is valid for any distribution of space charge and for any possible time evolution (Mora-Seró *et al.*, 2006b). Furthermore, Equation (1.45) allows for the simulation of SPV signals by modeling systems of differential equations or by following, for example, large numbers of random walks of individual charge carriers (Chapter 4).

Separation of photogenerated charge carriers in space can have different origins and can occur in different regions of a sample. The SPV is a superposition of the contributions from all photogenerated charge carriers separated in space. In this sense, the center of charge approach includes the superposition principle. Furthermore, the superposition principle allows for the separate analysis of different mechanisms leading to separation and relaxation of photogenerated charge carriers.

1.4.4.   *SPV in the picture of a parallel plate capacitor*

Equation (1.45), which describes the center of charge approach, corresponds to the potential of a parallel plate capacitor with a thickness d_{sep} and a dielectric constant ε_r which is charged by the charge N_Q (Figure 1.19(a)). The values of d_{sep} and N_Q depend on numerous parameters of photoactive materials and processes, which can be studied by SPV techniques, for example, as a function of time, illumination, temperature and ambient (see Chapter 2).

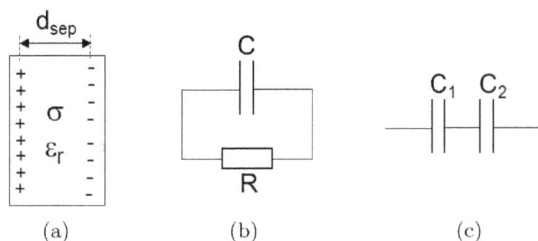

Figure 1.19. (a) Schematic diagram of charge separated in space across a distance d_{sep} in a sample with a given relative dielectric constant (ε_r) and a conductivity (σ), (b) corresponding equivalent circuit with a capacitor and a shunt resistance and (c) two capacitors connected in series as a superposition of two regions in a sample with charge separated in space (shunt and series resistances are omitted).

A photoactive material is also characterized by the σ. Therefore, the relaxation of photogenerated charge separated in space is limited by a shunt resistance defined by the bulk conductivity and/or the conductivity over a barrier, the area of the sample and d_{sep} (Figure 1.19(b)).

Photogenerated charge carriers can be separated in space in different regions of a sample, for example, at buried interfaces. In such a case, the different regions with photogenerated charge separated in space act as capacitors connected in series the potentials of which are added (superposition, Figure 1.19(c)).

Equation (1.45) allows for the analysis of the sensitivity of SPV measurements. For this purpose, the SPV is plotted as a function of d_{sep} for different values of N_Q in Figure 1.20. The value of d_{sep} can vary over a wide range from the order of 0.1 nm to several micrometers and more. For example, light-induced polarization or charge separation within a monolayer of organic molecules causes very short values of d_{sep}. In layers of nanoparticles such as semiconductor QDs, charge can be separated over several nanometers. In SCRs of semiconductors, charge can drift across distances depending on doping (for example, d_{sep} is of the order of 200 nm for a surface SCR of crystalline silicon doped by 10^{16} cm^{-3}).

The value of N_Q can vary over many orders of magnitude from values as low as 10^6 cm^{-2} or less to values of the order of 10^{13}–10^{14} cm^{-2}. The dependence of SPV on d_{sep} is illustrated in Figure 1.20 for values of N_Q ranging from 10^6 to 10^{14} cm^{-2}. For comparison, the density of surface atoms is of the order of 10^{14}–10^{15} cm^{-2}. For example, for a value of N_Q of 10^6 cm^{-2} and d_{sep} equal to 6 nm, the SPV is about 100 nV. Such a low value can be measured with a lock-in amplifier (see Chapter 2) and gives

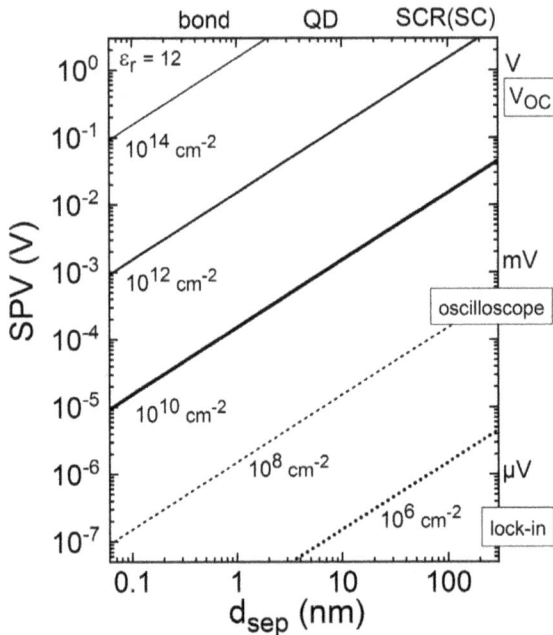

Figure 1.20. Dependence of SPV on d_{sep} for different areal densities of charge carriers separated in space for $\varepsilon_r = 12$. Bond, QD and SCR(SC) denote the range of the length of a chemical bond, the size of a semiconductor QD and the extension of a SCR in a doped semiconductor absorber for solar cells, respectively. V_{OC}, oscilloscope and lock-in are assigned to characteristic values of the open-circuit voltage in solar cells and to the limits of the sensitivity of SPV measurements with an oscilloscope and a lock-in amplifier, respectively.

an impression about the limit for the sensitivity of SPV measurements. The SPV is about $1\,mV$ for N_Q and d_{sep} equal to 10^{10} cm^{-2} and $6\,nm$, respectively. For comparison, $1\,mV$ is of the order of the sensitivity for transient SPV measurements at short times. For solar cells, the SPV shall be of the order of $1\,V$, which can be reached for charge separation across SCRs of semiconductors and values of N_Q of the order or larger than 10^{12} cm^{-2}.

The recombination of charge separated in space is often limited by transport. Limitation by transport can be expressed by resistances in equivalent circuits (see, for example, Figure 1.18(b)). Incidentally, SPV measurements allow for the investigation of transport-related phenomena without the need for extraction of photogenerated charge carriers at contacts as in measurements of the photocurrent.

Furthermore, SPV measurements are only sensitive to regions in a sample where photogenerated charge carriers can access charge separation, i.e. to d and/or the diffusion length of photogenerated charge carriers.

1.5. Mechanisms Leading to SPV Signals

1.5.1. *Charge separation across space charge regions*

Mobile charge carriers are separated in electric fields due to drift. Regions of built-in electric fields, such as pn-junctions, serve as charge-selective contacts in solar cells. Surface SCRs (see, for example, Sze (1981)) and corresponding built-in electric fields arise due to capture of charge carriers at surfaces of semiconductors. The charge captured at a semiconductor surface is compensated by the charge in the SCR. Usually, semiconductor surfaces are depleted due to the capture of majority charge carriers at surface states (for an n-type doped semiconductor see Figure 1.21(a)). At first glance, an SCR is characterized by its extension (d_{SCR}) and by the built-in potential or band bending (φ_0). Under illumination, the built-in potential is reduced to φ_{ill} due to separation of photogenerated charge carriers (Figure 1.21(b)).

The value of d_{SCR} can be approximated by taking into account φ_0, the dielectric constant and the density of ionized donors (N_D^+) or acceptors (N_A^-):

$$d_{\text{SCR}} \approx \sqrt{\frac{2 \cdot \varepsilon_o \cdot \varepsilon_r \cdot \varphi_0}{q \cdot N_{D(A)}^{+(-)}}} \tag{1.46}$$

As an example, d_{SCR} is plotted in Figure 1.22 for different values of φ_0 and N_D^+. Depending on doping and φ_0, d_{SCR} ranges between the nanometer and micrometer range. For example, d_{SCR} is about 200 nm for φ_0 and N_D^+ equal to 0.5 V and 10^{16} cm^{-3}, respectively. Under weak illumination, d_{SCR} can be considered as the charge separation length.

Drift across an SCR is the fastest process of separation of photogenerated charge carriers in the conduction and valence bands of a crystalline semiconductor. The transit time (t_{tr}) can be roughly estimated from the values of d_{SCR} and φ_0:

$$t_{\text{tr}} \approx \frac{d_{\text{SCR}}^2}{\mu_e \cdot \varphi_0} \tag{1.47}$$

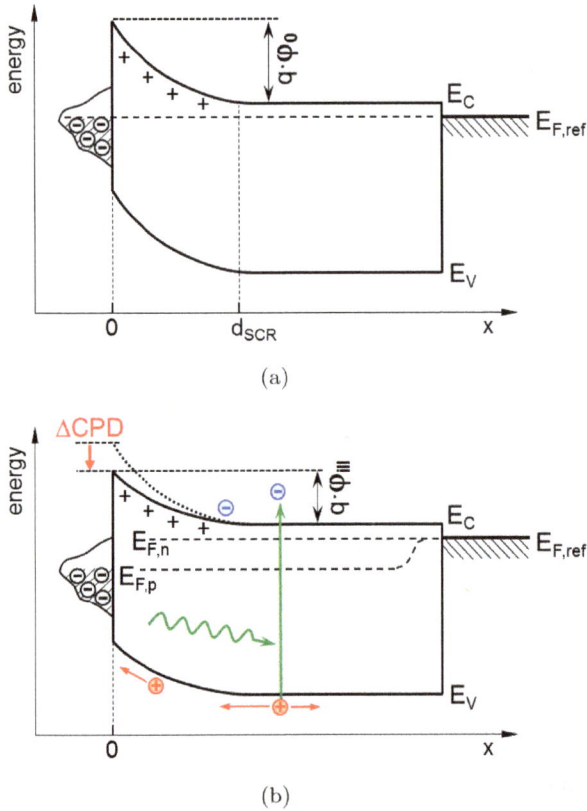

Figure 1.21. Schematic diagram of a surface SCR of an n-type doped semiconductor with surface states in (a) thermal equilibrium and (b) under moderate illumination. The exchange of charge between the SCR and surface states is neglected. $q \cdot \varphi_0$ and $q \cdot \varphi_{\mathrm{ill}}$ denote the surface band bending in thermal equilibrium and under illumination, respectively.

The value of t_{tr} is of the order of ps for a crystalline semiconductor. Capture and emission of charge carriers takes much longer time due to dissipation of energy by phonon cascades. Therefore, the change of the charge in surface states can be neglected at very short times.

At an ideal pn-junction, the photovoltage is given by the Fermi-level splitting, i.e. by the difference between the Fermi energies of electrons ($E_{F,n}$) and holes ($E_{F,p}$), which is equivalent to the difference between φ_{ill} and φ_0 at short times. The same approach can be applied to the SPV caused by charge separation across a surface SCR at a semiconductor

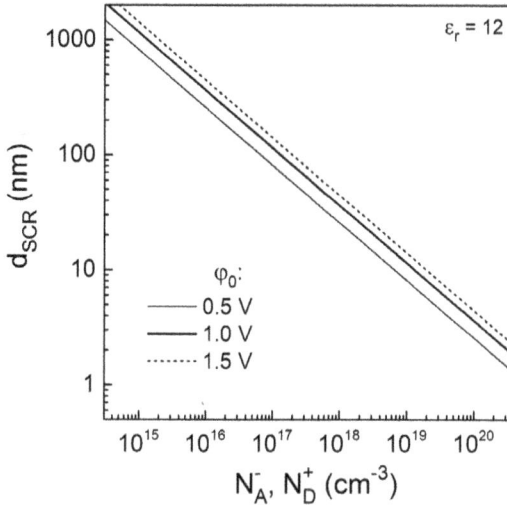

Figure 1.22. Dependence of the extension of a surface SCR on the density of ionized acceptors (N_A^-) or donors (N_D^+) for φ_0 equal to 0.5, 1.0 and 1.5 V (thin solid, thick solid and dashed lines, respectively).

surface (SPV_{SCR}):

$$\text{SPV}_{\text{SCR}} = \frac{E_{F,n} - E_{F,p}}{q} = \varphi_{\text{ill}} - \varphi_0 \tag{1.48}$$

Integration of the Poisson equation by using the Boltzmann equations for the densities of free electrons and holes in a non-degenerate semiconductor results in a solution for the SPV_{SCR}. Following Garrett and Brattain (1955) and Johnson (1958), the implicit equation connecting the density of photogenerated charge carriers (Δn) with the densities of charge carriers in thermal equilibrium (n_0 and p_0), φ_{ill} and φ_0 can be written in the following form (see for more detailed explanations also Kronik and Shapira (1999), and references therein):

$$\Delta n = \frac{p_0 \cdot [e^{-y_0} - e^{-y}] + n_0 \cdot [e^{y_0} - e^{y}] + (p_0 - n_0) \cdot (y_0 - y)}{e^{y} + e^{-y} - 2} \tag{1.49}$$

where y_0 and y are equal to $q \cdot \varphi_0 / (k_B \cdot T)$ and $q \cdot \varphi_{\text{ill}} / (k_B \cdot T)$, respectively.

The analysis of Equation (1.49) gives an impression about the dependence of SPV_{SCR} on Δn, φ_0 and T. As an example, Figure 1.23 shows SPV_{SCR} as a function of Δn or Δp for differently doped c-Si. At lower values of Δn or Δp, SPV_{SCR} increases by 0.06 V per decade (for 300 K), similarly

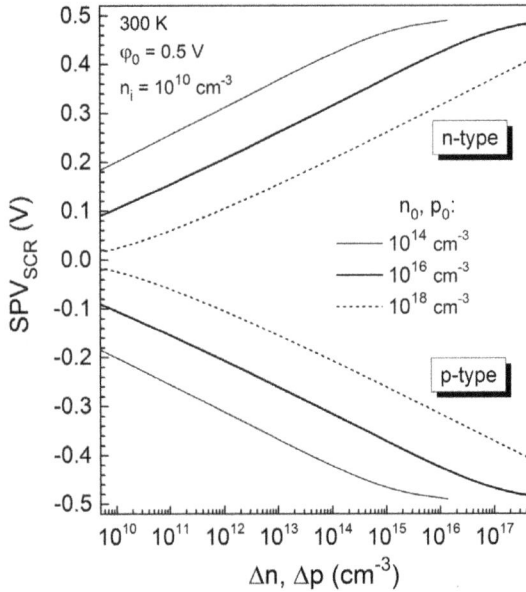

Figure 1.23. Dependence of SPV$_{\mathrm{SCR}}$ on Δn or Δp for n- and p-type doped c-Si ($n_i = 10^{10}\,\mathrm{cm}^{-3}$) with $\varphi_0 = 0.5\,\mathrm{V}$ and $n_0(p_0)$ equal to 10^{14}, 10^{16} and 10^{18} cm^{-3} (thin solid, thick solid and dashed lines, respectively).

as for the forward current of an ideal diode. The value of SPV$_{\mathrm{SCR}}$ tends to saturate at φ_0 when Δn or Δp become larger than the density of dopants by about 10 times. If Δn or Δp and n_0 or p_0 are known, Equation (1.48) can be used for the measurement of φ_0.

Equation (1.49) represents an idealized approximation for measurements at very short times. In reality, the distribution of charge separated in space is influenced by changes of the occupation of defect states at and/or near the surface due to charge transfer. The distribution, the type, the absorption cross-section and the capture cross-sections for electrons and holes of defect states may be considered in simulation tools for semiconductor devices. The simulation of the relaxation of photogenerated charge separated across a surface SCR is challenging for this reason. For relaxation, photogenerated charge carriers separated across a surface SCR have to overcome the barrier. With regard to thermionic emission and tunneling, the relaxation time depends exponentially on the barrier. At first glance, the barrier is reduced by a value of $q \cdot$ SPV$_{\mathrm{SCR}}$. Therefore, the decay time of the relaxation of SPV$_{\mathrm{SCR}}$ depends on the SPV$_{\mathrm{SCR}}$ itself

(Hamers and Cahill, 1991). However, models are still missing for a precise description of the separation of photogenerated charge carriers and their relaxation across a surface SCR. There is even no precise model for SPV at a Si/SiO_2 interface, which is the most studied interface.

1.5.2. *Diffusion and Dember photovoltage*

Inhomogeneous photogeneration of mobile electrons and holes takes place in semiconductor crystals for absorption of light with E_{ph} significantly larger than E_g. Diffusion of mobile charge carriers is driven by gradients of the concentrations of electrons and/or holes. Usually, electrons diffuse faster than holes, i.e. the diffusion constant of electrons (D_e) is larger than that of holes (D_h). This leads to a separation of photogenerated electrons and holes in space (Dember, 1932). The corresponding SPV is called the Dember photovoltage (U_D).

For estimating the order of magnitude of U_D, an expression for U_D will be derived for moderate photogeneration in an n-type doped semiconductor, in which Δn and Δp are much larger than the density of holes in thermal equilibrium (p_0) but much smaller than the density of electrons in thermal equilibrium (n_0) (Frenkel, 1933; Tauc, 1957):

$$p_0 \ll \Delta n = \Delta p \ll n_0 \tag{1.50}$$

The diffusion currents of electrons $(I_{diff,e})$ and holes $(I_{diff,h})$ have opposite signs due to the negative and positive charge of diffusing electrons and holes, respectively:

$$I_{diff,e} + I_{diff,h} = q \cdot D_e \cdot \frac{d(\Delta n)}{dx} - q \cdot D_h \cdot \frac{d(\Delta p)}{dx} \tag{1.51}$$

An electric field appears between diffusing electrons and holes. The electric field causes drift currents of electrons $(I_{drift,e})$ and holes $(I_{drift,h})$. The electric field can be written as the negative gradient of the Dember photovoltage:

$$I_{drift,e} + I_{drift,h} = q \cdot n \cdot \mu_n \cdot \frac{dU_D}{dx} + q \cdot p \cdot \mu_h \cdot \frac{dU_D}{dx} \tag{1.52}$$

In SPV measurements, the total current is equal to 0:

$$0 = I_{diff,e} + I_{diff,h} + I_{drift,e} + I_{drift,h} \tag{1.53}$$

The Einstein equation connects the diffusion coefficient of electrons or holes ($D_{e,h}$) with their mobility ($\mu_{e,h}$):

$$D_{e,h} = \frac{k_B \cdot T}{q} \cdot \mu_{e,h} \tag{1.54}$$

The combination of Equations (1.51)–(1.54) results in the following equation:

$$\frac{dU_d}{dx} = \frac{k_B \cdot T}{q} \cdot \frac{\mu_e - \mu_h}{n \cdot \mu_e + p \cdot \mu_h} \cdot \frac{d(\Delta n)}{dx} \tag{1.55}$$

Equation (1.55) can be simplified by introducing the definition of the conductivity (Equation (1.20)). The value of U_D can be obtained by solving Equation (1.55) under consideration of Equation (1.20), whereas σ_0 and σ_{ill} denote the conductivity in thermal equilibrium and under illumination, respectively:

$$U_D = \frac{k_B \cdot T}{q} \cdot \frac{\mu_e - \mu_h}{\mu_e + \mu_h} \cdot \int_{\sigma_0}^{\sigma_{ill}} \frac{d\sigma}{\sigma} \tag{1.56}$$

Integration of Equation (1.56) and consideration of Equation (1.20) and condition (1.50) give the expression for U_D:

$$U_D = \frac{k_B \cdot T}{q} \cdot \frac{\frac{\mu_e}{\mu_h} - 1}{\frac{\mu_e}{\mu_h} + 1} \cdot \ln\left[1 + \frac{\Delta n}{n_0} \cdot \left(1 + \frac{1}{\frac{\mu_e}{\mu_h}}\right)\right] \tag{1.57}$$

The value of U_D is a linear function of temperature and depends on the mobility ratio (μ_e/μ_h) and on the ratio between the density of photogenerated charge carriers and the density of majority charge carriers in thermal equilibrium ($\Delta n/n_0$). As an example, Figure 1.24 shows the dependence of U_D on Δn for μ_e/μ_h equal to 3.5 and different values of n_0.

Equation (1.57) shall be considered for the analysis of large signal SPV measurements (see, for example, Heilig *et al.* (1979)). Furthermore, Equation (1.57) can be used for the independent calibration of Δn.

Typical values of μ_e/μ_h are 3.7 for c-Si (Jacoboni *et al.*, 1977) or 27 for GaAs (Blakemore, 1982). For moderately doped silicon ($n_0 = 10^{14}\,\mathrm{cm}^{-3}$), U_D can be as a large as 70 mV ($\Delta n \approx 10^{16}\,\mathrm{cm}^{-3}$). Values of Δn can be, for example, of the order of 10^{15} or $10^{13}\,\mathrm{cm}^{-3}$ for c-Si or GaAs solar cells under constant illumination at AM1.5 (air mass 1.5 corresponds to a power of sunlight of $100\,\mathrm{mW/cm^2}$), respectively (see, for example, Dittrich (2018)). Therefore, U_D can be neglected for solar cells under operation at AM1.5. In contrast, U_D can become large in insulators and can therefore become important in photocatalytic systems based on undoped semiconductors with relatively large band gaps.

The Dember photovoltage (Equation (1.57)) does not contain information explicitly about the charge separation length of diffusing electrons and holes. With increasing time, Δn and Δp decrease due to recombination which causes a reduction of U_D. However, a long-lasting increase of SPV signals in time was observed in transient SPV measurements for materials with very low conductivity (Timoshenko *et al.*, 2000). This phenomenon was ascribed to the so-called independent diffusion which causes an increase of the charge separation length in time and therefore of SPV signals in transient measurements in photoactive materials with long dielectric relaxation times.

1.5.3. Charge separation by asymmetric charge transfer and injection across surfaces and interfaces

At surfaces and interfaces of photoactive materials, chemical bonds are distorted and modified in comparison to the bulk over up to several monolayers. This causes the formation of electronic states at and near surfaces and interfaces different to those in the bulk. Surface states are

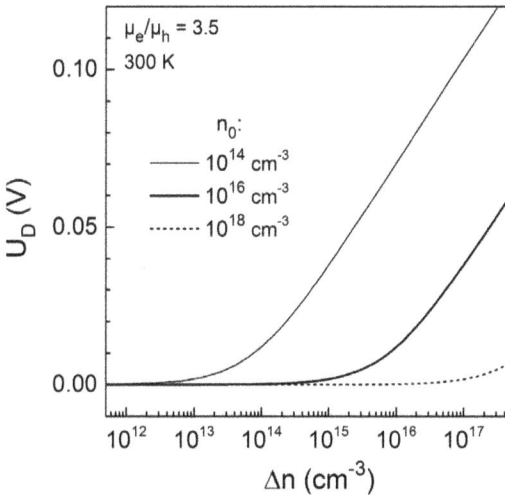

Figure 1.24. Dependence of the Dember photovoltage on Δn for an n-type semiconductor with μ_e/μ_h equal to 3.5 and n_0 equal to 10^{14}, 10^{16} and 10^{18} cm^{-3} (thin solid, thick solid and dashed lines, respectively).

usually localized. The density of electronic states in distorted surface regions of semiconductors or photocatalysts is usually rather high and disorder plays an important role. Surface states can be delocalized in the sense of, for example, extended states in a nanoparticle deposited on the surface of a photoactive material. Surface states can also be introduced by the dedicated deposition of chemical species, such as sensitizing organic molecules or clusters of catalysts, at surfaces of photoactive materials.

Photogenerated charge carriers can be directly transferred from the bulk of a photoactive material into surface states. A transfer of a photogenerated electron can consist, for example, by a step of resonant tunneling from the conduction band into a localized surface state followed by local relaxation and/or trapping at a surface state with lower energy (Figure 1.25(a)). On the contrary, an electron can be excited in a surface state or a dye molecule and injected from the excited state into the conduction band of a semiconductor by resonant tunneling (Figure 1.25(b)). Incidentally, injection means the transfer of charge carriers into delocalized states. Injection can be as fast as tens on fs, for example, for injection of electrons from dye molecules into semiconductor electrodes (Eichberger and Willig, 1990). Injection is the main mechanism for photogeneration of mobile charge carriers and local charge separation in dye-sensitized solar cells (O'Regan and Grätzel, 1991).

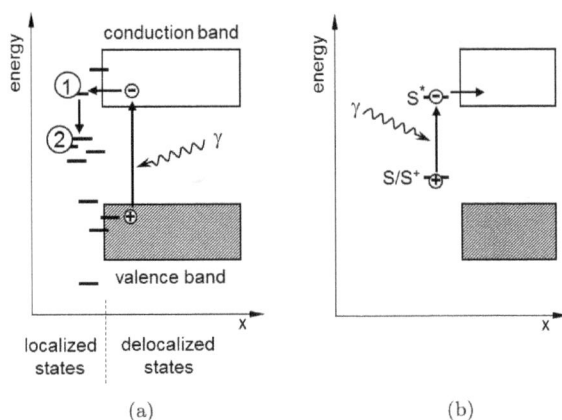

Figure 1.25. (a) Transfer of a photogenerated electron from the conduction band in a semiconductor into a localized state in the distorted surface region consisting of a tunneling (1) and a relaxation or trapping (2) steps and (b) injection of a photoexcited electron from a molecular state into the conduction band of a semiconductor.

Depending mainly on the occupation of surface states and the distance between localized surface states and delocalized states in the bulk, the charge transfer or injection rates can be very different. In the past, surface states were distinguished by fast and slow states in relation to their interaction with delocalized states in a semiconductor (Many and Goldstein, 1965).

Asymmetric transfer of photogenerated charge carriers from a bulk into surface states means preferential trapping of positive or negative charge due to different transfer rates. This leads to separation of photogenerated charge carriers in space and therefore to SPV signals. Asymmetric charge transfer is crucial, for example, for photocatalysis. SPV techniques are very useful for the investigation of surface states due to the high sensitivity and selectivity in the direction of charge transfer. Under illumination, the densities of electrons and holes are increased near the surface of a photoactive material, i.e. photogenerated electrons and holes can be both transferred into distorted surface regions. In following steps, photogenerated charge carriers transferred into distorted surface regions can undergo differently further trapping, de-trapping and hopping transport. This opens additional opportunities for the investigation of surface states by SPV techniques, such as transient SPV spectroscopy (see also Chapter 5).

The dielectric constants can be rather different in the bulk and in distorted or sensitized surface regions. The reduction of ε_r during charge transfer across an interface leads to an increase of the SPV with regard to Equation (1.45). It can be useful to define an effective charge separation length (d_{eff}).

$$d_{\text{eff}} = \frac{d}{\varepsilon_r} \tag{1.58}$$

A detailed description of the interaction between surface and bulk states and the precise assignment of surface states to chemical bond configurations is challenging since variations in coordination, tunneling, dielectric environment and local barriers shall be taken into account. Furthermore, the implementation of localized surface states and delocalized bulk states into common simulation tools is not trivial due to different simulation concepts.

1.5.4. *Charge separation due to dissociation of excitons*

An exciton is a quasiparticle describing a bound state between a mobile electron and a mobile hole. Excitons play an important role, for example, in organic semiconductors with conjugated molecules (see, for example, Barford (2013)) or in QDs (see, for example, Murray *et al.* (1993)). Excitons are electrically neutral. Therefore, excitons cannot be directly detected by SPV. Excitons can be investigated by SPV only after their dissociation followed by separation of the positive and negative charge carriers in space.

In organic semiconductors, excitons dissociate at donor–acceptor heterojunctions (Figure 1.26(a)). The energy of the LUMO is higher in the conjugated donor molecules than in the acceptor molecules. The wave functions of an exciton in the donor and of the electrons in the acceptor molecules partially overlap and a CTS (see, for example, Vandewal (2016)) can be formed. The interaction between an exciton and a CTS can lead to a weakening of the exciton binding energy so that excitons can dissociate and electrons can be transferred into acceptor molecules (Figure 1.26(b)).

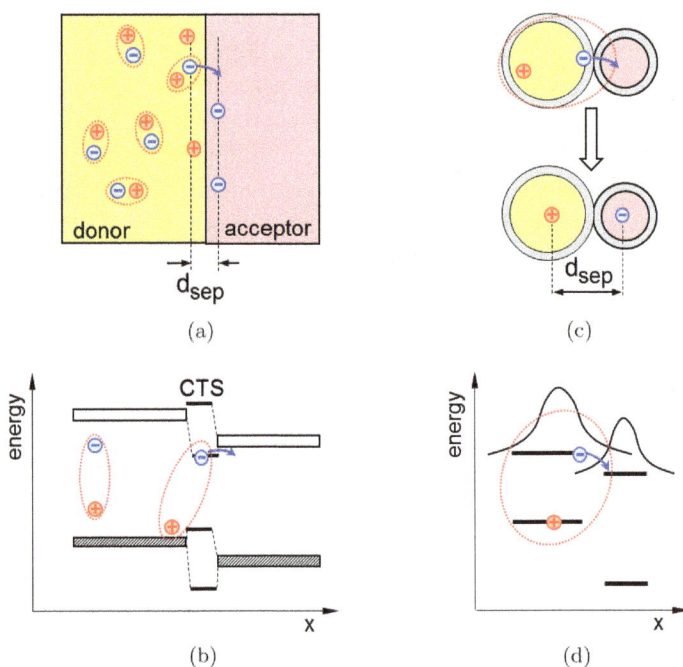

Figure 1.26. (a) Schematic diagram of the dissociation of excitons at an interface between two layers of conjugated donor and acceptor molecules (b) via the formation of CTS at the donor–acceptor heterojunction and (c) of the dissociation of an exciton in a semiconductor QD (d) via tunneling of the electron through the shells of surfactant molecules at a type II heterojunction formed between neighbored QDs.

This process can be very efficient and was successfully applied in organic solar cells (concept of the bulk heterojunction, Yu *et al.* (1995)).

Colloidal QDs consist of an inorganic semiconductor core and a shell of organic surfactant molecules for stabilizing the semiconductor core (see, for example, Murray *et al.* (1993)). In a QD, the extension of the wave function of an exciton is limited by the size of the nanoparticle. The wave function of an exciton in a QD can "leak out" depending on the thickness of the shell of organic surfactant molecules (see, for example, Guyot-Sionnest *et al.* (2003)). This makes excitons in QDs extremely sensitive to surface treatments. At interfaces between QDs, the wave function of an exciton can partially overlap with the wave functions of electronic states in a neighbored semiconductor QD (Figure 1.26(c)) so that the exciton can dissociate by tunneling of the electron into a neighbored QD (type II QD heterojunction, Figure 1.26(d)).

The dissociation of excitons is also possible at interfaces between layers of conjugated molecules or QDs and surfaces with appropriate donor or acceptor states, for example, at metal oxides or other semiconductors. Furthermore, excitons also dissociate in strong electric fields.

1.5.5. *Light-induced change of surface dipoles by polarization, adsorption and desorption of molecules*

A light-induced change of the CPD between the sample and reference electrodes is also possible without photogeneration of free charge carriers or dissociation of excitons. For example, the polarization of organic molecules can change under illumination. Organic molecules, for example, can contain rather different moieties enabling polarization and intramolecular transfer of excited charge carriers.

Layers and monolayers of organic molecules with a high orientation can be deposited on different substrates by making use of specific anchoring groups and self-assembling (Ulman, 1996). The oriented deposition of organic molecules gives the opportunity to study photoinduced polarization or intramolecular charge transfer (Figure 1.27) by SPV techniques.

The shape or configuration of relatively small organic molecules is rather rigid. Under illumination, electrons can be excited and partially attracted to one of the moieties of a molecule. This causes a change of the surface dipole (Figure 1.27(a)) and therefore of ΔCPD. The resulting SPV signal can be detected by SPV techniques. However, one has to keep in mind that a corresponding charge separation length is not related to

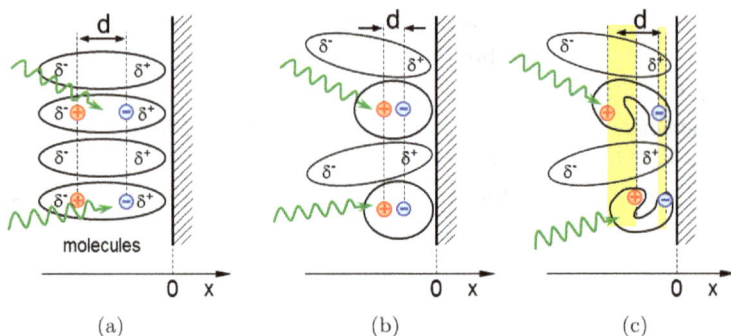

Figure 1.27. (a) Schematic diagram of a layer of highly oriented rigid organic molecules which can be polarized under illumination due to intramolecular charge transfer and (b) of molecules which can change the shape under illumination by shrinking or (c) folding. Positive and negative polarization are denoted by δ^+ and δ^-, respectively.

the separation of photogenerated free charge carriers but to a shift in polarization.

More complex organic molecules such as proteins may undergo deformation (Figure 1.27(b)) or folding (Figure 1.27(c)) under the action of electrostatic charge or illumination. This can additionally change the dipoles of molecules under illumination.

At surfaces, photons can activate chemical reactions and processes leading to adsorption (Figure 1.28(a)) or desorption (Figure 1.28(b)) of molecules. The change of chemical bonds at a surface and/or the change of the coverage of a surface cause strong changes in surface dipoles. In such a case, SPV signals are not related to the separation of mobile charge carriers in space but to a non-reversible variation of the CPD.

An activation of chemical reactions, adsorption and desorption of molecules can happen at practically any surface. This has consequences for the organization of SPV experiments. The control of the ambient atmosphere can especially play an important role. Depending on a given surface and experiment, it can be reasonable to perform SPV measurements in air (for example, metal oxides), inert gas atmosphere (for example, conjugated polymers), high vacuum (measurements at low temperature) or ultra-high vacuum (reconstructed surfaces). Furthermore, chemical bonds in organic molecules can be cracked under UV light. Therefore, it is useful to reduce the exposition to UV light of sensitive samples and surfaces and to repeat SPV measurements in order to distinguish non-reversible and reversible ΔCPD. Incidentally, the sensitivity of the ΔCPD to adsorption

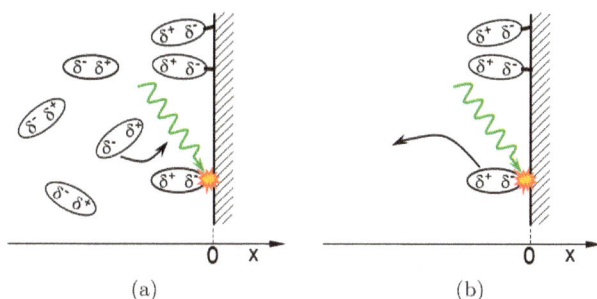

Figure 1.28. (a) Schematic diagram of the change of a surface dipole due to the activation of adsorption and (b) desorption of molecules by photons.

and desorption of molecules may be applied for gas sensing and non-reversible SPV signals can be used for dedicated studies of photochemical reactions at surfaces.

1.6. Summary

Separation of photogenerated charge carriers in space causes SPV signals. Photogeneration means the excitation of free charge carriers in a photoactive material by light absorption. SPV signals depend on the fundamental processes of light absorption, recombination and charge transport (Figure 1.29).

In solar cells and photocatalysts, photogenerated electrons and holes are usually excited by fundamental band-to-band absorption in semiconductor absorbers. Photogeneration of free charge carriers is also possible by excitation of electrons or holes from defect states. Electrons and holes can form bond states resulting in excitonic absorption.

Tunneling-assisted absorption takes place for transitions between occupied and unoccupied electronic states the wave functions of which overlap in space, for example, (i) between donor and acceptor molecules, (ii) between valence and conduction bands for photon energies below the band gap of semiconductors in high electric fields (Franz–Keldysh effect), (iii) at interfaces between metals and semiconductors or between different semiconductors (internal photoemission) or (iv) between surface states and the valence or conduction bands of a semiconductor. Tunneling-assisted absorption across interfaces leads directly to separation of charge carriers in space.

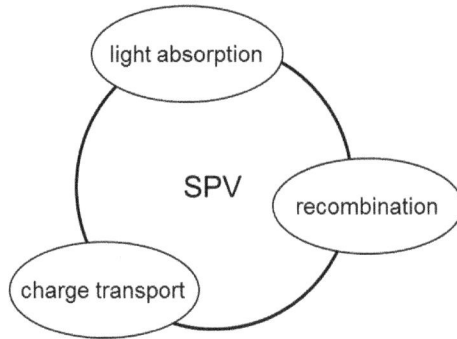

Figure 1.29. Fundamental processes on which SPV signals depend.

The total density of photogenerated charge carriers is limited by recombination processes. The basic recombination mechanisms are radiative, Auger, SRH and surface recombination.

Transport of positive and negative charge carriers to different regions in space causes disturbances of space charge. Electrical conductivity causes the equilibration or relaxation of disturbances in a space charge. The relaxation of a disturbance in a space charge can be dominated by the conductivity in the bulk. The bulk conductivity depends on the density of mobile charge carriers and their mobility. The dielectric relaxation time limits the relaxation of the disturbance of a space charge in a bulk material. Depending on the material and the transport mechanism, the dielectric relaxation time can vary over a huge range.

The relaxation of a disturbance of a space charge can also be limited by transport over barriers. Corresponding relaxation times or RC time constants can vary over many orders of magnitude depending on the barrier height, temperature (thermionic emission) or the width of the barrier (tunneling).

The consideration of the relaxation time of the disturbance of a space charge gives an impression about the orders of magnitude of sample parameters (conductivity, barrier height, width of a barrier layer) which can be investigated by SPV measurements.

SPV signals are measured as the light-induced change of the CPD between a sample and a reference electrode (ΔCPD). The center of charge approach and the superposition principle are very useful for the simulation and analysis of SPV signals. SPV signals can be described in terms of a parallel plate capacitor whereas the thickness of the capacitor corresponds

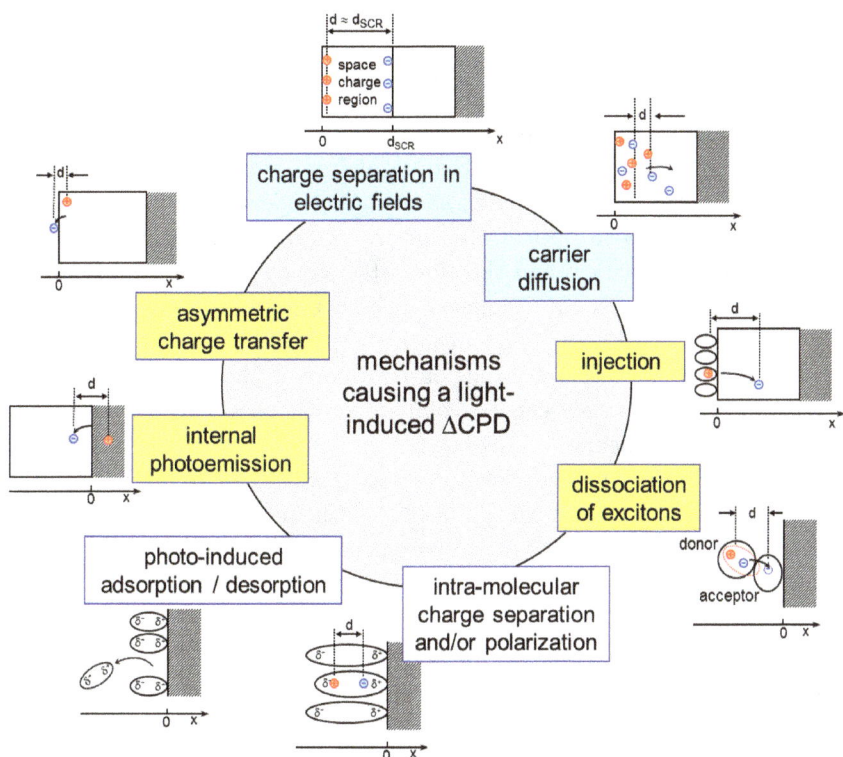

Figure 1.30. Overview of mechanisms causing a light-induced ΔCPD, i.e. SPV signals.

to the charge separation length, i.e. the difference between the centers of positive and negative charges. SPV signals can be measured for areal densities of charge carriers as low as 10^6 cm^{-2} and charge separation lengths as low as the length of chemical bonds.

SPV signals can be caused by different mechanisms of charge separation (Figure 1.30). Separation of photogenerated charge carriers takes place in electric fields (separation in SCRs) and in regions with gradients of the density of photogenerated charge carriers (diffusion and Dember effect). Internal photoemission and injection across interfaces, asymmetric transfer of photogenerated charge carriers across surfaces and dissociation of photogenerated excitons in combination with asymmetric charge transfer lead also to SPV signals.

Furthermore, light-induced changes of the polarization of molecules and intramolecular charge separation also result in a ΔCPD. In addition, the possibility of non-reversible changes in the CPD due to light-induced surface chemical reactions and adsorption and desorption of molecules shall be considered in SPV measurements.

SPV signals are directly related to photogeneration, i.e. to optical absorption resulting in photogeneration of free charge carriers and their separation in space. In this sense, spectral-dependent SPV measurements can be compared with optical absorption spectra. The density of photo-generated charge carriers and therefore SPV signals are limited by recom-bination processes, i.e. SPV signals can be compared with recombination measurements which are complementary to SPV measurements. Charge separation and relaxation depends on transport of photogenerated charge carriers, i.e. SPV signals can be compared with photocurrent and other transport measurements.

SPV measurements are locally resolved with regard to the charge separation length and/or the diffusion length of photogenerated charge carriers. The resolution of SPV measurements in space can be very high in relation to surfaces and interfaces across which charge separation is caused, for example, by injection or internal photoemission.

Due to the complexity of phenomena and mechanisms involved in the formation and relaxation of SPV signals, the interpretation of SPV signals can be rather puzzling. Therefore, dedicated experiments are needed for the investigation of certain properties of photoactive materials and/or mechanisms of, for example, optical absorption, photogeneration, recombination and charge transport.

Chapter 2

Measurement Techniques of Surface Photovoltage Signals

2.1. Introduction

The principles of surface photovoltage (SPV) measurements are based on (i) nulling of the light-induced change of the electrostatic potential between sample and reference electrodes with an external voltage (Kelvin probe techniques), (ii) out-coupling of the light-induced potential drop across a parallel plate capacitor with high-impedance buffers (fixed capacitor techniques), (iii) keeping constant the light-induced change of the current of low energetic electrons with an external voltage in ultra-high vacuum (UHV) (electron beam technique) and (iv) measuring the light-induced shift of photoelectron emission spectra in UHV (photoelectron emission spectroscopy). In a Kelvin probe, the ac current between a vibrating mesh and the sample surface or the ac electrostatic force between a vibrating tip and the sample surface are nulled. The highest resolution in space is reached for SPV measurements with a Kelvin probe force microscope (KPFM). Kelvin probes are sensitive to any light-induced change of the contact potential difference. Modulated and transient SPV measurements in the fixed capacitor arrangement provide the highest sensitivity and the widest range in time starting from nanoseconds, respectively. The parameters of operational amplifiers in high-impedance buffers, the adjustment of the measurement configuration and dedicated measures for the reduction of noise are important for transient SPV measurements over many orders of magnitude in time. Kelvin probes and fixed capacitors can be applied for SPV measurements in vacuum, gas atmospheres and electrolytes. In photoelectron emission spectroscopy, the incorporation of very short light pulses from synchrotrons into pump–probe experiments enables the correlation of SPV signals with changes in surface bond configurations at times up to the picosecond range.

2.2. Kelvin Probe Measurements

2.2.1. *Measurement of contact potential differences with Kelvin probes*

A configuration for SPV measurements includes a sample electrode and a reference electrode. The sample electrode is contacted or coated with the photoactive material. The work functions of the sample and reference electrodes are denoted by W_s and W_{ref}, respectively. When both electrodes are electrically connected, electrons flow from the electrode with the lower work function to the electrode with the higher work function until the same Fermi energy is reached at both electrodes. As a consequence, a charged capacitor is formed between both electrodes (Figure 2.1(a)).

The difference between W_{ref} and W_s is defined as the CPD (Equation (1.33)). The electrostatic potential (U_{es}) between the reference and sample electrodes corresponds to the difference between the vacuum energies of the reference and sample electrodes, $E_{vac,ref}$ and $E_{vac,s}$, respectively:

$$q \cdot U_{es} = E_{vac,ref} - E_{vac,s} \tag{2.1}$$

The electrostatic potential is equal to the CPD if both electrodes are shunted. The U_{es} can be changed by applying a bias potential (V_b) between both electrodes (Figure 2.1(b)):

$$U_{es} = CPD + V_b \tag{2.2}$$

The principle of a Kelvin probe is to adjust the U_{es} to 0 by varying V_b. The electrostatic potential between two electrodes is 0 if the vacuum energies are well aligned:

$$U_{es} = 0: \quad CPD = -V_b \tag{2.3}$$

The adjustment of U_{es} to 0 is probed with an indicating parameter (Figure 2.2).

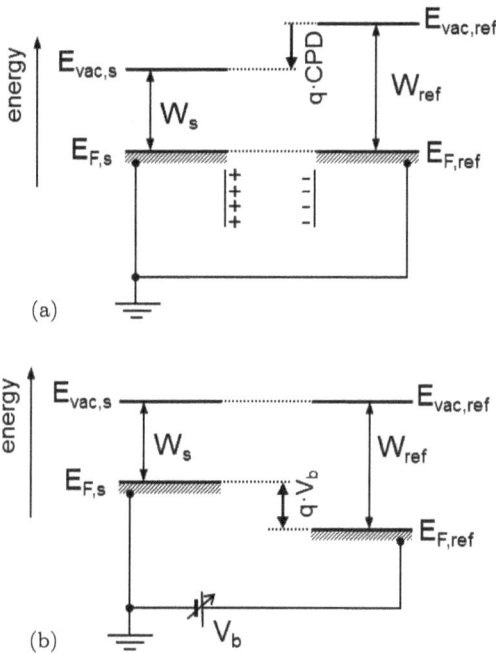

Figure 2.1. Simplified schematic diagram of the work functions of (a) a shunted reference (W_{ref}) and sample (W_s) electrodes and (b) of both electrodes connected with an external voltage source. The photoactive material on the sample electrode is omitted for clarity. In a Kelvin probe, the electrostatic potential between the sample and reference electrodes is adjusted to 0 (or nulled) with a bias potential.

A Kelvin probe allows for the fundamental investigation of the CPD since it connects a bias potential directly with the value of the CPD. The so-called null method was applied for the first time by Lord Kelvin for the measurement of the CPD between different metals at different temperatures and in different ambient including vacuum (Kelvin, 1898).

The adjustment of U_{es} to 0 can be detected by making use of the charge at the capacitor. The capacitance between both electrodes (C) depends on their distance (z). Therefore, the charge at the capacitor is a function of z:

$$Q(z) = (\text{CPD} + V_b) \cdot C(z) \qquad (2.4)$$

The charge can be adjusted to 0 by varying V_b. In this case, the CPD is equal to the $-V_b$:

$$Q = 0: \quad \text{CPD} = -V_b \qquad (2.5)$$

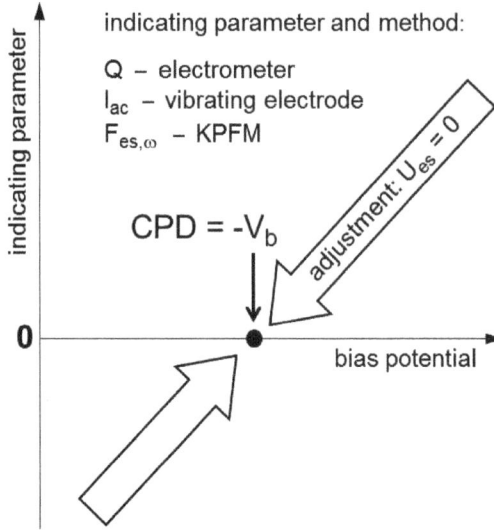

Figure 2.2. Principle of CPD measurements with the null method, i.e. adjustment of the electrostatic potential to 0 with a bias potential, by using charge at a capacitor (Q), an alternating current through a vibrating macroscopic electrode (I_{ac}) or an electrostatic force between the sample and a tip $(F_{es,\omega})$ as indicating parameters.

Lord Kelvin adjusted Q, as the indicating parameter for the null method, to 0 by using a highly sensitive quadrant electrometer under variation of the distance between the electrodes by hand (Kelvin, 1898), as shown in Figure 2.3(a). In the experiments of Lord Kelvin, the diameter of the electrodes was of the order of 10 cm and the distance between both electrodes was varied over several mm to several cm.

The distance between two electrodes can be changed periodically with a vibrating electrode (Figure 2.3(b)). As a consequence, if U_{es} is not 0, an alternating current (I_{ac}) flows due to the change of C (Zisman, 1932):

$$I_{ac} = (CPD + V_b) \cdot \frac{dC(t)}{dt} \tag{2.6}$$

In a Kelvin probe based on a vibrating electrode, I_{ac} is the indicating parameter for adjusting V_b to 0:

$$I_{ac} = 0 : \quad CPD = -V_b \tag{2.7}$$

The modulated electrostatic force (F_{es}) between the sample surface and a tiny reference electrode (tip) is used as the indicating parameter in

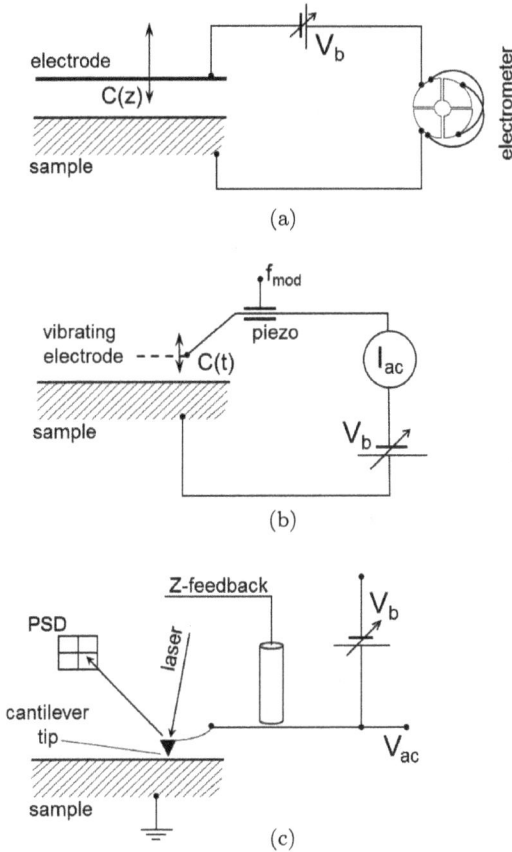

Figure 2.3. Schematic diagrams of the principle null methods of Kelvin probes based on the indicating parameters of charge, alternating current and electrostatic force ((a)–(c), respectively). PSD denoted the position-sensitive detector.

a Kelvin probe force microscope (KPFM), (Nonnenmacher *et al.*, 1991). F_{es} increases with decreasing distance. The tiny reference electrode is fixed at a vibrating cantilever. The piezoelectric effect is applied for the precise adjustment of very short distances between the sample and the reference electrodes in KPFM.

In KPFM, a bias potential and a modulated potential (V_{ac}) are applied to the sample or to the tiny reference electrode, which vibrates in front of the sample surface at a very short distance (Figure 2.3(c)). The deflection of the tip is measured with a position sensitive detector (PSD) by using the reflected spot of a laser.

The potential at the tip is given by (ω_{ac} is the angular modulation frequency):

$$V_{tip} = V_b + V_{ac} \cdot \sin(\omega_{ac}t) \tag{2.8}$$

The electrostatic force at the tip is given by (For details, see also, for example, Melitz *et al.*, 2011):

$$F_{es} = -\frac{1}{2} \cdot \frac{\partial C(z)}{\partial z} \cdot (\text{CPD} + V_{tip})^2 \tag{2.9}$$

Equation (2.9) can be separated into the DC electrostatic force and the electrostatic forces modulated at ω_{ac} and $2\omega_{ac}$. The part of the electrostatic force modulated at ω_{ac} can be written as

$$F_{es,ac} = -\frac{\partial C(z)}{\partial z} \cdot (\text{CPD} + V_b) \cdot V_{ac} \cdot \sin(\omega_{ac}t) \tag{2.10}$$

The signal related to $F_{es,ac}$ is measured with the PSD and a lock-in amplifier and adjusted to 0 by varying V_b.

$$F_{es,ac} = 0 : \quad \text{CPD} = -V_b \tag{2.11}$$

The KPFM allows for the investigation of the CPD with high resolution in space of several nanometers since the reference electrode is fabricated as a tiny tip and distances can be precisely adjusted.

Photogenerated charge carriers separated in space cause a dipole layer. A dipole layer on a sample electrode causes a change of the work function of the sample electrode. Therefore, the bias potential, at which U_{es} is 0, changes from V_b in the dark to V_b' under illumination due to the appearance of the light-induced change of the CPD (ΔCPD, Figure 2.4).

$$U_{es} = 0 : \quad \text{CPD} + \Delta\text{CPD} = -V_b' \tag{2.12}$$

The value of the SPV is given by the difference between V_b' and V_b:

$$\text{SPV} = V_b' - V_b = -\Delta\text{CPD} \tag{2.13}$$

2.2.2. *SPV measurements with a macroscopic Kelvin probe*

For SPV measurements with a Kelvin probe, the vibrating electrode can be fabricated from a fine gold mesh through which the sample is illuminated (Besocke, delta phi). The diameter of the vibrating electrode can be of the order of several millimeters. The distance between the vibrating electrode and the surface of the sample is of the order of 0.5–1 mm. The vibrating

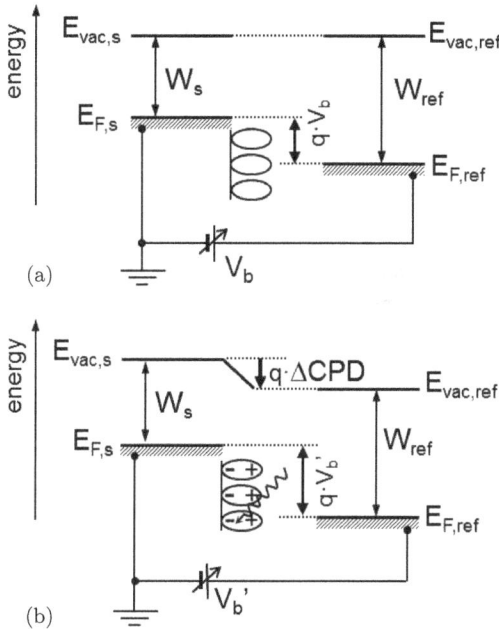

Figure 2.4. Simplified energy schemes of a reference electrode and a sample electrode with a (a) photoactive layer in the dark and (b) under illumination. Under illumination, the CPD changes by ΔCPD. A bias potential is applied between both electrodes in order to adjust the electrostatic potential to 0.

electrode can be driven, for example, with a piezo crystal at a modulation frequency equal to the resonance frequency of a reed connected with the reference electrode (modulation frequency between 100 and 200 Hz; Besocke and Berger, 1976). The sensitivity and the response time of CPD measurements with a Kelvin probe based on a vibrating electrode are of the order of 0.1 mV and several seconds, respectively. The sensitivity of CPD measurements with a Kelvin probe based on a vibrating electrode depends on the quality of the electrical ground, on the mechanical quality of the vibrating electrode, on the lock-in amplifier for the measurement of I_{ac} and on the feedback unit for adjusting I_{ac} to 0.

SPV measurements with a Kelvin probe are sensitive to any fast or slow processes that lead to the separation of photogenerated charge carriers in space. In addition, one has to take into account the fact that measurements with a Kelvin probe are also sensitive to any other change of the CPD, caused, for example, by the adsorption or desorption of molecules

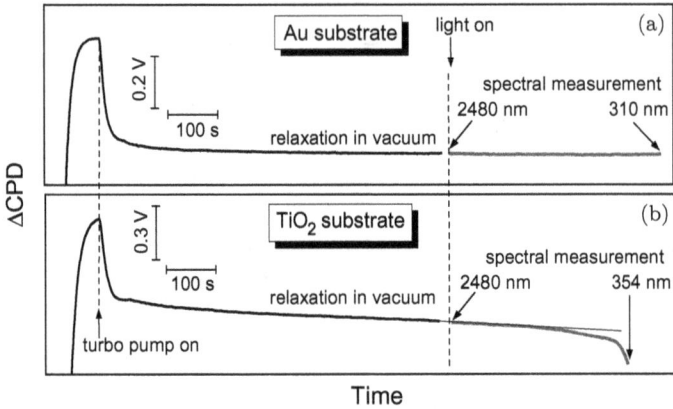

Figure 2.5. Examples for the evolution of ΔCPD during relaxation in vacuum and under illumination with decreasing wavelength for (a) gold and (b) TiO$_2$ electrodes. The wavelength of the exciting light was continuously changed from near-infrared to UV during illumination.

from a surface or by chemical reaction at a surface, including also the reference electrode. Therefore, SPV measurements with a Kelvin probe can sensitively depend on the history of illumination and/or adsorption and desorption of molecules. This shall be considered for the regime of spectral-dependent SPV measurements.

As an example, Figure 2.5 shows the time dependence of ΔCPD of a gold electrode and of a TiO$_2$ layer during evacuation to high vacuum and under monochromatic illumination at wavelengths decreasing in time from near-infrared to ultraviolet (UV). After switching on the Kelvin probe in the rough vacuum, ΔCPD increased and tended to saturate for the gold electrode within tens of seconds, but not for the TiO$_2$ layer. After switching on the turbo-molecular pump, the ΔCPD signals decreased similarly within the first tens of seconds for both samples. At longer time, ΔCPD tended to saturate for the gold electrode but not for the TiO$_2$ layer.

The ΔCPD of the gold electrode remained unchanged under illumination. For the TiO$_2$ layer, ΔCPD followed the relaxation in vacuum under illumination with near-infrared light. Under illumination with visible light, ΔCPD of the TiO$_2$ layer started to become more negative faster than under relaxation in vacuum, i.e. a component appeared in the ΔCPD signal which was related to separation of photogenerated electrons toward the sample electrode. Under illumination with UV light, ΔCPD of the TiO$_2$ layer started to become more negative faster than under illumination with visible light, i.e. an additional process (fundamental absorption in TiO_2) appeared which also caused the separation of photogenerated electrons toward the sample electrode.

The example shown in Figure 2.5 demonstrates the strong dependence of the relaxation of ΔCPD signals on the nature of the sample. As a consequence, a baseline, as known from other spectroscopic methods, is often not well defined for spectral-dependent SPV measurements with a Kelvin probe. It is recommended to wait with the measurement of SPV spectra until ΔCPD is relaxed or, if this takes too long, to repeat several spectral-dependent measurements under identical conditions, including an identical time interval between the measurements of two spectra.

Similar to the relaxation of ΔCPD signals, the response time of SPV measurements with a Kelvin probe can be rather slow. Figure 2.6 shows examples for the increase and decrease of SPV signals after switching on and off the light, respectively. For a $Cu(In,Cu)S_2/In_2S_3$ layer system sandwiched between two ohmic contacts (one contact was floating), the increase and decrease of the SPV signals were within the resolution time of the system. For CdS deposited onto a SnO_2:F layer, in contrast, the increase

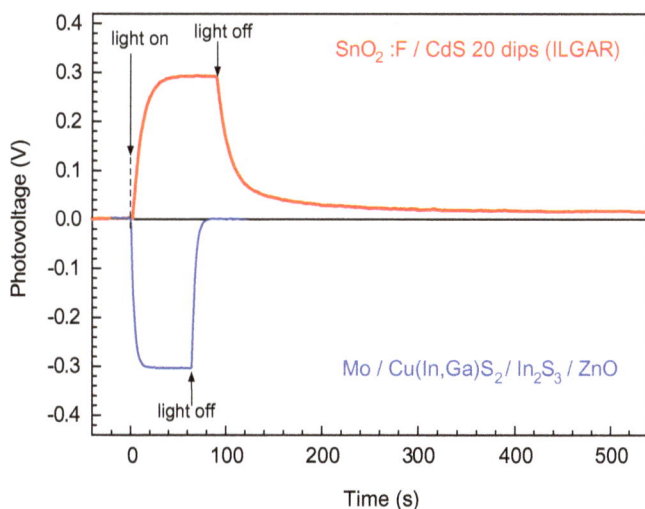

Figure 2.6. Example for the evolution and relaxation of SPV signals measured with a Kelvin probe based on a vibrating gold mesh for a CdS layer deposited onto SnO_2:F (red line) and of a $Mo/Cu(In,Ga)S_2/In_2S_3/ZnO$:Al structure (blue line). Excitation was performed with green light (530 nm). The area of the $Cu(In,Ga)S_2/In_2S_3/ZnO$:Al was reduced to the area of the vibrating electrode. The ZnO:Al contact was floating.

and decrease of SPV signals were rather slow. Incidentally, a slow increase and decrease of signals form a usual case in the SPV measurements with Kelvin probes.

A slow increase or decrease of SPV signals shall be considered in the measurement of SPV spectra with Kelvin probes. This will be illustrated by using the following simple model in which the SPV signal is proportional to the evolution of the density of photogenerated charge carriers (Δn) in time, i.e. the charge separation length is constant. The relaxation time (τ) can be considered as a dielectric relaxation time or lifetime. A generation spectrum normalized to $G^{\max}$($G(h\nu)$, $h\nu$ denotes the photon energy) with a characteristic increase at a band gap of 1.6 eV and an exponential decrease at higher photon energies, which is characteristic for blackbody radiation, is considered:

$$\frac{d(\Delta n)}{dt} = G - \frac{\Delta n}{\tau} \tag{2.14}$$

In spectral-dependent measurements, the wavelength (or photon energy) is varied stepwise in time (Figure 2.7(a)). The index i is used for the generation at the $h\nu_i$:

$$G_i = G_i[h\nu_i] \tag{2.15}$$

Furthermore, several values of the SPV signal are averaged at a given G_i (denoted by the index k). Therefore, $\Delta n_{i,k}$ corresponds to one measurement event. The time interval between two measurement events is given by Δt, which is of the order of several seconds in SPV measurement with a Kelvin probe. The discretization of Equation (2.14) results in the following expression:

$$\Delta n_{i,k} = \Delta n_{i,k-1} + G_i \cdot \Delta t - \Delta n_{i,k-1} \cdot \frac{\Delta t}{\tau} \tag{2.16}$$

For a number of m averages, the value of Δn_i is given as

$$\Delta n_i = \frac{1}{m} \cdot \sum_{k=1}^{m} \Delta n_{i,k} \tag{2.17}$$

The time interval for the change of $h\nu(\Delta t_i)$ can be significantly shorter than the relaxation time. Figure 2.7(b) shows the time-dependent values of $\Delta n/G^{\max}$ following from the simple model with $\tau/\Delta t_i = 10$ for scans with increasing and decreasing $h\nu$ in time.

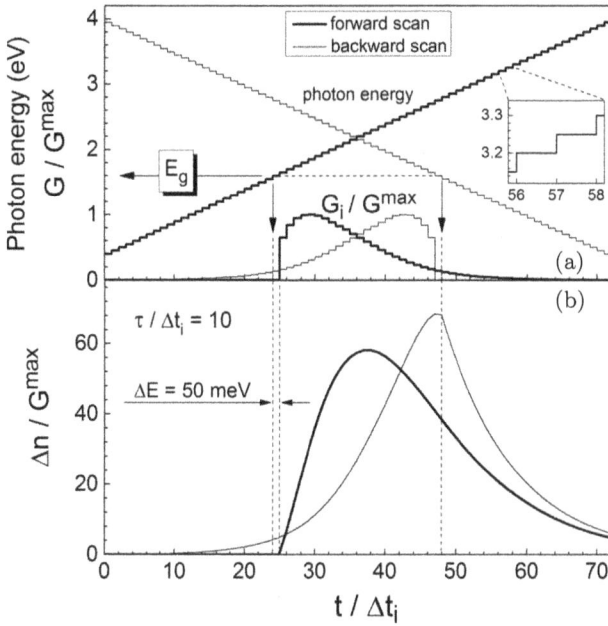

Figure 2.7. Example for a regime of spectral-dependent measurements with forward (increasing photon energy, thick lines) and backward (decreasing photon energies, thin lines) scans of photon energies and hypothetic generation (a) and for the corresponding densities of photogenerated charge carriers following the model after Equations (2.18) and (2.19) (b).

For the forward scan, i.e. for $h\nu_i$ increasing in time, the band gap can be well measured as an onset within the spectral resolution. This is not the case for the backward scan, i.e. for $h\nu_i$ decreasing in time. Therefore, the spectral-dependent SPV measurements with a Kelvin probe shall start at the lowest $h\nu$ at which photogeneration is still absent and continued at increasing $h\nu$.

The variation of Δt_i may cause some changes in measured SPV spectra. Figure 2.8 shows forward (Figure 2.8(a)) and backward (Figure 2.8(b)) scans of $\Delta n/G^{max}$ for values of Δt_i shorter and longer than τ (generation spectrum as in Figure 2.7). In all forward scans, the $h\nu$ at which the band gap appeared is identical. In contrast, the determination of E_g is not well defined in all backward scans. Therefore, spectral-dependent SPV measurements with a Kelvin probe in forward scans are well suitable for the precise measurement of onsets of photogeneration, such as band gaps or defect states in the band gap.

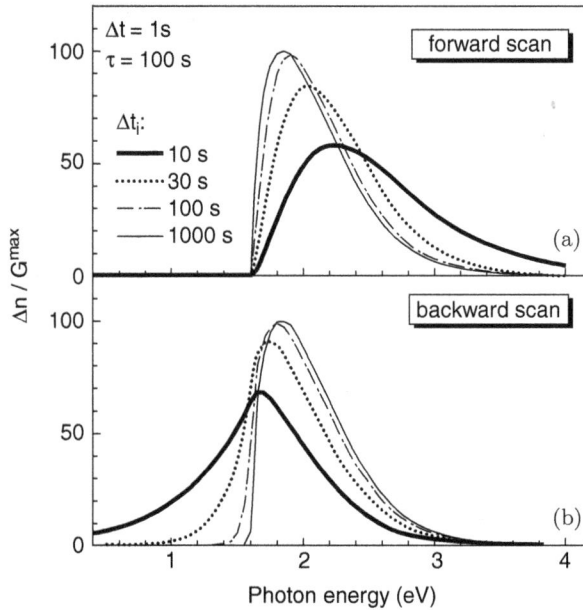

Figure 2.8. Spectra of $\Delta n/G^{max}$ simulated following the model after Equations (2.18) and (2.19) for (a) forward and (b) backward scans. The relaxation time was set to 100 s and the values of Δt_i were 10, 30, 100 and 1000 s (thick solid, dotted, dashed and thin solid lines, respectively).

With increasing Δt_i, the maximum of the spectra shown in Figure 2.8 shifted toward lower or higher photon energies for forward or backward scans, respectively. Therefore, peak positions cannot be well analyzed as known for conventional spectroscopy and one has to be very careful with the comparison of peaks in SPV spectra measured with a Kelvin probe.

Processes with opposite directions of charge separation can contribute to SPV signals. Furthermore, SPV signals can depend very differently on the light intensity, relaxation and photon flux. Therefore, SPV spectra are usually not normalized to the photon flux. As exclusion, SPV spectra can be well normalized to the photon flux if the mechanisms of charge separation, transport and recombination do not change with photon energy and if the SPV signal is proportional to the photon flux (low signal case), which shall be verified in the experiment.

The light-induced change of the CPD can be investigated with Kelvin probes on samples immersed in electrolytes (Bastide *et al.*, 1999). For this purpose, the back contact of the sample electrode has to be sealed from the

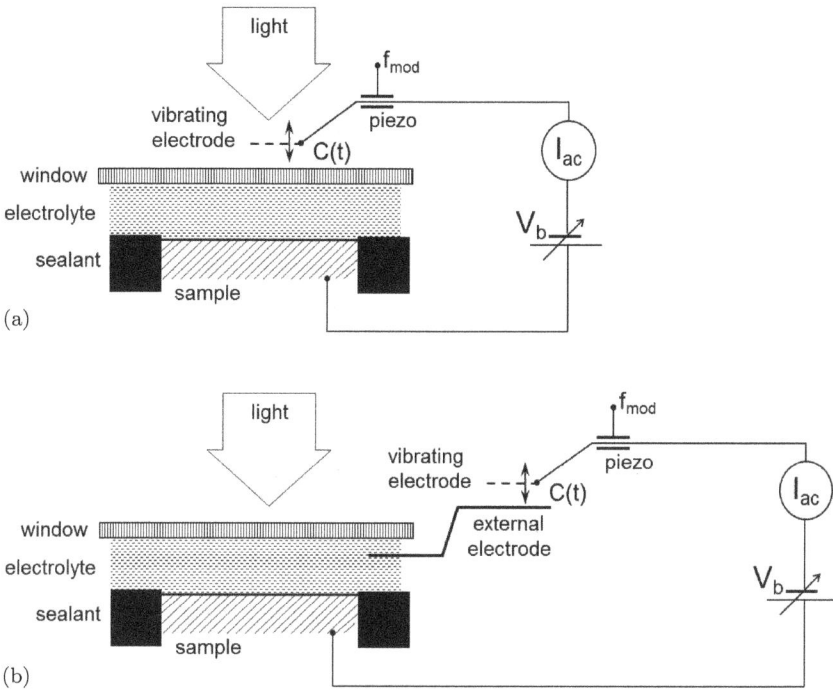

Figure 2.9. Schematic configurations for the measurement of the SPV on a sample immersed in an electrolyte with a Kelvin probe based on a vibrating mesh. The vibrating electrode can be placed directly (a) in front of the sample with the window or (b) in front of an external electrode contacting the electrolyte.

electrolyte (Figure 2.9). The vibrating reference electrode can be placed in front of a window sealing the cell with the sample and the electrolyte (Figure 2.9(a)). The sample can also be kept as a floating electrode (Rühle *et al.*, 2004) and the electrolyte can be contacted with an external metal electrode immersed into the electrolyte at one end (Figure 2.9(b)).

Dipole layers are usually formed at interfaces between electrolytes and solid matter due to polarization. In electrochemistry, such dipole layers are called Helmholtz double layers. It shall be taken into account that additional polarization appears at interfaces between an electrolyte and a window as well as between an electrolyte and a metal electrode. Furthermore, additional polarization is also caused by the dielectric properties of a window. The light-induced ΔCPD can be measured precisely with a Kelvin probe for samples immersed in electrolytes if the additional polarization phenomena are independent of illumination.

2.2.3.   *SPV measurements with Kelvin probe force microscopy*

In 1982, Binnig and Rohrer invented the scanning tunneling microscope, allowing for extremely high spatial resolution in surface analysis (STM, Binnig and Rohrer, 1982). Distances can be varied in STM measurements over extremely short regions by using piezo ceramics. A STM combines the extremely precise control of distances with the exponential dependence of a tunneling current between a metal surface and a sharp metal tip close to the surface. Instead of measuring a tunneling current, the force between a tip and a sample surface is measured in atomic force microscopy (AFM). In KPFM (see also the review of Melitz *et al.*, 2011), the tip is coated with a metal to that an electrostatic potential can be applied.

The force between two objects at short distances can be described by the Lennard-Jones potential (see, for example, Lennard-Jones, 1931) as shown in Figure 2.10(a). At extremely short distances, the potential is dominated by repulsion due to the interaction of electron orbitals. At a bit longer distances of the order of 0.5 nm, the potential between two objects is attractive due to induced electric dipoles (Van der Waals forces). The range of distances resulting in attractive forces is used in KPFM.

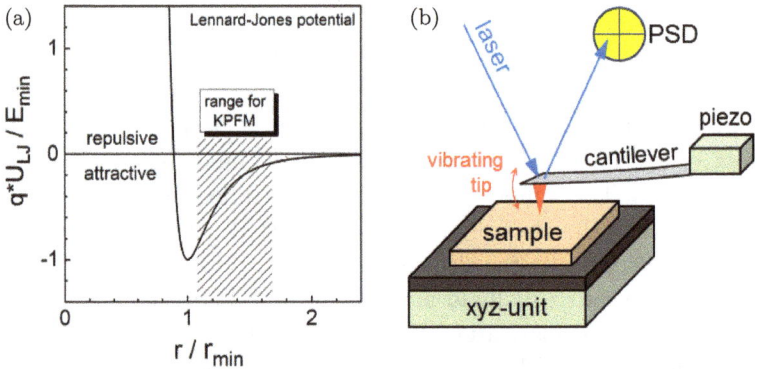

Figure 2.10. Dependence of the normalized Lennard-Jones potential on the normalized distance (repulsive potential $\sim (r_{min}/r)^{12}$, attractive potential $\sim (r_{min}/r)^{6}$). (a) The region of attractive force is applied for AFM and KPFM measurements. (b) Principle of AFM. A sharp tip vibrates at a close distance to the sample surface. The tip is fixed at the cantilever. The vibration is excited with a piezo crystal. The motion of the cantilever is detected by the deflection of a reflected laser spot on a position-sensitive detector (PSD). The sample is mounted on a piezo ceramic xyz-unit.

The change of the small electrostatic force between a tiny tip and a surface is detected in AFM with a vibrating cantilever (Figure 2.10(b)). The mechanical deformation of a cantilever, which is a tiny paddle made from crystalline silicon, is sensitive to extremely small forces or changes of forces. In KPFM, the change of the electrostatic force between the tip and a surface is detected by the variation of the amplitude of the deformation (amplitude modulated KPFM or AM-KPFM) or of the resonance frequency of the cantilever (frequency modulated KPFM or FM-KPFM). The deformation of the cantilever is measured with a reflected laser spot by using a PSD. In KPFM, the tip at the cantilever is coated with a metal and serves as the reference electrode in spatially resolved SPV spectroscopy (SRSPS) measurements. Depending on the modulation mode of the KPFM, the resolution of the measurement of CPD is of the order of a millivolt.

The measurement principle of the AM-KPFM is explained in Figure 2.11. The tip is vibrating in front of the sample surface (Figure 2.11(a)). The oscillation frequency of the cantilever is close to its resonance angular frequency ($\omega_{res} + \Delta\omega$). The resonance frequency is of the order of tens of kilohertz. The tip is scanned over the sample surface (Figure 2.11(b)). The amplitude of the oscillation of the cantilever remains constant for a constant distance between the tip and the sample surface. The distance between the tip and the sample surface can change depending on the topology. For example, the oscillation amplitude increases if the distance between the tip and the sample surface increases. In AFM, the amplitude is kept at a constant value (set point A_0) by adjusting the distance between the tip and the sample surface with the z-control. The topography is given by the xy-map of the z-values.

The work function of the sample surface can locally change, for example, due to local charging (Figure 2.11(b)). In KPFM, V_b and $V_{AC} \cdot \sin(\omega_{ac} \cdot t)$ are applied to the tip or to the sample. For increasing the sensitivity, ω_{ac} is chosen at the second resonance angular frequency of the cantilever ($2 \cdot \omega_{res}$, of the order of several hundreds of kilohertz). V_{ac} is of the order of $100 \, mV$. Incidentally, the V_{ac} should be as low as possible since it can influence the charge distribution at the sample surface.

When the charge distribution across the sample surface changes during the scan of the tip, $F_{es,ac}$ and therefore the vibration of the cantilever will change (Figure 2.11(c)). The value of $F_{es,ac}$ is nulled or set to a constant value (set point) by adjusting V_b. The value of V_b is equal to the CPD if $F_{es,ac}$ is exactly nulled. The distribution of the CPD is given by the xy-map of the V_b-values as shown schematically in Figure 2.11(d).

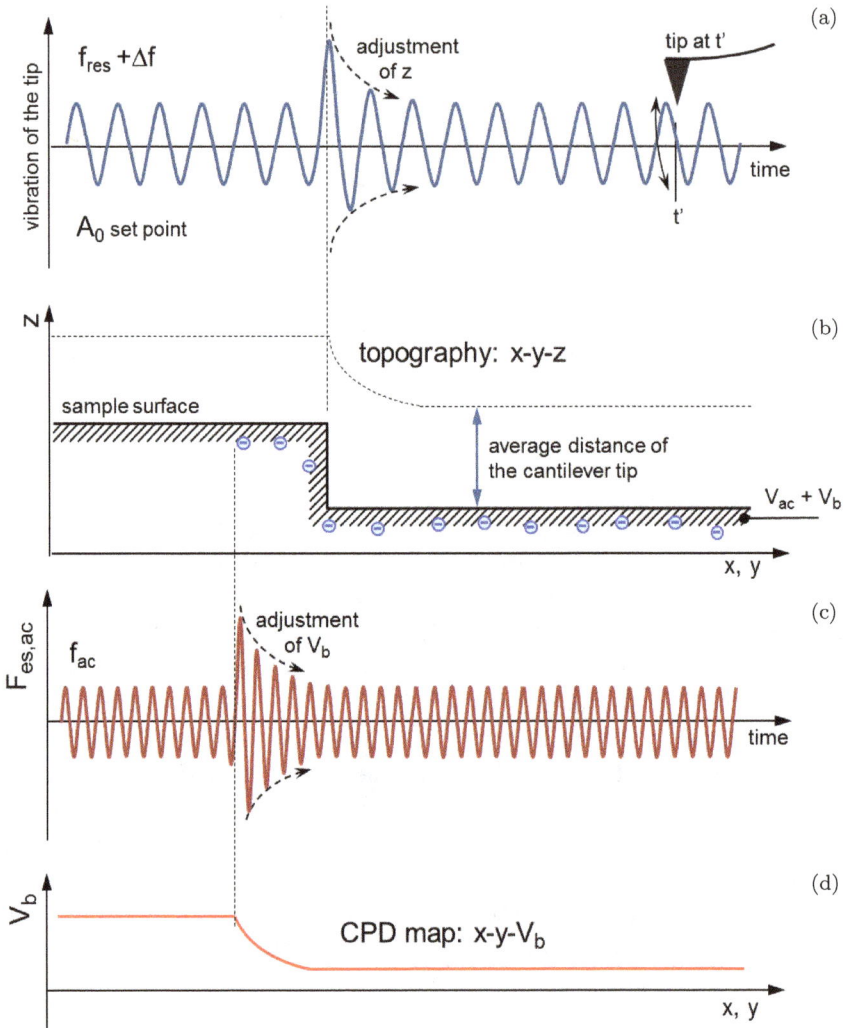

Figure 2.11. Principle of the amplitude modulated KPFM including the adjustment of the vibration amplitude of the cantilever to (a) a constants value by changing the distance between the sample surface and the tip at the cantilever, (b) cross-section of the sample surface with a step and a partially distributed negative charge with the schematic map of the topography, (c) modulated electro static force which is kept constant or nulled by adjusting V_b applied at the sample and (d) schematic map of the CPD signal which is equal to $-V_b$ if $F_{es,ac}$ is nulled.

The FM-KPFM works in a rather similar way as AM-KPFM whereas a constant frequency is defined as the set point (see Glatzel and Sadewasser, 2017 for more details). Incidentally, FM-KPFM is sensitive to the force gradient and allows for a higher spatial resolution then AM-KPFM. FM-KPFM is usually applied in UHV.

SPV maps with high spatial resolution can be obtained by plotting the difference of CPD maps under illumination and in the dark. Of course, the sample surface shall be screened from stray light of the laser used for the measurement of the deflection of the cantilever in order to avoid its influence on SRSPS measurements.

Setups for measuring SRSPS require the control of numerous parameters related to the topology (AFM), CPD (KPFM) and ΔCPD (SRSPS). Figure 2.12 shows a setup for the measurement of modulated SRSPS in ambient atmosphere with an AM-KPFM. The deflection of the laser spot is measured with the PSD. The PSD signal is amplified, converted into a potential and fed into a demodulator to separate the parts of the PSD signal related to the mechanical ($\omega_{res} + \Delta\omega$) and electrical ($\omega_{ac}$) modulations. Separate frequency generators provide the excitation potential for the piezo drive of the cantilever (V_{exc}) and the potential for the modulation of the electrostatic force.

The PSD signals related to mechanical and electrical modulation are measured with separate lock-in amplifiers. The outputs of the lock-in amplifiers are fed into separate feedback loops for keeping A_0 constant and for nulling $F_{es,ac}$ by adjusting the potential of the z-piezo drive (V_z) and V_b. The potentials V_{ac} and V_b are added and applied at the sample electrode.

V_b is additionally modulated by the SPV signal. A lamp with a mechanical chopper for modulation (angular frequency ω_{mod}) and a monochromator for setting the wavelength (λ) are used for the excitation of the SPV signal. The period of the modulated light shall be much shorter than the feedback time for the adjustment of V_b. The modulated SPV signal is measured with a third lock-in amplifier. For measuring spatially resolved maps, the potentials of the xy-piezo drives (V_x and V_y) are controlled. For the interested reader, more explanations and details about AFM, KPFM and measurement modes can be found in Glatzel and Sadewasser (2018).

The resolution of modulated SRSPS signals is of the order of $100\,\mu V$ (Streicher *et al.*, 2009, for FM-KPFM). The incident angle of the light

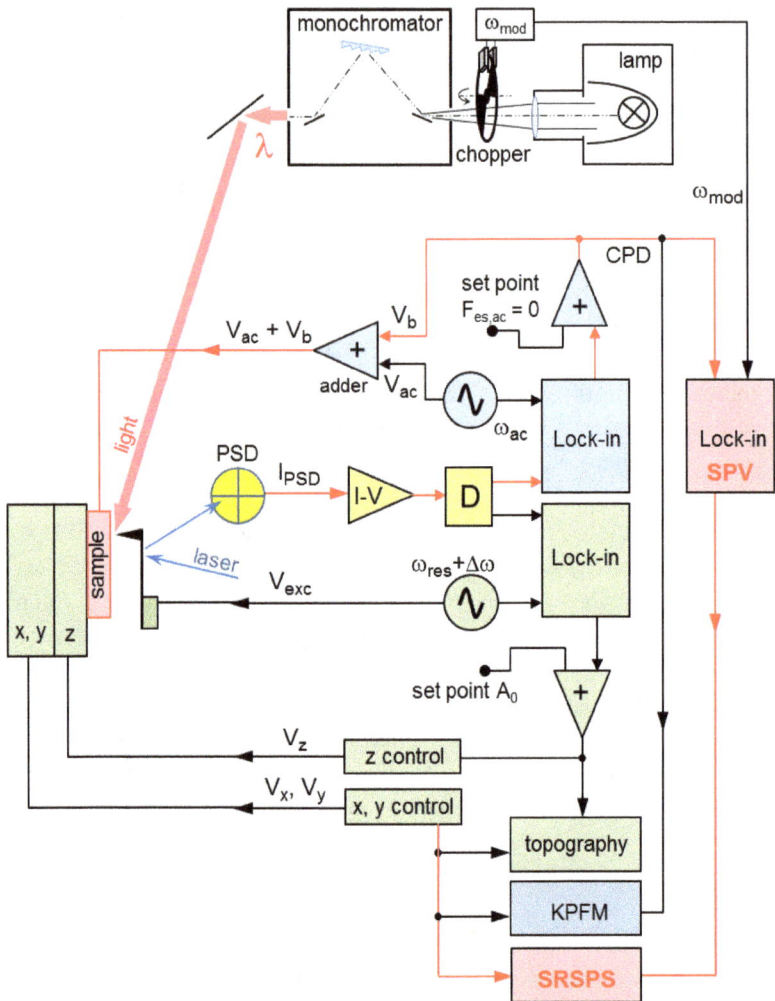

Figure 2.12. Setup for the measurement of the modulated SRSPS in the mode of AM-KPFM. The elements with the greenish, blueish and reddish backgrounds mark the parts for mapping the topography, CPD and SPV, respectively. The elements with the yellowish background include the PSD for measuring the deflection of the laser spot, the current to potential converter ($I - V$) and the frequency demodulator (D) separating the modulated signals related to $\omega_{res} + \Delta\omega$ (topography) and ω_{ac} (CPD). The potentials V_x and V_y and V_z are applied to the xy- and z-piezo drivers, respectively. The cantilever is excited with the potential V_{exc} at frequency $\omega_{res} + \Delta\omega$. The electrostatic force is modulated with the potential V_{ac} at frequency ω_{ac} and shifted by the bias potential (V_b). The set points for the feedback loops for the topography and the CPD are given by the vibration amplitude of the cantilever (A_0) and by the condition of nulling $F_{es,ac}$, respectively. The light for excitation of the SPV is modulated with an optical chopper at the frequency ω_{mod}. The red connections mark the signal lines for the measurement of the SRSPS.

exciting the SPV is very flat. Therefore, depending on the topology of a sample, partial shading can significantly influence SRSPS (Streicher *et al.*, 2009). SPV with a KPFM can be performed over a large area as shown, for example, at a pn-junction of SiC (Gysin *et al.*, 2015). Furthermore, SPV with a KPFM can provide information about the local charge separation at photocatalytic crystals of $BiVO_4$ and the role of co-catalysts (Zhu *et al.*, 2017). Incidentally, a limiting factor for broad application of SRSPS is that the measurement of SPV spectra with high resolution in space and the mapping of SPV signals with KPFM are rather time-consuming (of the order of hours for one measurement).

2.3. The Fixed Capacitor Arrangement

2.3.1. *Capacitive out-coupling of SPV signals*

In SPV measurements with fixed capacitors, the reference and sample electrodes are fixed and form a measurement capacitor (C_m — capacitance of the measurement capacitor). The reference and sample electrodes are connected via a measurement resistance (R_m). In the dark, the Fermi energies of the reference and sample electrodes are equilibrated by R_m. Figure 2.13 shows a simplified energy scheme (a) and the simplest equivalent circuit for

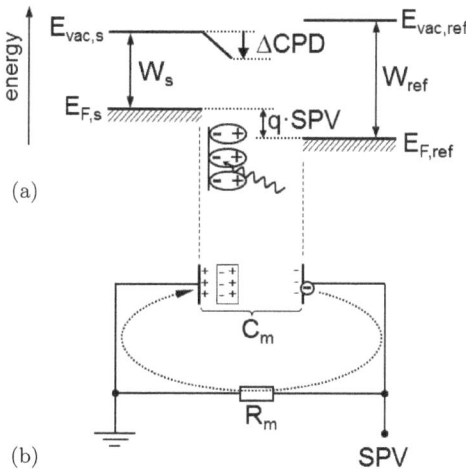

Figure 2.13. Energy scheme of (a) a reference electrode and a sample electrode with a photoactive layer under illumination and (b) simplified equivalent circuit for capacitive out-coupling of SPV signals.

SPV measurements with a fixed capacitor (b). Due to inductance, the SPV charges the measurement capacitor. The corresponding voltage drop can be directly measured. This gives the opportunity for fast SPV measurements. The charged measurement capacitor discharges within the $R_m \cdot C_m$ time constant. Therefore, SPV signals can be measured at times shorter than $R_m \cdot C_m$.

Capacitive out-coupling means the measurement of the voltage drop caused by SPV signals across C_m. Oscilloscopes and lock-in amplifiers are used as measurement devices for transient and modulated SPV measurements, respectively, with fixed capacitors. A charged capacitor discharges via a resistance. The internal resistance of a measurement device (load resistance, R_{load}) acts as a shunt. For very fast transient SPV measurements, R_{load} is set to $50\,\Omega$. Therefore, an additional device, which matches the impedance of the measurement capacitor with the impedance of the measurement device, is required for capacitive out-coupling of SPV signals. The device required for capacitive out-coupling is called high-impedance buffer.

A high-impedance buffer (Figure 2.14) transfers a potential dropping at a very high impedance (input potential, U_{in}) to a potential dropping at a much lower impedance (output potential, U_{out}). For an ideal high-impedance buffer, the values of the SPV, U_{in} and U_{out} are equal (Figure 2.14(a)):

$$U_{out}(t) = U_{in}(t) = \text{SPV}(t) \tag{2.18}$$

Real high-impedance buffers (Figure 2.14(b)) are based on the operational amplifiers (OPAs) with a very low bias current (I_{bias}). The equivalent circuit of a real high-impedance buffer contains an ideal high-impedance buffer, a current source, an input capacitance (C_{in}), an input resistance (R_{in}) and an output or source resistance (R_{source}). For conventional OPAs, I_{bias}, C_{in} and R_{in} are of the order of several pA, pF and TΩ, respectively. For SPV measurements in the range of nanosecond, R_{load} and R_{source} shall be equal in order to avoid reflections of the signal caused by impedance mismatch. Therefore, the value of R_{source} is set to $50\,\Omega$. R_{source} and R_{load} form a resistive voltage divider for the measured potential ($U_{meas}(t)$):

$$U_{meas}(t) = \frac{R_{load}}{R_{load} + R_{source}} \cdot U_{out}(t) \tag{2.19}$$

I_{bias} flows through R_{in} and causes an offset potential (U_{offset}). Due to the very high R_{in}, U_{offset} would be high and limited in practice by

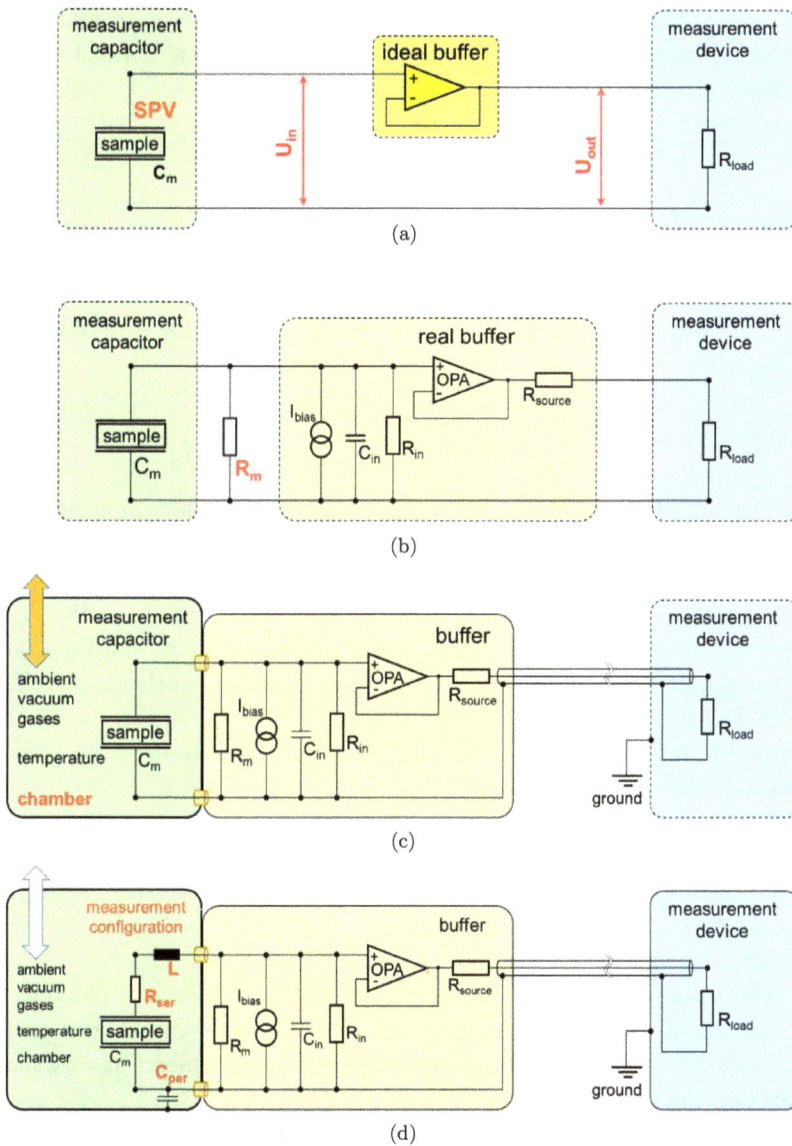

Figure 2.14. Equivalent circuits for the capacitive out-coupling of a potential (a) with an ideal high-impedance buffer, (b) with a real high-impedance buffer, (c) with a high-impedance buffer integrated into a measurement setup and (d) equivalent circuit including elements of the measurement configuration, i.e. an inductance (L, caused, for example, by connecting wires), a series resistance (R_{ser}, caused, for example, by the TCO of the reference electrode) and a parasitic capacitance (C_{par}, caused, for example, by electrical feed-through).

the potential of the power supply of the OPA. Therefore, an additional resistance, much lower than R_{in}, shall be connected in parallel to R_{in} in order to limit U_{offset}. The resistance limiting U_{offset} is called the measurement resistance (R_m):

$$U_{offset} = I_{bias} \cdot \frac{R_m \cdot R_{in}}{R_m + R_{in}} \tag{2.20}$$

The value of R_m is usually of the order of $1 - 100\text{G}\Omega$. Therefore, U_{offset} is of the order of several millivolts for conventional high-impedance buffers.

The bias current charges C_m in time. This causes a drift potential ($U_{drift}(t)$):

$$U_{drift}(t) = U_{offset} \cdot \left[1 - \exp\left(-\frac{t}{R_m \cdot C_m} \right) \right] \tag{2.21}$$

The drift can influence the baseline during SPV measurements with fixed capacitors. It is recommended to start SPV measurements after U_{drift} reached the saturation value of U_{offset}.

For practical reasons, R_m is often assembled together with the high-impedance buffer and connectors for the power supply of the OPA and for the input and output signals into one compact shielded box. The whole device is also called buffer (Figure 2.14(c)). A buffer can be directly connected, for example, to chambers for temperature-dependent SPV measurements in controlled ambient.

The equivalent circuit of a real measurement configuration contains additional elements aside R_m and C_m. For example, electrical feed-through, sheet resistances of electrodes and connecting wires cause parasitic capacitances (C_{par}), series resistances (R_{ser}) and inductivities (L), respectively (Figure 2.14(d)).

The values of C_{par}, R_{ser} and L are of the order of several pF, tens of Ω and tens of nanohenry, respectively. Therefore, SPV signals can excite damped oscillations ($U_{LCR}(t)$) influencing SPV measurements at very short times (see the following paragraph).

C_{par}, C_{in} and C_m act together as a capacitive voltage divider for U_{in}. Furthermore, C_m discharges via R_m with the time constant $R_m \cdot C_m$. The elements in the equivalent circuit including the buffer and the measurement configuration limit also the resolution time of U_{out}. In the simplest case, a resolution time can be described by a logistic growth function with an onset time constant (τ_{on}).

The signal at the output of a high-impedance buffer is a superposition of SPV(t), $U_{LCR}(t)$ and U_{drift}, which are limited by τ_{on}, $R_m \cdot C_m$ and the capacitive voltage divider:

$$U_{out}(t) = \frac{\exp\left(-\frac{t}{R_m \cdot C_m}\right)}{1 + \exp(-\frac{t}{\tau_{on}})} \cdot \frac{C_m}{C_m + C_{in} + C_{par}}$$
$$\cdot [SPV(t) + U_{LCR}(t) + U_{drift}(t)] \qquad (2.22)$$

Depending on the time range or on the modulation period, measured SPV signals can significantly deviate from the SPV signals resulting directly from the spatial separation of photogenerated charge carriers in a sample. This shall be taken into account for the analysis of SPV signals measured with fixed capacitors. The SPV signal undisturbed by the measurement configuration and the high-impedance buffer are also called the source function of the SPV signal. Furthermore, the incomplete discharge of C_m via R_m in the dark can cause, due to the influence of the electric field of remaining photoinduced charge, additional uncertainties in modulated SPV measurements or in transient SPV measurements with averaging of transients.

High-impedance buffers can be applied for any SPV or photovoltage measurements independently of the nature of the fixed capacitor. A fixed capacitor can be formed, for example, by implementing an insulator between the sample and reference electrodes, by the space charge region in a solar cell (measurement of the so-called V_{OC} decay, i.e. the relaxation of the open circuit voltage in time) or by a Helmholtz layer at a semiconductor/electrolyte interface (Figures 2.15(a)–2.15(c), respectively).

For conventional SPV measurements, a reliable fixed capacitor is formed by pressing the reference and sample electrodes one on the other whereas an insulator is placed between both electrodes. Mica (muscovite) is a well-suitable insulating mineral with a layer structure. Muscovite is transparent over a wide spectral range and thin sheets with a thickness of tens of micrometer can be peeled off along the Van der Waals planes. Usually, a mica sheet is much thicker than the charge separation length. Therefore, the value of C_m is mainly given by the thickness of the mica sheet. Taking into account a relative dielectric constant of mica of about 8 and a diameter of the reference electrode of several millimeter, the value of C_m is of the order of 10–100 pF. Instead of mica, other insulators can also be used for forming the measurement capacitor.

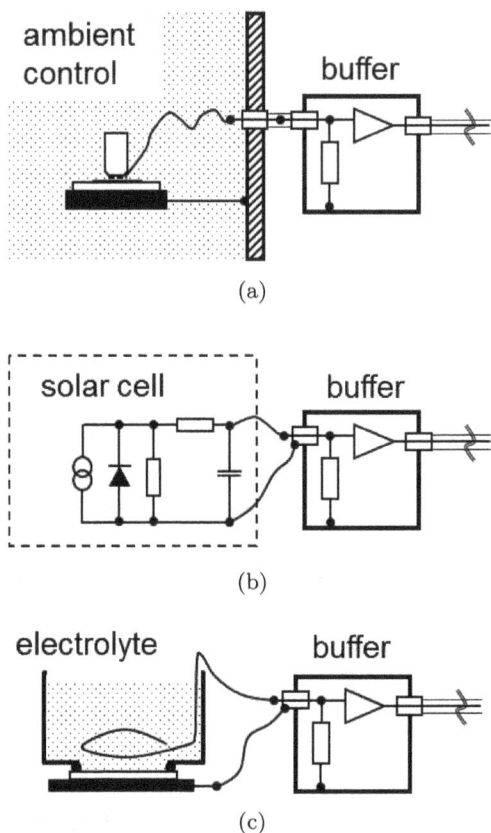

Figure 2.15. Application of high-impedance buffers for (a) SPV measurements under ambient control, (b) V_{OC}-decay measurements of solar cells and (c) photovoltage measurements in electrolytes.

Figure 2.16 shows two electrode configurations for well reliable SPV measurements on silicon wafers (after Heilig (1974), Figures 2.16(a)–2.16(c)) and on relatively small samples, thin films or powders (after Figures 2.16(d)–2.16(f)). The reference electrodes consist of a thin conductive metal oxide such as a SnO_2:F layer deposited onto a glass plate shaped afterwards (Figure 2.16(b)) or onto the phased side of a quartz cylinder shaped before deposition (Figure 2.16(f)).

For the characterization of the surface band bending of silicon wafers and its dependence on a bias potential (Heilig, 1974), the c-Si wafer is contacted with a metallic needle at the backside which has been etched

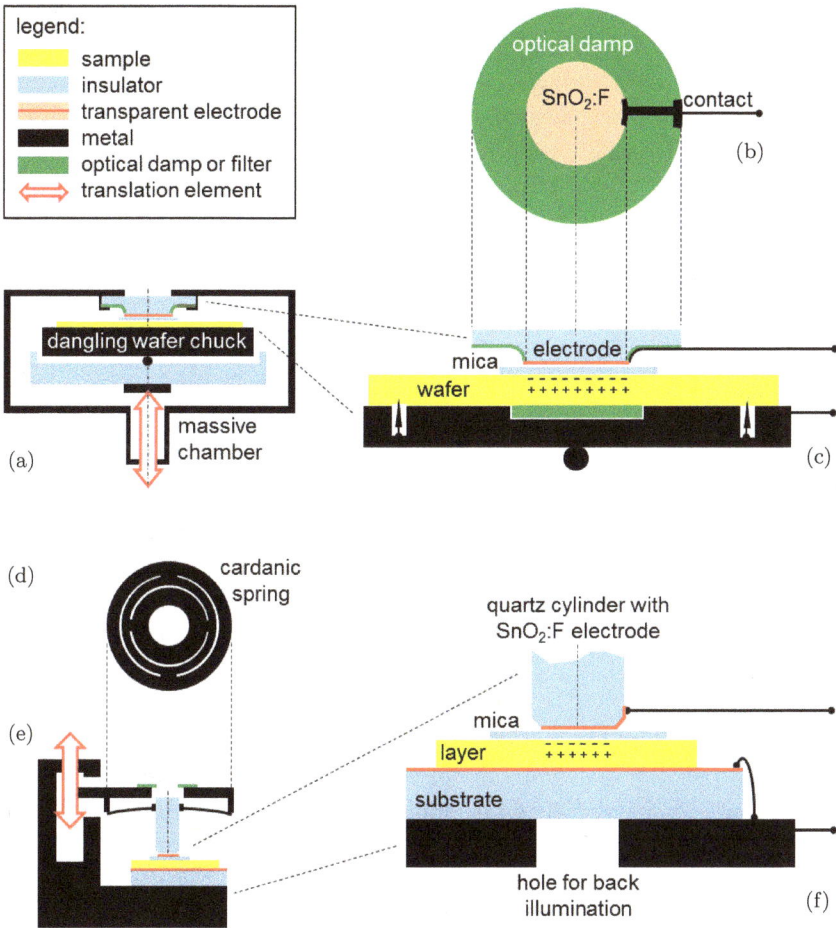

Figure 2.16. Electrode configurations for SPV measurements with fixed capacitors on large wafers ((a)–(c), electrode design after Heilig, 1974) on small samples and thin films ((d)–(f)). Side views: (a), (c), (e) and (f). Top view of the large electrode: (b), top view of the cardanic spring for smart pressing: (d).

in HF (hydrofluoric acid) solution before in order to remove silicon oxide. The needle contact is placed closer to the edge of the wafer. An optical damp is placed underneath the measurement capacitor in order to avoid an influence of light reflected at the sample electrode (single pass of light). The reference electrode is contacted with an evaporated gold stripe bonded to the external contact wire. The part of the glass sheet which is not covered by the electrode is coated with a black light absorbing layer in order to avoid uncontrolled illumination.

Relatively small samples, powders and thin films can be investigated in the configuration of the measurement capacitor with a quartz cylinder. This relatively small configuration allows for its implementation into vacuum (Duzhko *et al.*, 2001b) or UHV chambers (Parvan *et al.*, 2018).

2.3.2. Time range of SPV measurements with fixed capacitors

The resolution time of SPV measurements can be limited by the bandwidth of the measurement device, by the bandwidth of the high-impedance buffer, by the duration time of the exciting light pulse and by the measurement configuration of the experimental setup with the sample and the electrode. The resolution time is limited by the resolution time of the slowest component. The bandwidth of fast oscilloscopes is 1 GHz or higher. The bandwidth of OPAs with a low bias current is of the order of hundred(s) of MHz. For example, the bandwidths and I_{bias} of the OPA656, OPA659 and LTC6268 are 500, 650 and 350 MHz and 2 pA, 50 pA and 2 fA, respectively. Therefore, resolution times of about 0.3–1 ns can be realized with high-impedance buffers.

SPV transients are excited with short light pulses. Historically, short light pulses of the duration time of several microseconds were generated with a spark discharge (see, for example, Johnson, 1957; Heilig, 1968; Heilig *et al.*, 1979). In the 1960s, short powerful laser pulses were generated for the first time by varying the effective reflectivity at the end of a ruby rod with a Kerr cell (McClung and Hellwarth, 1961). Now, aside pulsed dye lasers, tunable Q-switched Nd:YAG lasers with optical parametric amplifiers are used for the generation of short laser pulses with duration times of a couple of ns in a broad range of wavelength. Light pulses with duration times of the order on 1–100 ns can be also generated with pulsed laser diodes. Incidentally, the application of ultra-short laser pulses is not really useful for SPV measurements with fixed capacitors due to the limited resolution time.

SPV transients can excite damped oscillations in LCR circuits. The natural frequency (ω_0) and the decay constant (δ) of an LCR circuit are given by

$$\omega_0 = \frac{1}{\sqrt{L \cdot C}} \tag{2.23}$$

$$\delta = \frac{R}{2 \cdot L} \tag{2.24}$$

The period of the natural frequency (T_0) is given by

$$T_0 = 2 \cdot \pi \cdot \sqrt{L \cdot C} \tag{2.25}$$

The function of the damped oscillations excited by an SPV transient is given by

$$U_{LCR}(t) = A_{LCR} \cdot \sin\left(\sqrt{\omega_0^2 - \delta^2} \cdot t - \varphi_{LCR}\right) \cdot \exp(-\delta \cdot t) \tag{2.26}$$

where A_{LCR} and φ_{LCR} denote the amplitude and phase angle of the damped oscillations, respectively.

For a thin wire of a length of 1 cm, L is about 10 nH (see, for example, Rosa, 1908). The wire connecting the reference electrode with the high-impedance buffer is several cm long. Therefore, the value of T_0 is of the order of 1–10 ns for fixed capacitor arrangements. The damping time is the reciprocal damping constant and can amount to values between several hundreds of picoseconds and tens of nanoseconds. Therefore, the LCR circuit of the measurement configuration can strongly influence the measurement of SPV transients at times up to tens of nanoseconds. On the other side, the influence of the LCR circuit on measured SPV transients can be usually neglected at times longer than 20–30 ns.

As an example, Figure 2.17 shows a superposition of a fast SPV transient (increase and decrease within about 1 ns and 10 ns, respectively) and U_{LRC} for L and C equal to 60 nH and 5 pF and R equal to 25, 90 and 125 Ω, respectively. For R equal to 25 Ω, damped oscillations can be well distinguished for about five periods up to times between 15 and 20 ns. For higher damping at R equal to 90 Ω, only one damped oscillation can be seen within about 5 ns. This damped oscillation causes an overshoot at times within about 1 ns and an undershoot at times between about 1 and 2.5 ns. The damping at R equal to 125 Ω is close to maximum and a damped oscillation cannot be detected whereas a sharp overshoot at very short times decays within 1 ns. Therefore, slight differences in the LCR circuit can cause relatively large differences in SPV transients at short times.

In practice (see also Chapter 4), the influence of the LCR circuit of the measurement configuration on SPV transients can be reduced by reducing L and by adjusting R to a short damping time. Short damping times can be achieved for underdamping when the squared damping constant gets close to the squared natural frequency.

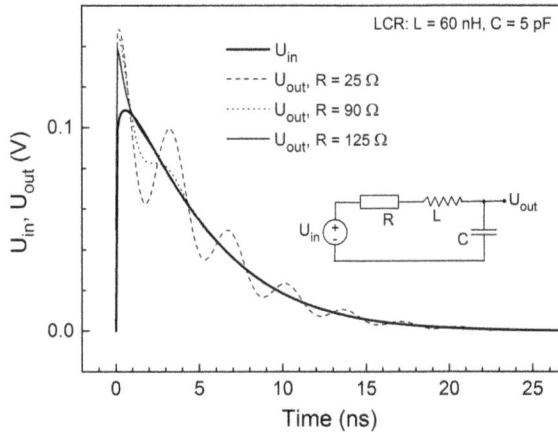

Figure 2.17. Superposition of an SPV transient (thick solid line) and damped oscillations caused by LCR circuits typical for an electrode configuration with a 6-cm long wire ($\approx$10 nH/cm) connecting the front electrode with the buffer and different values of R (25, 90 and 125 Ω, dashed, dotted and thin solid lines, respectively).

Sharp overshoots in SPV transients at very short times can be caused by the measurement configuration. For the unambiguous interpretation of SPV transients at very short times, it is recommended to vary the values of L and/or R in the LCR circuit in order to separate several damped oscillations and obtain the values of L, C and R. Then the source function of the SPV transient can be deduced.

At long times, the range of transient SPV measurements is limited by the $R_m \cdot C_m$ time constant. The value of C_m cannot be changed in a wide range. Therefore, the time range can only be varied significantly by varying R_m in accordance with I_{bias}.

As an example, Figure 2.18 compares a laser pulse and the source functions of a short and a long retarded to more than 100 s SPV transient with the $U_{drift}(t)$ and the exponential decays given by the $R_m \cdot C_m$ time constants for R_m and I_{bias} of 10 GΩ and 2 pF, 1 GΩ and 50 pF and 500 TΩ and 2 fA, respectively. The measurement of SPV transients up to 1–10 ms is well defined at moderate constant offsets reached within about 100 ms–1 s for the lower values of R_m and moderate values of I_{bias}. Related transient SPV measurements with conventional high-impedance buffers (R_m up to tens of gigaohms) are very robust and reliable within about $\leq$100 μV–1 mV in the time range between about 10 ns and 100 ms at a constant offset.

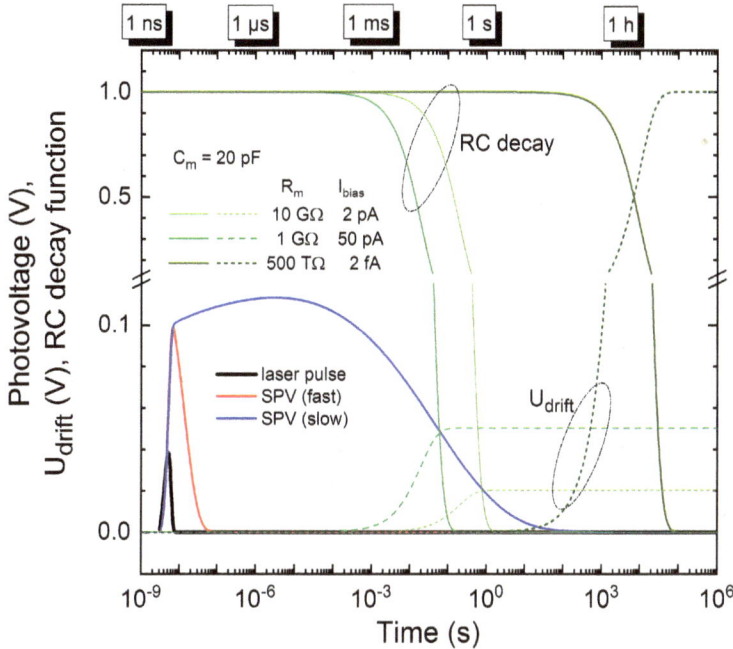

Figure 2.18. Example for a short laser pulse (black line) and source functions of a fast (red line) and very slow and retarded (blue line) SPV transients in comparison with U_{drift} (green dashed lines) and exponential decays for the $R_m \cdot C_m$ (green solid lines) for R_m and I_{bias} equal to 10 GΩ and 2 pA (green thin lines), 1 GΩ and 50 pA (green medium lines) and 500 TΩ and 2 fA (green thick lines), respectively. The logarithmic time scale ranges from 1 ns to 10^6 s. For clarity, the onset of the laser pulse was shifted in time to 3 ns.

For the measurement over 12 orders of magnitude in time from 1 ns to 10^3 s (Dittrich *et al.*, 2017a), R_m is extremely high and U_{drift} increases within hours to values of the order of 1 V. Therefore, at very long times, a drift cannot be avoided in measurements with fixed capacitors and high-impedance buffers. The drift shall be taken into account in the analysis of SPV transients at very long times. Furthermore, the long-lasting drift has consequences for the organization of the experiment since the measurement capacitor shall be discharged from time to time or the drift has to be compensated with an additional device. In general, the drift at very long times can be measured and is relatively well defined by the performance of the corresponding OPA. However, R_{in} and I_{bias} can fluctuate over longer times so that uncertainties may occur.

2.3.3. *Reduction of noise in transient SPV measurements*

The extremely high measurement resistance makes transient SPV measurements very sensitive to sources of electrical or electromagnetic noise and disturbances. For example, a fluctuation of the current by only 10 fA results in a fluctuation of the SPV signal by 1 mV for R_m equal to 100 GΩ. Therefore, effective measures for the reduction of electrical and electromagnetic noise and disturbances are required for highly sensitive transient SPV measurements in a wide range.

Sources of noise and disturbance are summarized in Figure 2.19. Due to electromagnetic induction and balancing, local fluctuations of the electric potential of the ground between separately grounded components result in fluctuations of the current and vice versa. Ground loops are therefore a major source of noise and disturbances in transient SPV measurements. In order to avoid ground loops, only the measurement device, i.e. the oscilloscope, is connected with the common electrical ground in the laboratory. Rechargeable batteries are recommended as isolated power supplies for high-impedance buffers. Furthermore, the measurement

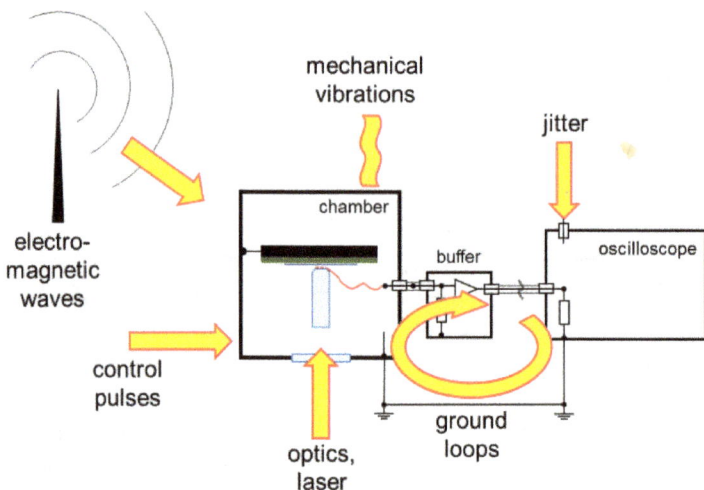

Figure 2.19. Sources of noise and disturbances in transient SPV measurements. Ground loops, electrical control pulses from equipment in the laboratory, electromagnetic waves, fluctuations in electrical (jitter) and/or optical components and mechanical vibrations can cause noise and disturbances in SPV transients.

chamber shall be electrically disconnected from vacuum pumps, optical tables, vacuum gauges and other equipment.

Correlated disturbances (or correlated noise) are well reproducible and can appear in certain time domains of SPV transients. Correlated disturbances can be caused, for example, by electromagnetic pulses from the power supply of the laser. Correlated disturbances cannot be reduced by averaging. Therefore, correlated disturbances shall be measured separately under dark condition (dark transient) and subtracted from measured SPV transients (correction to the dark transient).

Uncorrelated disturbances (or uncorrelated noise) are not reproducible. Uncorrelated disturbances can be caused, for example, by radio waves (high frequency) or electrical signals of other equipment in the laboratory (kHz range). Usually, a vacuum chamber is sufficient for electromagnetic shielding. However, optical windows and insulating O-rings are leaks for the penetration of electromagnetic waves and pulses. For reducing the influence of electromagnetic waves and pulses in transient SPV measurements, the area of optical windows shall be as small as possible, regions of unshielded O-rings have to be shunted and it is recommended to keep the wire connecting the reference electrode and the feedthrough to the buffer as short as possible.

A jitter of trigger pulses, i.e. fluctuations in time of switching on the measurement events, can disturb averaged SPV transients at very short times. Therefore, trigger pulses must be very well synchronized with the laser pulses. A low jitter of trigger pulses can be achieved, for example, by combining a beam splitter with a fast photodiode and a so-called splitter amplifier giving a large signal with a very sharp increase at a well-defined threshold related to the intensity of the laser pulse.

Mechanical vibrations cause noise due to changes in optical alignments and/or C_m. Therefore, the optical system and the measurement chamber must be mechanically decoupled from vibrating equipment such as rotary pumps. As an optimum, the laser, the optical system and the measurement chamber are fixed at the same damped optical table and the propagation of mechanical vibrations, for example, via vacuum tubes or hoses, is interrupted by implementing massive dampers. Incidentally, this may not be trivial in large complex setups including, for example, UHV chambers and equipment.

For corrections and analysis of transients, it is reasonable to set the baseline of a transient to 0 at the time before the laser pulse is switched on

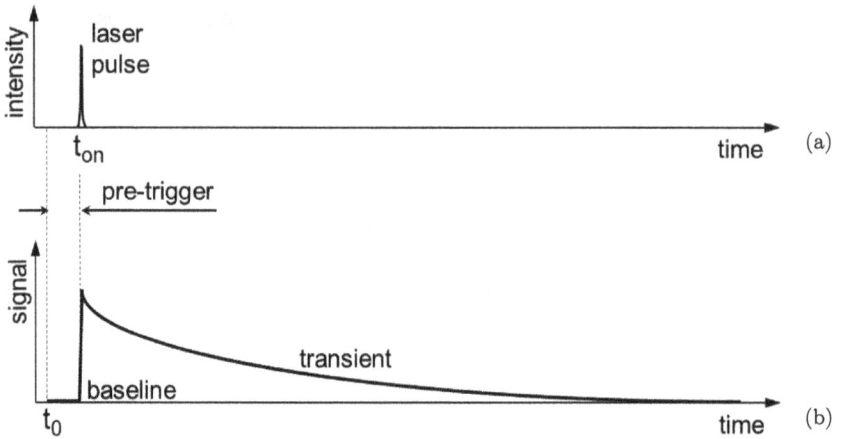

Figure 2.20. Schematic diagram of (a) exciting laser pulse and (b) corresponding SPV transient. For determining the baseline, a pre-trigger time $t_{on} - t_0$ has to be set on the oscilloscope (t_{on} and t_0 denote the time when switching on the laser and when starting the measurement, respectively).

(baseline correction). For this purpose, several data points are measured before switching on the laser pulse (t_{on}). Therefore, the measurement of a transient is the time t_0 before t_{on}. The time interval between t_o and t_{on} is called the pre-trigger time (Figure 2.20). After the baseline correction, the SPV transient can be corrected to the capacitor voltage divider.

Transient measurements on a logarithmic time scale can be used for reducing noise with increasing time. In the following, the procedure of the logarithmic readout and averaging in time is explained in more detail (Dittrich *et al.*, 2008). A logarithmic readout means that the time interval between registered samples increases exponentially in time (Figure 2.21). For this purpose, a certain number of a sample on board of the digital oscilloscope (NS_{board}) has to be related to the number of a certain sample to be registered (NS_{reg}) by an exponential function:

$$NS_{board} = int \left[A1 \cdot \exp \left(\frac{NS_{reg}}{A2} \right) \right] + A3 \qquad (2.27)$$

Incidentally, at the beginning of a transient, a range with a linear readout of 100–1000 samples is required for pre-triggering and for registration of the fastest process in the transient.

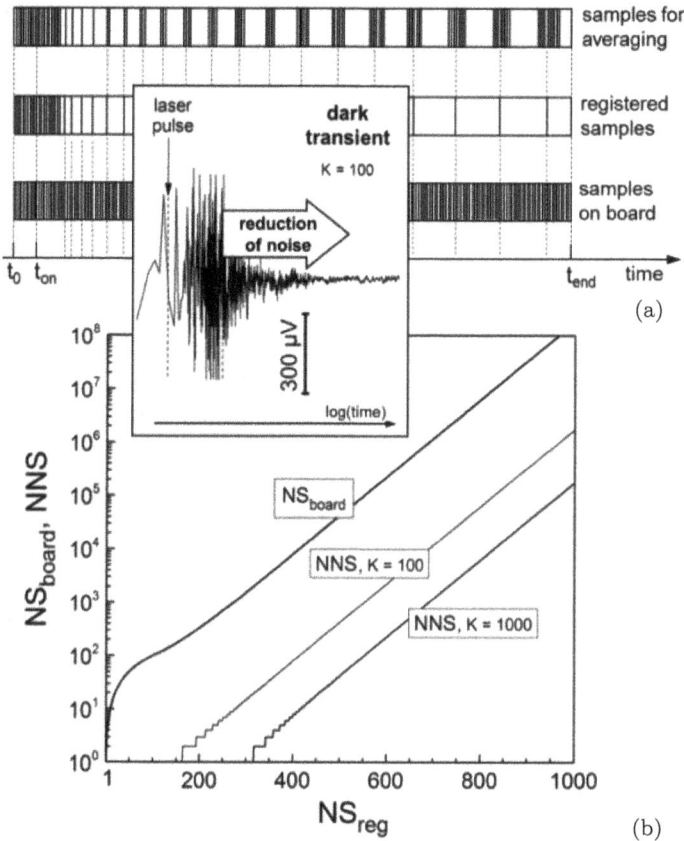

Figure 2.21. (a) Principle of a logarithmic readout and (b) example for the pattern of a logarithmic readout. A logarithmic readout means an exponential increase of the time interval between stored samples and an exponential increase of the number of samples for averaging around a stored sample. The inset shows a measured dark transient with logarithmic readout (see Dittrich *et al.*, 2008 for more details).

The pattern of the logarithmic readout (see also Figure 2.21(b)) can be adjusted by the numbers $A1$, $A2$ and $A3$ in Equation (2.27), which can be, for example, of the order of 10, 60 and 50, respectively (maximum number of samples on board: 10^8, 100 samples with a linear readout followed by about 800 registered samples with a logarithmic readout). By applying a logarithmic readout, the number of 10^7–10^8 samples in a measured transient can be reduced to about 10^3–10^4 registered samples in a transient ready for further analysis. In accordance with the logarithmic readout, the number of samples for averaging (NNS_{board}) can be increased in time:

$$NNS_{board} = \frac{int\left[A1 \cdot \exp\left(\frac{NS_{reg}}{A2}\right)\right] + A3}{K} \qquad (2.28)$$

The parameter K in Equation (2.28) has to be chosen in such a way that no information is lost, i.e. that the averaging procedure does not disturb the shape of the measured transient. A reliable number of NNS_{board} is of the order of 100–1000 times below the given NS_{board} related to a given NS_{reg}. An example for a pattern of a logarithmic readout is depicted in Figure 2.21(b) for K equal to 100 and 1,000. As an example, the reduction of noise in time by applying a logarithmic readout and Equation (2.28) is demonstrated for the measurement of a dark transient in the inset of Figure 2.21. By applying the logarithmic readout, the noise was reduced from hundreds of microvolt at shorter times to tens of microvolt at longer times.

Sometimes, uncorrelated disturbances cannot be completely eliminated in a setup. Then, the application of twin electrodes with high-impedance twin buffers can be useful in order to eliminate the influence of uncorrelated disturbances in SPV transients (Figure 2.22).

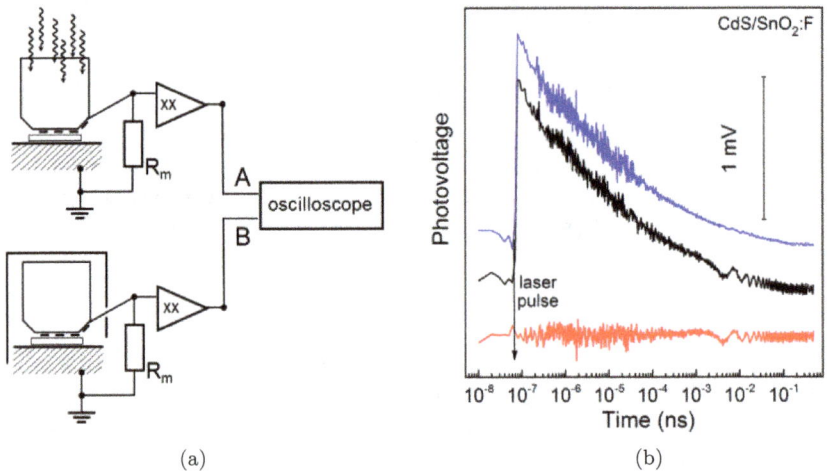

(a)   (b)

Figure 2.22. Principle (a) and example (b) of transient SPV measurements with twin electrodes and twin buffers. Illuminated (back line) and dark (red line) transients are measured under identical conditions. The difference of both transients (A – B, blue line) is free of disturbances (see Dittrich *et al.*, 2008 for more details).

In this configuration, one transient is measured under illumination and another transient is measured in the dark under identical conditions and at the same time. The difference of both transients is free from uncorrelated (and correlated) disturbances (Dittrich *et al.*, 2008).

2.3.4. *Modulated SPV measurements*

Due to the lock-in technique, the highest sensitivity of SPV measurements is reached for modulated measurements in the fixed capacitor arrangement. In contrast to the measurements with Kelvin probes, only those SPV signals are measured, which can follow the modulation frequency (f_{mod}), i.e. the response of processes with relaxation times much longer than about $1/f_{mod}$ (modulation period, T_{mod}) is filtered out.

A lock-in amplifier combines a phase-sensitive switch and an integrator. In the simplest case, the sign of the incoming signal is, first, switched (not inverted/inverted) with respect to f_{mod} and the phase shift between the signal and a reference signal and, second, integrated afterwards in time. Therefore, only the modulated part of the signal is amplified whereas noise signals at frequencies other than f_{mod} are nulled out. The operations performed by a lock-in amplifier can be implemented with analog devices (Michels and Curtis, 1941). Lock-in amplifiers can also be realized with software (digital lock-in, Wang, 1990). Instead of analyzing inverted and non-inverted signals, the signal can be multiplied by a sine wave of the modulation frequency in digital lock-in amplifiers. In modern lock-in amplifiers, a digital signal processing chip is used for the lock-in analysis. The use of digital signal processing has the advantage that the influence of higher harmonics can be eliminated and that the signals can be multiplied with the exact sine wave.

For reducing the noise in modulated SPV measurements (δ_{SPV}), the integration time constant of the lock-in amplifier (τ_{int}) can be increased and/or averaging over a certain number of readouts (N_{avg}) can be implemented:

$$\delta_{SPV} \sim \frac{1}{\sqrt{\tau_{int}}}; \sim \frac{1}{\sqrt{N_{avg}}} \qquad (2.29)$$

Well-suited values of f_{mod}, τ_{int} and N_{avg} are, for example, 23 Hz, 0.3 s and 10, respectively. This results in time intervals for the measurement of one point in modulated SPV spectra of the order of 1–10 s. Therefore, a modulated SPV spectrum can be measured within about 10–20 min.

Double phase lock-in amplifiers are used in modulated SPV measurements in order to get information about fast and slow components in SPV signals in relation to the modulation period (T_{mod}). Double phase lock-in means that there are two separate channels for which the functions controlling the switch of the sign before integration are shifted by $T_{mod}/4$, i.e. by 90°, to each other (sin and cos in digital processing). The principle of the double phase lock-in is explained in Figure 2.23 for modulated light (Figure 2.23(a)) and a positive SPV signal increasing when switching on the light and decreasing when switching off the light (Figure 2.23(b)).

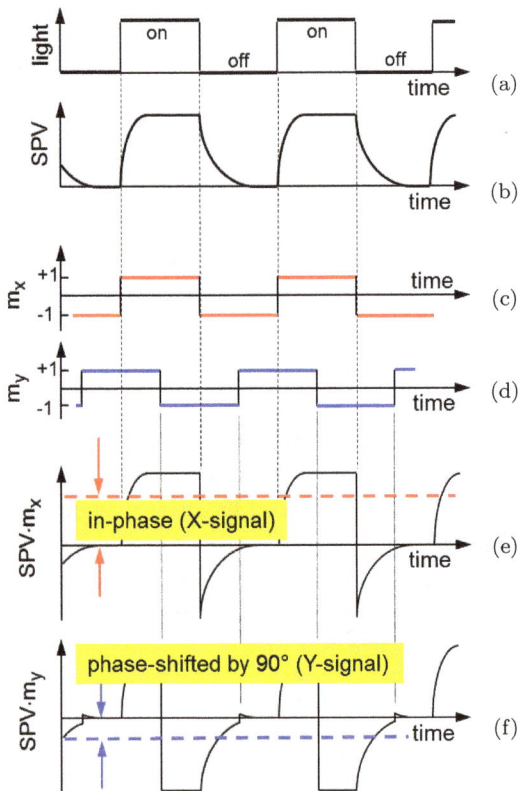

Figure 2.23. (a) Modulated light intensity, (b) idealized SPV response, (c) simplest functions for the formation of the in-phase and (d) phase-shifted by 90° signals, and products of the SPV response and (e) m_x or (f) m_y functions. The dashed lines in (e) and (f) mark the in-phase (X) and phase-shifted by 90° (Y) signals resulting from the integration, respectively.

The switch of the sign of the SPV signal before integration means that the SPV signals are multiplied with a periodic function of the same f_{mod} which is $+1$ (not inverted) or -1 (inverted). The phase angle (φ) is equal to $0°$ or $180°$ at the times when the light is just switched on or off, respectively. The functions switching from -1 to $+1$ and from $+1$ to -1 at φ equal to 0 or $-90°$ and $180°$ or $90°$, respectively, are denoted by m_x (Figure 2.23(c)) or m_y (Figure 2.23(d)). In lock-in amplifiers with digital processing, m_x and m_y are given by $\sin(2\pi \cdot f_{mod} \cdot t)$ and $\cos(2\pi \cdot f_{mod} \cdot t)$, respectively.

The SPV signals multiplied by m_x or m_y and integrated in time are called the in-phase (X-signal, Figure 2.23(e)) and phase-shifted by $90°$ (Y-signals, Figure 2.23(f)).

$$X = \int SPV(t) \cdot m_x(t) \cdot dt \qquad (2.30')$$

$$Y = \int SPV(t) \cdot m_y(t) \cdot dt \qquad (2.30')$$

A positive or negative X-signal means that electrons are separated preferentially toward the sample or reference electrodes, respectively.

The response of modulated SPV signals can be very fast in relation to T_{mod} (Figure 2.24(a)), i.e. the SPV signal follows directly the modulated light. In this case, the X-signal reaches the maximum and the Y-signal is equal to 0 (Figures 2.24(b) and 2.24(c)). In contrast, the response of a modulated SPV signal can be very slow in relation to T_{mod} (Figure 2.24(d)), i.e.

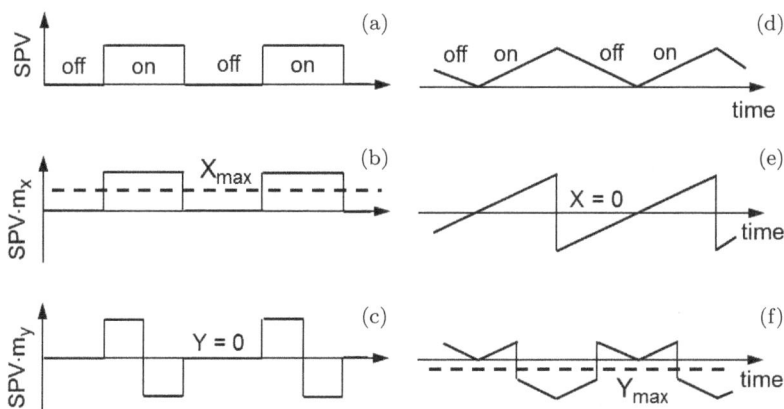

Figure 2.24. Idealized fast (a) and slow (d) SPV responses and corresponding products of the SPV response and the m_x ((b) and (e)) and m_y ((c) and (f)) functions.

the SPV signal increases linearly when the light is switched on and decreases linearly when the light is switched off. In this case, the X-signal is equal to 0 and the Y-signal reaches the maximum (Figures 2.24(e) and 2.24(f)). The investigation of the frequency dependence of the X- and Y-signals can provide information about relaxation times of charge separated in space.

Trapping of photogenerated charge carriers can strongly influence modulated charge separation. For example, the density of free photogenerated charge carriers is usually much higher than the density of trapped charge carriers during illumination. Therefore, free charge carriers separated in space dominate usually the SPV signal during illumination. After switching off the light, free charge carriers recombine much faster than trapped charge carriers. Therefore, trapped charge carriers separated in space often dominate the SPV signal after switching of the light.

Instead of using X- and Y-signals, modulated SPV measurements can be represented in terms of the amplitude (R) and the phase angle (ϕ) which are defined as

$$R = \sqrt{X^2 + Y^2} \tag{2.31}$$

$$\varphi = \frac{180°}{\pi} \cdot \arctan\left(\frac{Y}{X}\right) \tag{2.32}$$

The range of the phase angle contains information about the direction of modulated charge separation in space. For example, a phase angle in the range between 270° and 0° (360°) corresponds to the separation of positive charge carriers toward the reference electrode.

SPV signals are a superposition of free and trapped charge carriers separated in space. Depending on the distribution in space, the nature and the occupation of trap states, the contribution of trapped charge carriers to SPV signals can be of the same or opposite sign as the contribution of the free charge carriers separated in space. As an example, the influence of trapping of photogenerated charge carriers at surface states on modulated SPV signals is illustrated in Figure 2.25. Incidentally, trapping can strongly influence the overall distribution of charge in modulated SPV measurements and can therefore also dominate the sign of X-signals.

If free and trapped charge carriers are separated toward the same direction (Figures 2.25(a) and 2.25(b)), the sign of the SPV signal does not change after switching off the light and the Y-signals are negative or positive for positive or negative X-signals, respectively. For those cases, the phase angles range between 270° and 0° or between 90° and 180°, respectively.

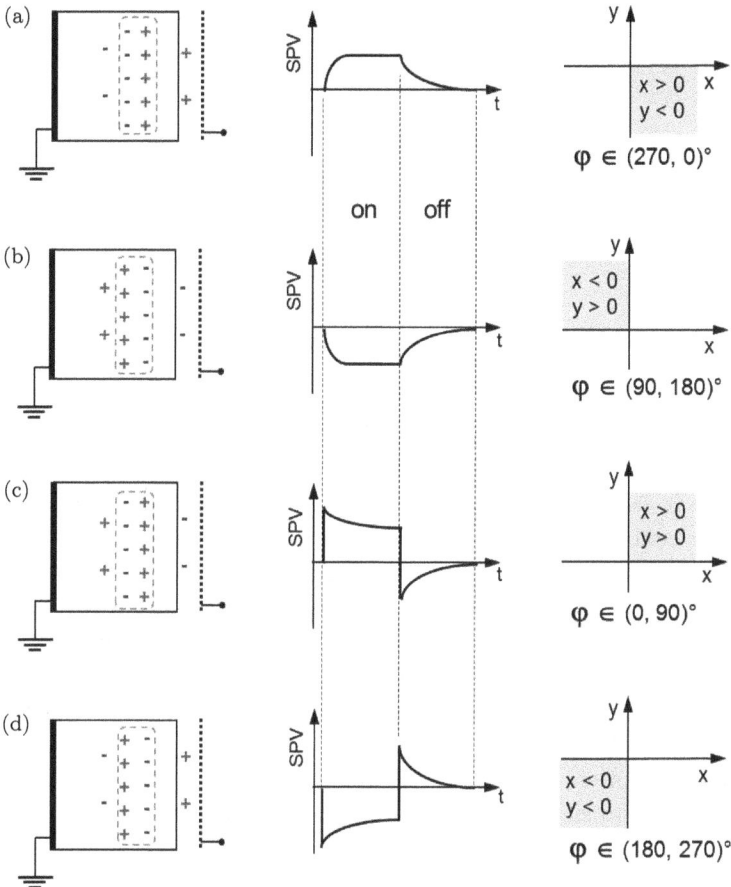

Figure 2.25. Illustration for the influence of trapping of charge carriers at surface states on modulated SPV signals. Positive and negative charges marked by the dashed squares denote those free charge carriers following fast recombination after switching off the light. (left) Schematic diagrams of the preferential separation of free and trapped charge carriers with the same ((a) and (b)) and opposite ((c) and (d)) directions of charge separation and of preferential separation of positive or negative charge toward the reference electrode during light on ((a), (c) or (b), (d), respectively). (middle) Idealized SPV signals during light on and light off. (right) Quadrants for corresponding X- and Y-signals and ranges of phase angels.

If free and trapped charge carriers are separated toward opposite directions (Figures 2.25(c) and 2.25(d)), the sign of the SPV signal changes after switching off the light and both the X- and Y-signals are positive or negative. The Y-signals are positive or negative for preferential trapping of

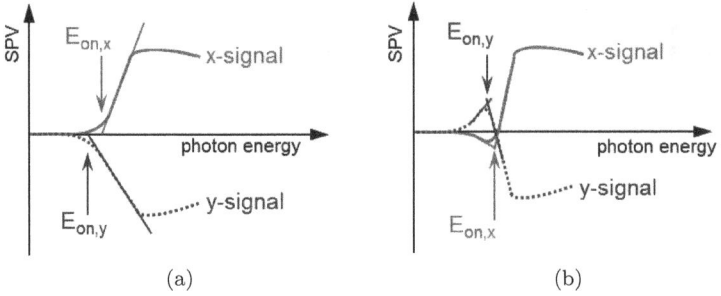

Figure 2.26. Schematic SPV spectra of X- and Y-signals for a semiconductor with a positive band bending and defect states (a) near the valence band edge in the bulk or (b) near the surface causing preferential trapping of holes. $E_{on,x}$ and $E_{on,y}$ denote the onset energies, which are close to the band gap of the semiconductor.

electrons or holes, respectively, at surface states. For those cases, the phase angles range between 0° and 90° or 180° and 270°, respectively.

Depending on the influence of trapping, X- and Y-spectra can be rather different and the analysis of modulated SPV spectra is often not straightforward. In the simplest case, transition energies can be described by onset energies of X- and Y-signals ($E_{on,x}$ and $E_{on,y}$, respectively, Figure 2.26). Onset energies can be close to, for example, an optical band gap but are not equal to it in most cases. Incidentally, the spectra of the Y-signals are often shifted to lower photon energies in comparison to the X-signals due to slower discharge of deeper defect states.

2.4. SPV Measurements with Electron Beams and Photoelectrons

2.4.1. *Spectral-dependent measurements at a constant electron beam*

The electron current (I_e) from an electron gun toward a sample surface depends on the potential between the cathode of the electron gun and the sample surface. Under illumination, this potential changes by the value of ΔCPD. Therefore, electron beams can be applied for the measurement of ΔCPD (Figure 2.27(a)). For this purpose, I_e is adjusted to a low constant value of the order of 0.1–1 μA by applying V_b. At constant I_e, the change of V_b caused by illumination is equal to ΔCPD:

$$\Delta\text{CPD} = (V_{b,\text{ill}} - V_{b,\text{dark}})|_{I_e=\text{const}} \qquad (2.33)$$

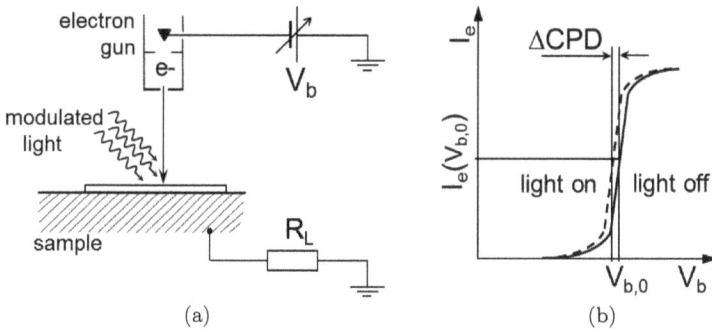

Figure 2.27. Principle of a measurement of ΔCPD with an electron beam.

The measurement of ΔCPD with an electron beam can be applied, for example, for mapping of SPV signals over relatively large areas (Henzler and Clabes, 1974). The lateral resolution of SPV measurements with an electron beam depends on the diameters of the electron and light beams and their alignment. The sensitivity of SPV measurements with electron beams is increased by modulating the light (Steinrisser and Hetrick, 1971).

For the modulated measurement of ΔCPD, an electron gun operating at low voltage is installed in UHV and directed toward the sample surface. The work function of the metal oxide of the cathode is relatively low and the cathode is indirectly heated for sufficient electron emission. The bias potential of the electron gun can be varied and well stabilized in a narrow range of several volts. The electron current is measured as the voltage drop at the load resistance (R_L, of the order of 100 kΩ). During the SPV measurement, the sample is illuminated with modulated monochromatic light and the modulated voltage drop at the load resistance is measured with a lock-in amplifier. V_b is adjusted to a value near the maximal change of the current in the current–voltage characteristic ($V_{b,0}$ in Figure 2.27(b)). The change of the voltage drop at R_L (ΔU_L) is given by R_L and the difference between I_e at $V_b + \Delta$CPD and I_e at V_b:

$$\Delta U_L = R_L \cdot (I_e(V_b + \Delta\text{CPD}) - I_e(V_b)) \tag{2.34}$$

The value of ΔCPD is much smaller than $V_{b,0}$. Therefore, ΔU_L is proportional to the slope of the current–voltage characteristic at $V_{b,0}$ and to ΔCPD:

$$\Delta U_L = R_L \cdot \left.\frac{\partial I_e}{\partial V_b}\right|_{V_b=V_{b,0}} \cdot \Delta\text{CPD} \tag{2.35}$$

Incidentally, instead of applying a potential source and a load resistance (Steinrisser and Hetrick, 1971), a constant current device can be used for modulated SPV measurements with electron beams (Clabes and Henzler, 1980).

Due to the modulation of the light and the application of a lock-in amplifier, the change of the CPD due to spectral-dependent illumination can be measured with a very high sensitivity of the order of 1 μV (Steinrisser and Hetrick, 1971). For comparison, this sensitivity is much higher than for SPV measurements with a vibrating Kelvin probe and is of a similar order as the sensitivity for modulated SPV measurements with a fixed capacitor in a vacuum chamber.

The application of electron beams with low energy and intensity for SPV measurements allows for the temperature- and spectral-dependent investigation of surface states at semiconductor surfaces under extremely clean conditions in UHV. This was nicely demonstrated, for example, on silicon crystals cleaved and conditioned in UHV (Kasupke and Henzler, 1976; Clabes and Henzler, 1980).

Surface contaminations are crucial for the investigation of surface states in UHV. As an additional advantage, the application of electron beams with low energy and intensity does not introduce surface contaminations during operation for many hours (Clabes and Henzler, 1980).

2.4.2. *SPV measurements with photoelectrons*

Electrons excited with photons that have energy larger than the surface work function of the sample electrode can leave the sample into vacuum. This process is called photoemission. The electrons emitted by photoemission are called photoelectrons. The measurement of the density of photoelectrons as a function of their kinetic energy is called photoemission spectroscopy (PES).

The mean free path of electrons (λ_{mfp}) in matter defines the information depth of photoelectrons. The value of λ_{mfp} depends on the energy of the electron (see Seah and Dench 1979 and references therein). Figure 2.28 depicts λ_{mfp} as a function of energy for solids in terms of nanometer and of monolayers for inorganic compounds, elements and organic compounds.

In the range between about 10 and 100 eV, the values of λ_{mfp} are of the order of 0.5–5 nm or 1–10 monolayers. The barrier for the emission of photoelectrons depends on the binding energy and on the surface work function of the material. Therefore, PES is extremely surface-sensitive and

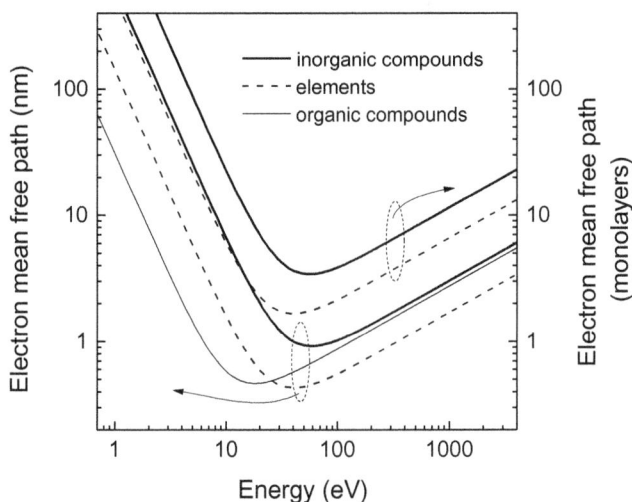

Figure 2.28. Energy dependence of the electron mean free path in organic compounds, elements and organic compounds (thick solid, dashed and thin solid lines, respectively) by using Equation (5) given in Seah and Dench (1979).

the kinetic energy of photoelectrons is sensitive to the CPD and its change. In addition, due to the extremely short λ_{mfp} in solids and short λ_{mfp} in gases, measurements of the emission of photoelectrons shall be performed under UHV conditions.

Figure 2.29 shows a schematic diagram for photoemission from states E_S in a layer on top of the sample electrode into vacuum. The kinetic energy of the photoelectron ($E_{k,0}$) is measured with an analyzer (see, for example, Henzler and Göpel, 1991). The kinetic energy of the photoelectron in the dark is measured with respect to the vacuum level of the reference electrode ($E_{k,\mathrm{dark}}$).

The kinetic energy of photoelectrons depends on the energy of the exciting photons and the energy of the electronic states from which the photoelectrons are excited. In PES, electrons excited from valence states (order of 1–10 eV) are distinguished from electrons excited from core levels (order of 10–1000 eV). Typical light sources for PES are, for example, helium lamps operating in the HeI (21.2 eV) and HeII (40.8 eV) modes, x-ray anodes, such as Mg (Mg Kα line, 1253.6 eV) or synchrotrons (<10–3000 eV). PES from valence states and from core levels are also called ultraviolet photoelectron spectroscopy (UPS) and x-ray photoelectron spectroscopy (XPS), respectively.

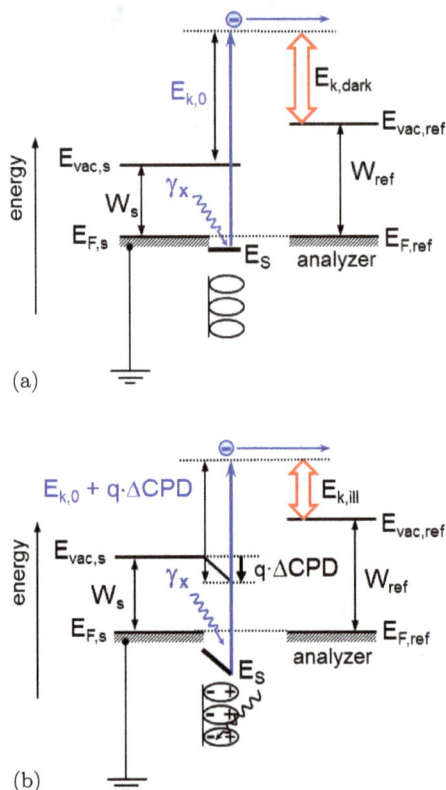

Figure 2.29. Photoelectron emission in (a) the dark and (b) under illumination. The photoelectron is excited by a photon (γ_x) from the state E_S. The kinetic energy of the photoelectron (E_k) is measured with respect to E_{vac} of the analyzer. The difference $E_{k,ill} - E_{k,dark}$ corresponds to the light-induced ΔCPD.

Under illumination, the kinetic energy of photoelectrons depends on ΔCPD. Hence, the SPV can be investigated by measuring E_k under illumination ($E_{k,ill}$) and in the dark ($E_{k,dark}$):

$$\Delta\text{CPD} = \frac{E_{k,ill} - E_{k,dark}}{q} \tag{2.36}$$

Incidentally, photons used for the excitation of photoelectrons can already cause a SPV. Furthermore, shifts in the spectra of E_k can also be induced by other phenomena, such as changes of the chemical environment of surface atoms. This has to be considered in the analysis of PES spectra. A rigid shift of all features related to core levels in PES spectra without changes in the line shapes under illumination is a clear indication for SPV whereas the shift is equal to the SPV (Margaritondo *et al.*, 1980). Similar shifts of valence band peaks, of the edge toward the Fermi energy and of the secondary electron onset under illumination also provide clear evidence for an SPV (Jaegermann, 1986).

Synchrotrons provide unique conditions for time-dependent SPV measurements by PES. Electrons are accelerated to a very high velocity move in a storage ring and emit synchrotron radiation of photons with energies from the infrared to the x-ray (Figure 2.30(a)). In a synchrotron, electrons move in the so-called bunches of a duration time of tens of picoseconds, whereas the minimum time between bunches is of the order of 2 ns (see also the reviews of Yamamoto and Matsuda, 2013; Neppl and Gessner, 2015).

For SPV measurements, laser pulses are synchronized with bunches for PES analysis whereas sophisticated triggering between the electronic master clock of the storage ring and the laser pulses is decisive (Widdra *et al.*, 2003). For very fast measurements, a delay line (Figure 2.30(b)) is applied in order to define the time values in a measured transient (pump-probe

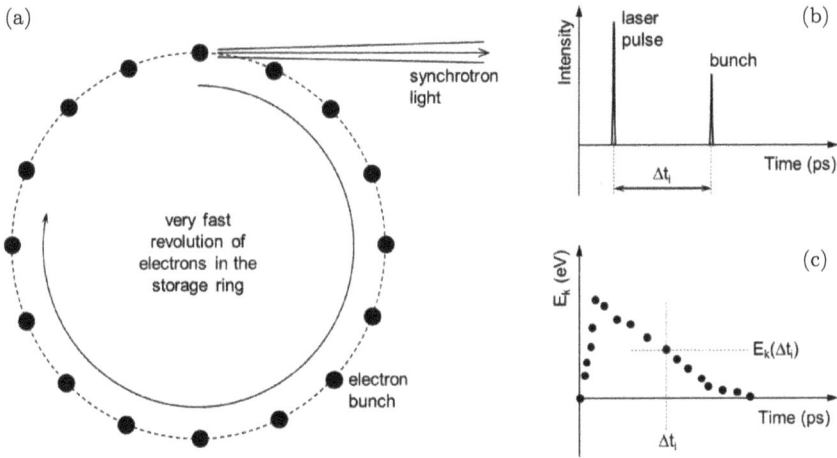

Figure 2.30. Principle of measuring very short SPV transients by photoelectron spectroscopy at a synchrotron. Short electron bunches move in a storage ring and emit very short pulses of synchrotron light (a) which is used for the excitation of photoelectrons from the sample surface. Short laser pulses used for photogeneration in the sample are synchronized with electron bunches. A tunable time delay between the laser pulses (pump beam) and electron bunches (probe beam) allows for probing the sample at different times after photogeneration (b). SPV transients are built from the shifts of the kinetic energy of photoelectrons at the different times (c).

principle, see also Figure 2.30(c)). At the same time, the PES measurement shall be faster than the time between subsequent bunches. A resolution time of 60 ps has been achieved for PES measurements at synchrotrons (Widdra *et al.*, 2003). Related systems were applied, for example, to the investigation of the SPV response by PES of $SiO_2(1.5\,nm)/Si(100)$ in UHV (Bröcker *et al.*, 2004), p-GaAs, ZnO/p-$Si(100)$ or a surface of ZnO sensitized with N3 dye molecules (Neppl *et al.*, 2014).

The resolution of the kinetic energy (ΔE_k) can be estimated for a given geometry of the spectrometer, for example, for a time-of-flight (TOF) spectrometer (Yamamoto and Matsuda, 2013):

$$\Delta E_k = \sqrt{\frac{8 \cdot E_k^2}{m_e} \cdot \frac{\Delta t}{d_{\text{flight}}}} \qquad (2.37)$$

where m_e, Δt and d_{flight} denote the electron mass, the time resolution and the flight distance, respectively. Therefore, the potential resolution in short SPV transients measured by PES at a synchrotron can be of the order of tens of millielectron volts if E_k, Δt and d_{flight} are of the order of tens of electronvolts, 100 ps and 1 m, respectively. While going to excitation and analysis with ultra-short (fs range) pulses (see, also Yamamoto and Matsuda, 2013), the energy resolution will be limited by the Heisenberg uncertainty principle. Therefore, the application of pulses with duration times shorter than 0.1 ps seems unreasonable for time-resolved SPV measurements with PES.

The information from very short SPV transients by XPS at a synchrotron is combined with additional information which can be obtained from XPS measurements. This gives the opportunity for direct correlation between processes of charge separation and changes of chemical bonds and surface species. It is possible, by using higher energetic photons, to analyze photoelectrons which are emitted from a surface in a chemical environment through an ultra-thin separating membrane into the UHV (Shavorskiy *et al.*, 2014).

2.5. Summary

The principles of SPV measurements are based on the following:

(i) nulling of an ac current or an ac electrostatic force between two vibrating electrodes by applying a bias potential (Kelvin probes),

(ii) out-coupling of potentials changing in time with fixed capacitors and
      high-impedance buffers,
(iii) adjusting of a constant electron current with a bias potential, and
(iv) shifts of peaks in photoelectron spectroscopy.

Each of these principles has specific requirements to equipment,
limitations and fields of application. SPV signals can be excited with cw,
modulated or pulsed light sources at variable wavelengths.

Figure 2.31 gives a schematic overview of the time ranges and sensitivities for the different classes of SPV techniques.

Macroscopic Kelvin probes are the most fundamental approach for
measuring SPV since the photoinduced change in the bias potential
for nulling the ac signals is directly related to the SPV and since no
additional, the SPV signals disturbing, interactions are present. Kelvin
probes are sensitive to any fast and slow processes leading to separation of
photogenerated charge carriers in space. The sensitivity is about 0.1–1 mV
for SPV signals measured with Kelvin probes. The resolution time of Kelvin
probes is limited to the order of 0.1–1 s.

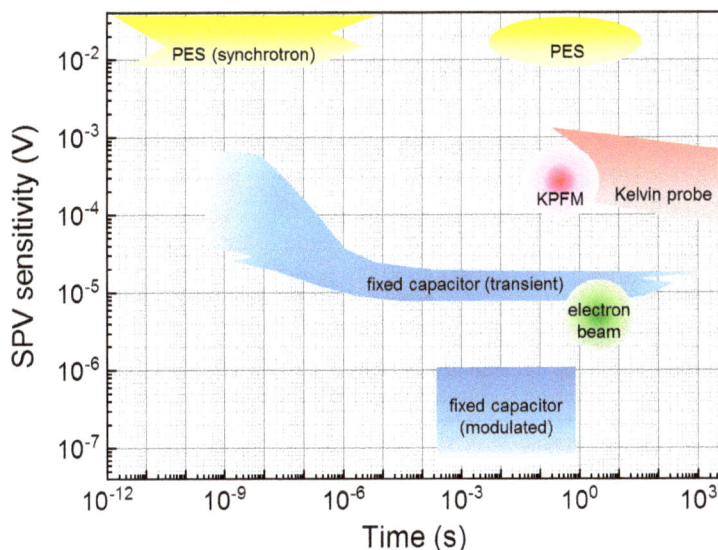

Figure 2.31. Overview of the time ranges and sensitivities for SPV measurements with
Kelvin probes (reddish), fixed capacitors (blueish), electron currents at low potential
(greenish) and PES and XPS at a synchrotron (yellowish).

Macroscopic Kelvin probes are very reliable and compact. For this reason, such electrodes can be well implemented into complex setups with variable ambient or liquid solutions.

Vibrating tiny tips fixed at vibrating cantilevers are used as reference electrodes in KPFM. The adjustment of the electrostatic force at a very short distance between the sample and reference electrodes allows for a high spatial resolution of the order of tens of nanometer in SPV measurements with KPFM. However, the electrostatic interaction between the sample surface and the tip can disturb the electronic states at a surface and therefore SPV signals.

SPV signals changing in time are required for the measurement with fixed capacitors and high-impedance buffers. In related SPV measurements, the reference potential may change due to some partial charging of the sample surface under repetitive modulated or pulsed illumination. This shall be taken into account when interpreting SPV signals.

High-impedance buffers are needed for matching the impedances of the fixed capacitor and of the measurement device. The input and output impedances of high-impedance buffers are of the order of $G\Omega$–$T\Omega$ and equal to $50\,\Omega$, respectively. The right choice of, for example, the lengths of connecting wires, electrical feedthroughs, insulators, shielding and input impedances is decisive for transient SPV measurements from the nanosecond to the hour range and for efficient reduction of noise. Measures such as a logarithmic readout or the measurement of difference transients can be useful for reducing the noise to the order of $10\,\mu V$ at longer times. In modulated SPV measurements with fixed capacitors, the noise can be further reduced to values below $100\,nV$.

Additional electronic interactions aside photogeneration and charge separation (Table 2.1) can disturb SPV signals. The application of electrical

Table 2.1. Sources for electronic interactions in SPV measurements with (macroscopic) Kelvin probes, microscopic Kelvin probes (KPFM), fixed capacitors, (low energy) electron beams, X-ray photoelectron spectroscopy and UV photoelectron spectroscopy.

	Light	V_b	V_{ac}	Electrons	X-rays	UV
Kelvin probe	X	X				
KPFM	X	X	X			
Fixed capacitor	X					
Electron beam	X	X		X		
PES (XPS)	X			X	X	
PES (UPS)	X			X		X

potentials has an influence on the population of trap states. The injection or emission of electrons changes the amount of charge in a sample. X-rays and (far) UV light cause photogeneration of additional mobile charge carriers. Incidentally, only SPV measurements with fixed capacitors do not require any additional interactions aside photogeneration and charge separation.

Electrodes for SPV measurements with fixed capacitors can be well adopted for measurements in practically any different ambient (UHV, vacuum, gas atmospheres, liquid solutions). This provides a very high degree of freedom for experimentalists. Furthermore, modulated SPV spectroscopy with fixed capacitors is, aside its high sensitivity, very reliable and relatively fast and therefore suitable for applications in research and development and for monitoring of technological processes.

The adjustment of a constant current of electrons with low energy with a bias potential allows for measurements of SPV signals over 3–4 orders of magnitude with a sensitivity of about 10 μV. This method can be applied, for example, for the investigation of surface states on ultra-clean surfaces in UHV. One has to keep in mind that interaction of electrons can cause changes in the surface. Incidentally, the influence of electrons with low energy on the surface is much smaller than for higher energetic electrons or X-rays.

PES is extremely surface-sensitive. The shift of peaks, for example, of core levels in XPS spectra, can be measured with an accuracy of the order of ten(s) of millielectron volts. Therefore, the resolution of the measurement of SPV signals by PES is of the order of 10–100 mV, which is the lowest resolution in comparison with the other techniques. Nevertheless, the analysis of the shift of peaks in XPS spectra is a very elegant method to separate SPV signals from chemical shifts and therefore to correlate SPV signals with changes in chemical bonds.

Synchrotrons give the unique opportunity to investigate extremely short SPV signals in pump–probe experiments with time-resolved XPS for analysis. Very short pulses of synchrotron radiation are caused by the short bunches of electrons (duration of about 20 ps) moving in the storage ring. The time delay between an exciting laser pulse and an analyzing bunch can be varied systematically and allows for the investigation of SPV transients at very short times. Such measurements take a relatively long time and demand extensive technical support. It is useful to apply the opportunities of a synchrotron for unique and sophisticated SPV experiments in fundamental research.

Chapter 3

Practical Aspects of Surface Photovoltage Measurements

3.1. Introduction

Practical experience is shared for different kinds of surface photovoltage (SPV) measurements. Basic setups for continuous, modulated and transient SPV spectroscopy are explained. Attention is paid to the intensity spectra and spectral resolution of monochromatic light sources, the role and suppression of stray light and the criterion of the low signal case for the characterization of band gaps and transition energies. SPV measurements can be performed on bulk semiconductors, thin and extremely thin photoactive layers and photoactive powders. Furthermore, a design is provided for the realization of a robust high-impedance buffer with a bandwidth of 500 MHz. An SPV measurement configuration contains capacitors and inductances. Therefore, SPV signals can excite oscillations which disturb SPV transients at times shorter than about tens of nanoseconds. Resolution times of 1–2 ns and/or sub-ns can be reached for transient SPV measurements by analog damping of oscillations and/or self-consistent numerical extraction of very short source transients, respectively. Parameters such as temperature, ambient atmosphere, bias potentials, electrolytes or light intensity can be varied during dedicated SPV experiments. The variation of parameters shall not disturb SPV signals. Therefore, abrupt changes of electric potentials and currents, photoinduced reactions at reference electrodes and coupling of control loops shall be avoided. Essential experimental measures are explained. Examples are shown for SPV measurements of a temperature-dependent band gap, an activation energy from transient SPV measurements, a pressure-dependent evolution of defects in nanoporous TiO_2, a bias potential dependence of large transient SPV signals in c-Si and diffusion lengths of photogenerated charge carriers obtained at constant modulated SPV signals in a c-Si crystal and $CH_3NH_3PbI_3$ powder.

3.2. Spectral-Dependent SPV Measurements

3.2.1. *Basic setups for continuous, modulated and transient SPV spectroscopy*

A basic setup for spectral-dependent SPV measurements consists of five general components (Figure 3.1). The monochromatic light source provides the monochromatic light of a given wavelength (λ). The light beam is formed and directed with the optical alignment system. The SPV signal arises in the measurement configuration and is measured with a measurement device. A computer is used for data transfer to the wavelength control of the monochromatic light source and for reading out data from the measurement device.

Monochromators with lamps are used as tunable monochromatic light sources for continuous and modulated illumination. Pulsed tunable lasers are applied for transient spectral-dependent SPV measurements.

In the following, aspects for SPV measurements with monochromators will be explained in more detail. In a monochromator, the light of a lamp is split into its spectral components by a dispersive element. Gratings or prisms are used as dispersive elements. The dispersion of a grating is constant and its dependence on temperature is weak due to low thermal

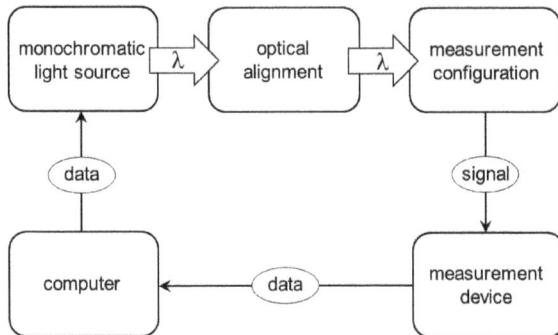

Figure 3.1. Principle block diagram for spectral-dependent SPV measurements.

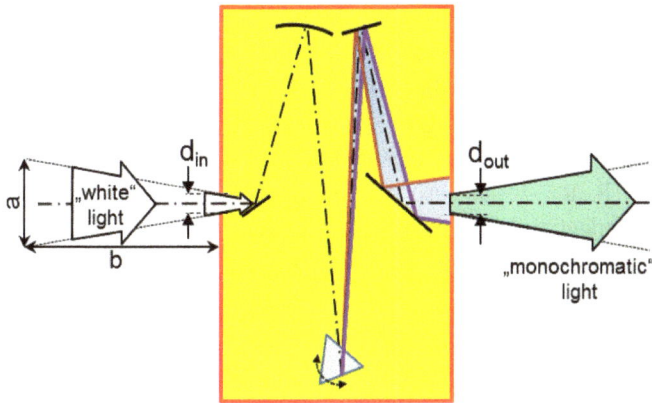

Figure 3.2. Schematic diagram of a prism monochromator. The ratio a/b is called the opening ratio. The widths of the entrance and exit slits are denoted by d_{in} and d_{out}, respectively.

expansion. In contrast, the dispersion of a prism depends on temperature. This shall be taken into account in SPV experiments, for example, with a quartz prism monochromator.

Light of a lamp enters a monochromator via the entrance slit and monochromatic light leaves a monochromator via the exit slit (Figure 3.2). The region of the entrance slit is illuminated homogeneously. Therefore, the intensity of light entering the monochromator is proportional to the width of the entrance slit (d_{in}). The intensity of monochromatic light leaving the monochromator is proportional to the width of the exit slit (d_{out}). The intensity of monochromatic light is proportional to the product of d_{in} and d_{out}. Usually, the values of d_{in} and d_{out} are changed symmetrically ($d_{in} = d_{out} = d_{slit}$).

A monochromator has a certain acceptance angle, which is also expressed by the opening ratio (a/b). The opening ratio is defined by the ratio between the diameter of a conus and its distance to the entrance slit at which light is available for spectral splitting. Light entering the monochromator at an angle larger than the acceptance angle is not available for spectral splitting. The opening ratio is given by the design of the optical components inside the monochromator.

The intensity of the light entering a monochromator is proportional to the squared opening ratio. A high opening ratio can be helpful for the characterization of samples with very low SPV signals. However, the intensity of stray light, i.e. non-monochromatic light leaving the monochromator, is also proportional to the squared opening ratio. For sensitive SPV measurements with a minimum number of artifacts, the intensity of stray light shall be as low as possible. Therefore, the application of a monochromator with a low opening ratio is recommended.

SPV spectra are often measured over a wide range of λ from the near infrared (NIR, 2.5–1.5 μm) to the ultraviolet (UV) regions (400–250 nm). For measurements in a wide spectral range with a grating monochromator, changes of gratings and order filters are demanded. This causes discontinuities in SPV spectra. Usually, such discontinuities cannot be eliminated in SPV spectra by normalization, as for transmission or photocurrent spectroscopies, since SPV signals are not necessarily proportional to the photon flux. In contrast, the application of a quartz prism monochromator allows for measurements of continuous spectra over a wide spectral range.

Halogen or xenon arc lamps (spectra in Figure 3.3) are usually used for continuous and modulated spectral-dependent SPV measurements. The

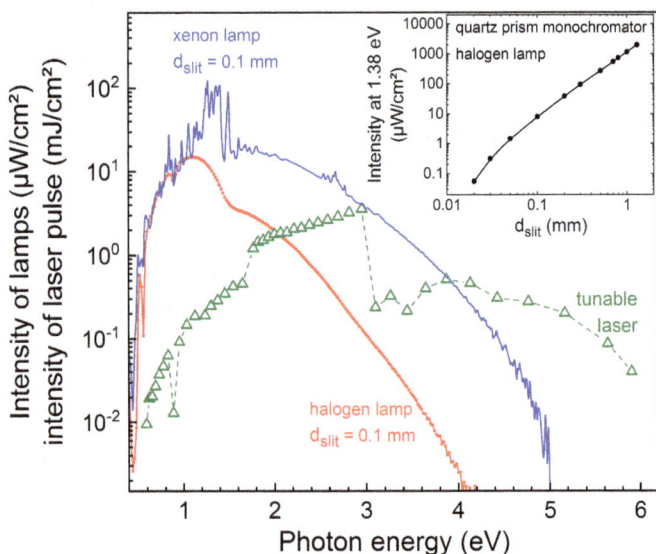

Figure 3.3. Spectra of monochromatic light sources consisting of a 100 W halogen lamp (red circles and line) or a 400 W xenon arc lamp (blue) with a quartz prism monochromator (SPM2) and of a pulsed tunable Nd:YAG laser (NT242) with a spectral cleaning unit and a fixed beam expander ($\times$3) (green dashed line and symbols). The inset shows the dependence of the intensity of the monochromatic light source with the halogen lamp on d_{slit}.

temperatures of the corresponding blackbody radiation are about 3,000 and 10,000 K. Therefore, halogen or xenon arc lamps are preferred for measurements in the NIR or UV regions, respectively. In addition, spectral lines in the NIR and around 460–500 nm of a xenon arc lamp shall be considered in the analysis of SPV spectra.

For transient SPV spectroscopy, a diode-pumped tunable Nd:YAG laser can be used as a pulsed monochromatic light source over a spectral range from about 2400 to 220 nm. An additional optical cleaning unit serves for suppression of harmonics of the fundamental laser line.

Figure 3.4 shows basic setups for continuous, modulated and transient SPV spectroscopy (Figures 3.4(a)–3.4(c), respectively). Continuous monochromatic light is directed onto a vibrating Kelvin probe (Figure 3.4(a)). Modulated or pulsed monochromatic light is directed onto the electrodes in the fixed capacitor arrangement (Figures 3.4(b) and 3.4(c), respectively). The light from the monochromator is collimated and slightly focused on the electrodes. The light intensity from the pulsed laser is reduced with a tunable beam expander and the spot size is limited with an additional aperture.

The electrodes and the sample are mounted in a shielded chamber with a quartz window. The piezo crystal and the vibrating gold mesh are connected with the Kelvin probe controller via electrical feedthroughs. For measuring modulated and transient SPV signals in vacuum and over wide ranges of gas pressures and temperatures, the high-impedance buffer is placed outside the chamber and connected with the reference electrode via a wire and an electrical feedthrough. The noise level of modulated SPV signals can be strongly reduced if the high-impedance buffer is powered by a battery and placed inside the shielded chamber.

For modulated SPV measurements, the light is modulated with a mechanical chopper, which can be placed in front of the entrance or behind the exit slit of the monochromator. Mechanical choppers are designed for a frequency range between several hertz and about 3–10 kHz. Incidentally, in order to keep a low noise level, f_{mod} should not be equal or close to the grid frequency or multiples of the grid frequency. For realizing a certain range of modulation frequencies with mechanical choppers, the number of segments of the chopper blade(s) is varied, which can also have some influence on the phase angle at the lock-in amplifier. The phase angle at the double-phase lock-in amplifier is calibrated, for example, with a silicon photodiode, the response of which is much faster than the modulation period. Light emitting devices (LEDs) can be applied for very precise

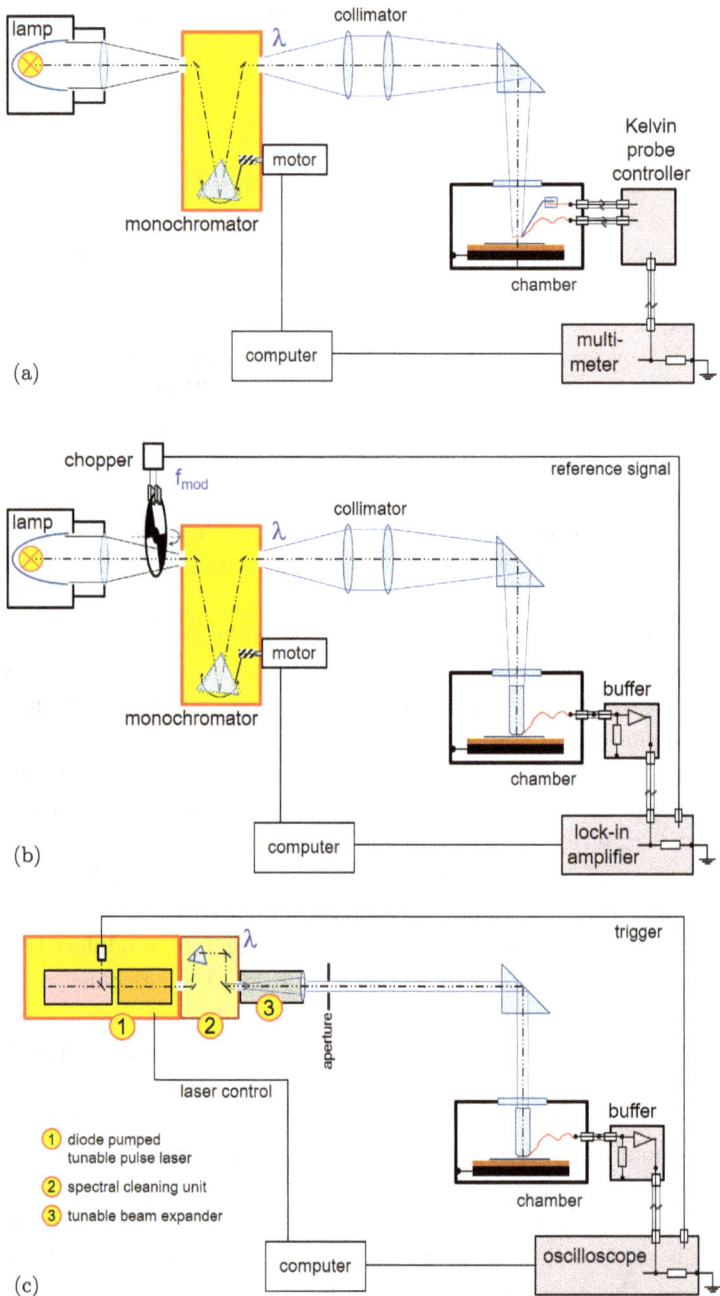

Figure 3.4. Schematic setups for (a) continuous, (b) modulated and (c) transient SPV spectroscopy.

modulated SPV measurements in a wide frequency range (mHz–MHz range) at fixed wavelengths.

SPV signals are measured with a multi-meter (continuous), a lock-in amplifier (modulated) or an oscilloscope (transient). A computer is used for reading out the SPV signals at a given wavelength. For reducing the noise, the number of averages (N_{avg}) can be increased and, for modulated measurements, the integration time constant of the lock-in amplifier (τ_{int}) can be increased:

$$\delta_{SPV} \sim \frac{1}{\sqrt{N_{avg}}} \tag{3.1}$$

$$\delta_{SPV} \sim \frac{1}{\sqrt{\tau_{int}}} \tag{3.2}$$

Time intervals for the measurement of one point in modulated SPV spectra are of the order of 1–10 s. Therefore, modulated SPV spectra can be measured within about 2–20 min.

Digital or sampling oscilloscopes are characterized by their maximum sampling rate (in units of samples per s; S/s) and maximum number of samples which can be registered in one transient. The maximum sampling rate and the maximum number of samples depend on the oscilloscope and are typically of the order of 100 MS/s–4 GS/s and 0.01 MS–1 GS, respectively.

3.2.2. *Spectral resolution of monochromatic light sources*

The intensity of the exciting light from a monochromator is proportional to the product of d_{in} and d_{out} (Figure 3.5(a)). The increase of the width of the slits causes an increase of the SPV signals but a reduction of the spectral resolution, i.e. an increase of the difference of the peaks of spectral lines which can be still distinguished in a spectrum ($\Delta\lambda$). For a precise characterization of, for example, the edge of the band gap of a semiconductor (Figure 3.5(b)) or the width of a spectral line (Figure 3.5(c)), the spectral resolution shall be high, i.e. $\Delta\lambda$ shall be low. This demand can be in conflict with a low SPV signal requiring a high intensity for spectral-dependent SPV measurements of a certain sample. Therefore, the application and/or interpretation of spectral-dependent SPV measurements can be limited by the intensity of SPV signals in accordance with the spectral resolution.

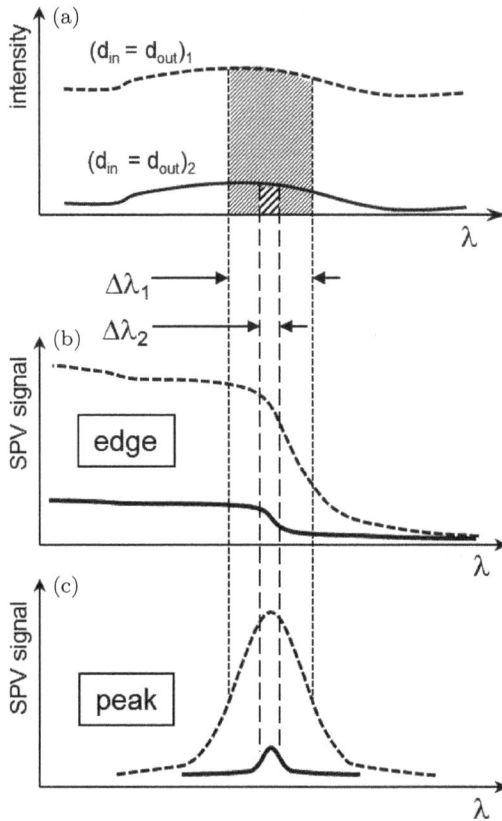

Figure 3.5. Schematic diagram of the intensity of monochromatic light for (a) broad or narrow slits of a monochromator (($d_{\mathrm{in}} = d_{\mathrm{out}})_1$ and ($d_{\mathrm{in}} = d_{\mathrm{out}})_2$, respectively) and (b) for corresponding SPV spectra near a band edge of a semiconductor or (c) a peak of a spectral line. $\Delta\lambda_1$ and $\Delta\lambda_2$ denote the spectral resolutions of the monochromator.

For further analysis, SPV spectra are usually presented as a function of photon energy ($h\nu$). For this purpose, the SPV spectrum shall be converted from a function of λ into a function of $h\nu$. The photon energy and wavelength can be converted using the following expression:

$$h\nu \cdot \lambda = \frac{h \cdot c}{q} = 1240 \cdot \mathrm{eV} \cdot \mathrm{nm} \tag{3.3}$$

Furthermore, one should keep in mind that a SPV spectrum is the dependence of the change of the SPV signal in relation to the interval of integration, i.e. the spectral resolution:

$$\frac{d(\text{SPV}(h\nu))}{d(h\nu)} = \frac{d(\text{SPV}(\lambda))}{d\lambda} \tag{3.4}$$

Therefore, $d\lambda$ or $d(h\nu)$ should be converted together with the axis of the spectrum. Incidentally, onset energies and relative changes of SPV signals are independent of the transformation of the values. Therefore, often only the wavelength is transformed into the photon energy when showing $\text{SPV}(h\nu)$ spectra.

The spectral resolution of the wavelength $(\Delta\lambda)$ is converted into the spectral resolution of the energy (ΔE):

$$\Delta E = \frac{h \cdot c}{q} \cdot \frac{\Delta\lambda}{\lambda^2} \tag{3.5}$$

For demonstration, $\Delta\lambda$ and ΔE of a quartz prism monochromator (SPM2, Carl Zeiss Jena) is depicted in Figure 3.6 for d_{slit} equal to 1.0 and 0.1 mm as a function of λ and of $h\nu$. As an example, $d(h\nu)$ is about 70 and 7 meV at a photon energy of 1 eV for d_{slit} equal to 1.0 and 0.1 mm, respectively. The characterization of exponential tails (E_t) is limited by the spectral resolution. Furthermore, steps of the order of 10–50 meV are

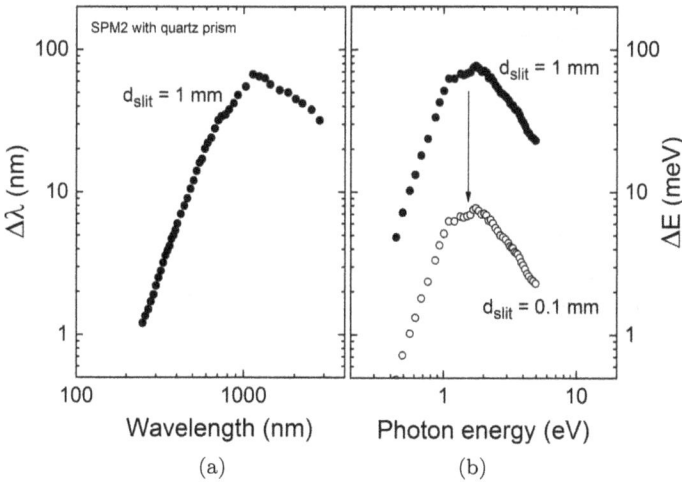

Figure 3.6. Spectral resolution of a quartz prism monochromator (SPM2) for d_{slit} equal to 1.0 and 0.1 mm as a function of (a) wavelength and (b) photon energy.

sufficient for the variation of $h\nu$ in spectral-dependent measurements with a quartz prism monochromator. Incidentally, the value of a band gap can be obtained with a precision higher than ΔE if including several values of SPV($h\nu$), which change weaker than the limitation by the spectral resolution, into the analysis with a fit function.

It is impossible to characterize samples with low SPV signals at a high spectral resolution. Nevertheless, the reduction of the spectral resolution to a range of around, for example, 50 meV enables modulated spectral-dependent SPV measurements on samples with very low signals. This is useful for the SPV analysis of photoactive materials with high band gaps and E_t larger than 50 meV, such as porous titania, and for the investigation of electronic states in the forbidden band gap of semiconductors, such as crystalline silicon or hybrid metal halide perovskites. A reduced spectral resolution of the order of 50 meV over a wide spectral range can be easily achieved with a quartz prism monochromator. Therefore, a quartz prism monochromator is very useful for the characterization of photoactive materials with very low modulated SPV signals.

Laser lines are usually very sharp in comparison to the dispersion of a monochromator. Therefore, a tunable pulsed laser based on a Nd:YAG laser with an optical parametric amplifier is practically "dispersion free" and allows for precise spectral-dependent SPV measurements. Furthermore, the spectral response of transient SPV signals can sensitively depend on the photon energy at which mobile charge carriers are excited. Related phenomena can be accurately studied by the transient SPV spectroscopy.

As an example for the role of the spectral resolution in SPV measurements, Figure 3.7 compares normalized modulated in-phase SPV spectra of moderately doped c-Si measured for different values of d_{slit} and a transient spectrum of highly n-type doped c-Si measured at 2 ns.

Due to the increasing spectral resolution with decreasing d_{slit}, the onsets of the spectra shifted toward shorter wavelengths with decreasing d_{slit}. For large values of d_{slit}, a precise measurement of the band gap of c-Si is not possible. For practical reasons, an onset energy (E_{on}) can be defined as the intersection of the slope in the inflection point of the spectrum with the x-axis. The corresponding values of E_{on} were about 1.03, 1.07, 1.08, 1.1 and 1.11 eV for d_{slit} equal to 1.0, 0.5, 0.2, 0.1 and 0.05 mm, respectively. For comparison, the band gap of c-Si is 1.12 eV at room temperature. The values of E_{on} can be close to E_g depending on the spectral resolution. Therefore, precision is lost on the price of high intensity at low spectral resolution.

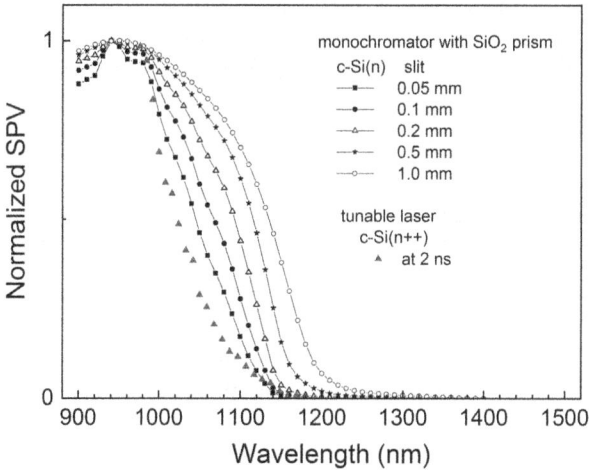

Figure 3.7. Normalized modulated in-phase SPV spectra near the band gap of moderately n-type doped c-Si measured with a quartz prism monochromator for d_{slit} equal to 0.05, 0.1, 0.2, 0.5 and 1.0 mm (filled squares, filled circles, open triangles, stars and open circles, respectively) and transient SPV spectrum of highly n-type doped c-Si measured at 2 ns with a tunable Nd:YAG laser with a spectral cleaning unit (filled triangles).

The measurement of modulated SPV signals is nearly impossible for highly doped c-Si even at very large values of d_{slit} due to the very short lifetime of photogenerated charge carriers. For low noise at short times, transient SPV spectra can be well measured for highly doped c-Si. For the given sample in Figure 3.7, the transient SPV spectrum of highly doped c-Si set on at a wavelength of about 1160 nm corresponding to a photon energy of about 1.07 eV. This value is almost very close to the band gap of highly doped c-Si (Overstraeten and Mertens, 1987).

If we look carefully at Figure 3.7, the modulated SPV spectra of c-Si measured with the monochromator were extended toward the longer wavelengths. Low SPV signals at photon energies far below the band gap might be interpreted as separation of photogenerated charge carriers excited from defect states. However, related SPV signals below the band gap of semiconductors can also be caused by stray light from the monochromator.

3.2.3. On the role of stray light

Stray light is defined as non-monochromatic light leaving a monochromator. Any monochromator has stray light. Depending on the construction of a monochromator and on its opening ratio, stray light is usually suppressed by 2–3 orders of magnitude in comparison to the monochromatic light. Incidentally, the quartz prism monochromator (SPM2, Carl Zeiss Jena) preferred by the authors of this textbook has an opening ratio of 1:8 enabling a relatively strong suppression of stray light at a relatively high intensity.

In moderately doped conventional semiconductors, SPV signals can be measured at intensities varied over many orders of magnitude. For example, SPV signals are of the order of several $100\,nV$ for a charge separation length of about $200\,nm$ and a density of photogenerated charge carriers of about $10^5\,cm^{-2}$ (see Chapter 1), which is 6–7 orders of magnitude below the density of the space charge. Therefore, stray light can cause strong artifacts in spectral-dependent SPV measurements.

The application of double and triple monochromators leads to a suppression of stray light by up to 5–6 or 7–9 orders of magnitude, respectively. However, the intensity of the monochromatic light might be rather low depending on the dispersion of a double or triple monochromator. Therefore, stray light is usually suppressed with optical long-pass filters in continuous and modulated spectral-dependent SPV measurements. An optical long-pass filter absorbs or transmits light with wavelengths below or above a cut-off wavelength, respectively. The cut-off wavelength shall be close to the band gap of the investigated semiconductor. Stray light is additionally suppressed by 5–6 orders of magnitude for wavelengths below the cut-off wavelength.

The influence of stray light is demonstrated in Figure 3.8 for on–off cycling on moderately n-type doped c-Si. In these measurements, the light was switched on and off at wavelengths of 1,050 and $1,100\,nm$ or $2,000\,nm$ corresponding to photon energies above or below the band gap, respectively. Without using a cut-off filter (Figure 3.8(a)), the SPV signals were 238 and $34\,mV$ at 1,050 and $2,000\,nm$, respectively. For the measurements with an optical filter cutting out photons with energies above $1.24\,eV$ ($1,000\,nm$, Figure 3.8(b)), the SPV signals were 235 and $17\,mV$ at 1,050 and $2,000\,nm$, respectively. For the measurements with an optical filter cutting out photons with energies above about $1.13\,eV$ ($1,100\,nm$ Figure 3.8(c)), the SPV signals were 220 and $5\,mV$ at 1,100 and $2,000\,nm$, respectively.

In the following, it is assumed that the SPV signals of the silicon wafer were caused by Fermi-level splitting in the bulk, i.e. surface recombination and asymmetric trapping are neglected. The Fermi-level splitting in the bulk is dominated by the ratio between the density of the photogenerated

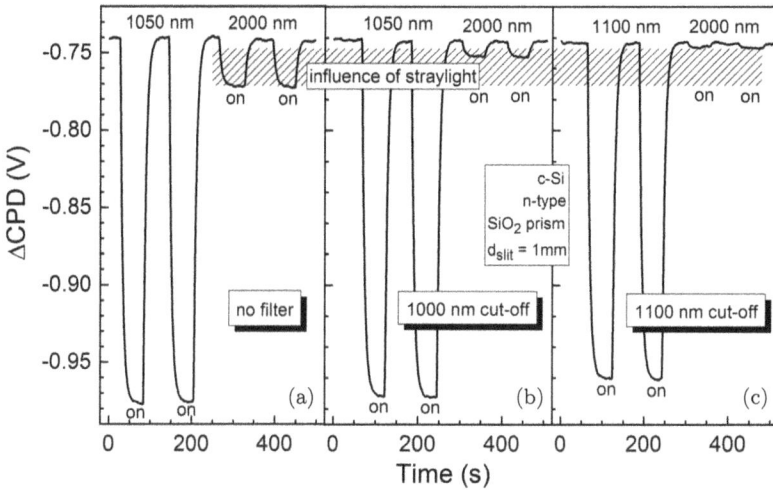

Figure 3.8. On–off cycles of ΔCPD of moderately n-type doped c-Si measured at 1,050 or 1,100 nm (two cycles in (a) and (b) or (c), respectively) and at 2,000 nm (two cycles) without a filter (a) and with long-pass filters with cut-off wavelengths at 1,000 and 1,100 nm ((b) and (c), respectively). The hatched area marks the influence of stray light.

charge carriers (Δp) and the density of the minority charge carriers in thermal equilibrium (p_0 for n-type doped c-Si). Therefore, the (steady-state) SPV signals can be approximated as

$$\text{SPV} \approx \frac{k_B \cdot T}{q} \cdot \ln\left(\frac{\Delta p}{p_0} + 1\right) \tag{3.6}$$

where q, k_B and T denote the elementary charge, Boltzmann constant and temperature, respectively.

It follows from Equation (3.6) that Δp was about 9,000 times larger than p_0 for absorption of light at 1,050 nm. For absorption of light at 2,000 nm without a cut-off filter, Δp was about 3–4 times larger than p_0, i.e. the stray light was suppressed by about 2,000–3,000 times in the monochromator. For absorption of light at 2,000 nm with a cut-off filter at 1,000 nm, Δp was as nearly equal to p_0, i.e. the stray light was additionally suppressed by about 3 times in comparison to the measurement without a cut-off filter. For absorption of light at 2,000 nm with a cut-off filter at 1,100 nm, Δp was about 5 times smaller than p_0. One shall keep in mind that stray light can be significant even in this case. The application of a filter with a cut-off below the band gap is recommended.

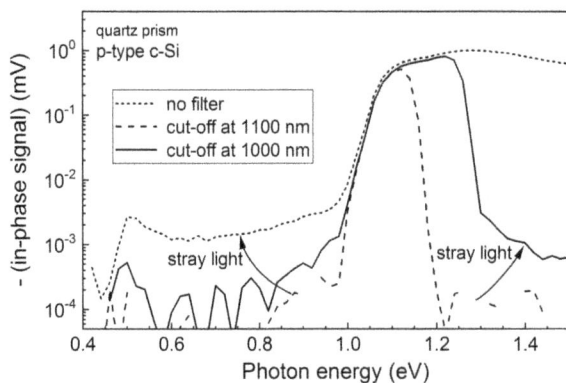

Figure 3.9. Modulated spectra of the negative in-phase SPV signals of moderately p-type doped c-Si measured without a filter (short dashed line) and with long-pass filters with cut-off wavelengths at 1,000 and 1,100 nm (dashed and solid lines, respectively).

The role of cut-off filters for the suppression of stray light is demonstrated for modulated SPV spectra of crystalline silicon in Figure 3.9. The filter cut-offs were chosen at 1,000 and 1,100 nm. For the filter cut-off at 1,000 nm, the SPV signals related to stray light were still above the level of noise. Furthermore, the SPV signals below the band gap and above the cut-off were of the same order of magnitude. Such a behavior can serve as a criterion for the influence of stray light on SPV spectra. For the filter cut-off at 1,000 nm, the SPV signals related to stray light were below the level of noise. Therefore, the doubtless characterization of defect states in the band gap of semiconductors by spectral-dependent SPV measurements can be rather challenging.

Lasers do not have stray light as monochromators. However, parasitic light at wavelengths different to that of a corresponding laser line can be present in laser beams. In transient SPV spectroscopy, such parasitic light is caused, for example, by pump light or plasma lines and harmonics of the fundamental laser line, which are incompletely filtered out. The parasitic light in transient SPV spectroscopy is suppressed by 5–6 orders of magnitude with an additional spectral cleaning unit behind the pulsed tunable laser.

3.2.4. *The low-signal case in SPV spectroscopy*

The characterization of transition energies, band gaps and exponentially distributed defect states close to the band gap of photoactive materials is an important application of SPV spectroscopy. For the analysis of the SPV spectra, a linear dependence of SPV signals on the light intensity or photon flux would be very useful. This can be the case at low SPV signals (so-called "low-signal case").

In analogy to the open circuit voltage (V_{OC}) of a solar cell, SPV signals are limited by a short circuit density (I_{SC}) and a parallel or shunt resistance (R_{par}). In the simplest case, the dependence of I_{SC} on V_{OC} can be written as

$$I_{SC} = I_0 \cdot \left[\exp \left(\frac{q \cdot V_{OC}}{k_B \cdot T} \right) - 1 \right] + \frac{V_{OC}}{R_{par}} \tag{3.7}$$

where q, k_B, T and I_0 denote the elementary charge, Boltzmann constant, temperature and diode saturation current density, respectively. The low-signal case is reached if the first term in Equation (3.7) is much lower than the second term. In the low-signal case, SPV signals are proportional to the photocurrent, i.e. to the photogeneration rate or photon flux.

As an example, Figure 3.10 shows SPV transients of highly n-type doped c-Si excited at 500 nm and different laser intensities. The SPV

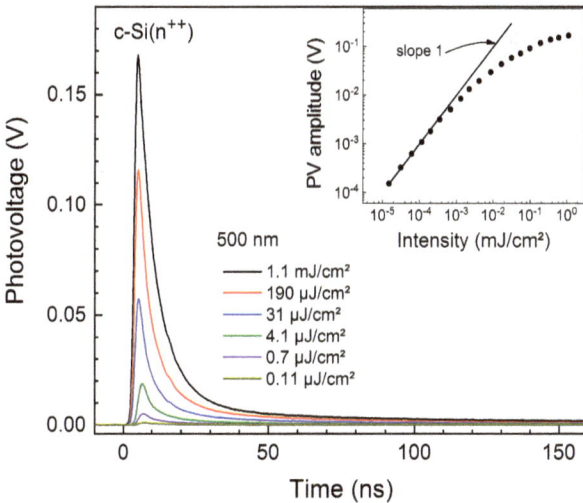

Figure 3.10. SPV transients of highly n-type dopes c-Si excited at 500 nm and laser intensities of $1.1 \, \text{mJ/cm}^2$, 190, 31, 4.1, 0.7 and $0.11 \, \mu\text{J/cm}^2$ (black, red, blue, olive, violet and dark yellow lines, respectively). The inset shows the intensity dependence of the maximum SPV signals. The duration time of the laser was about 3 ns at FWHM (full width at half maximum).

transients decayed within tens of nanoseconds, what is in agreement with the recombination lifetime in highly doped c-Si (Fossum *et al.*, 1983). At the highest laser intensity, the maximum SPV signal was about 170 mV. The dependence of the maximum SPV signals on the laser intensity is plotted in a double logarithmic scale in the inset of Figure 3.10. The SPV signals tended to saturate at the highest laser intensities. At the lowest laser intensities, the maximum SPV signals depended linearly on the laser intensity for values of the SPV of up to several millivolts. Therefore, for the given sample, transient spectral-dependent SPV measurements can be treated as low-signal cases for SPV signals up to 3–4 mV.

Photogenerated charge carriers can be separated toward opposite directions. This leads to different components in SPV signals. For the analysis of a SPV spectrum near transition energies or a band gap, one has to make sure that the mechanisms of charge separation, charge transport and recombination do not change within the analyzed spectral range.

The change of the sign of SPV signals gives clear evidence for a change of the mechanism(s) of charge separation and/or relaxation. In modulated SPV spectroscopy, a constant (or nearly constant) phase angle is a well-defined criterion for unchanged mechanisms of charge separation, charge transport and recombination. Therefore, modulated SPV spectra with a nearly constant phase angle and a linear dependence of the SPV signals on the light intensity can be well treated as low-signal cases.

Absorption coefficients are low near band gaps or close to transition energies. Consequently, the absorption lengths are usually larger or even much larger than the charge separation and/or diffusion lengths in a given photoactive material. Therefore, the photogeneration rates and the SPV signals are practically proportional to the absorption coefficient in such a case. The proportionality between low SPV signals and absorption coefficients can be a strong advantage of SPV spectroscopy in the low-signal case. However, one has to keep in mind that SPV spectroscopy is only sensitive to those absorption events which lead to photogeneration of mobile charge carriers and their separation in space. This is different to optical spectroscopy which is sensitive to any absorption event (Figure 3.11). Therefore, SPV measurements at low signals can probe absorption close to optical band gaps and absorption leading to excitation of trapped charge carriers. Molecular transitions can be probed if charge separation occurs within a layer of molecules or if electrons excited by light are injected into, for example, a metal oxide layer.

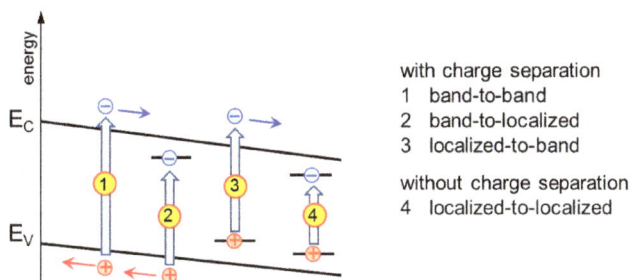

Figure 3.11. Electronic transitions in a semiconductor leading to charge separation (band-to-band, band-to-localized states and localized state-to-band, 1–3, respectively) and not leading to charge separation (localized state-to-localized state, 4) and which can therefore be probed by SPV spectroscopy in the low-signal case.

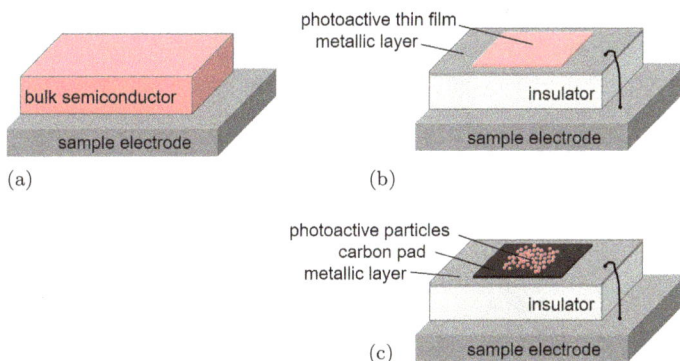

Figure 3.12. Arrangement of (a) a bulk semiconductor, (b) a photoactive thin film and (c) photoactive powder particles on the sample electrode for the investigation of transition energies by SPV spectroscopy.

3.2.5. *Examples for the investigation of transition energies*

Band gaps and transition energies can be investigated by SPV spectroscopy for bulk semiconductors, photoactive thin films and photoactive powders (Figure 3.12). Photoactive thin films can consist, for example, of photovoltaic absorber layers, layers of semiconductor quantum dots, films of conjugated polymers or sub-monolayers of organic molecules deposited onto a metal oxide. Powder samples can be prepared for SPV measurements, for example, by crumbling and fixing powder particles on a carbon pad which was fixed at the sample electrode before.

In the following, some examples will be demonstrated for the investigation of band gaps and electronic transitions in molecules by using transient, continuous and modulated SPV spectroscopy on c-Si and photoactive thin films of TiO_2 and $CuAlO_2$ and TiO_2 nanoparticles.

Crystalline silicon is the most studied semiconductor in general. SPV spectra of moderately doped c-Si can be well measured by modulated SPV spectroscopy. Layers of highly doped c-Si are applied for the formation of ohmic contacts in solar cell and electronic devices. For highly doped c-Si, the lifetime of minority charge carriers and the width of the space charge region are very short (of the order of 10 ns and 3 nm, respectively). For this reason, modulated SPV signals are extremely small for highly doped c-Si and the band gap cannot be measured with a high resolution since d_{slit} has to be increased very much. In transient SPV spectroscopy, SPV signals can be obtained at their maximum when the density of photogenerated charge carriers is at maximum. Therefore, the band gap of highly doped c-Si can be studied with high resolution by transient SPV spectroscopy.

The absorption coefficient increases by the power of 2 near the indirect band gap (E_{ig}) of c-Si. Figure 3.13 shows the square roots of the transient

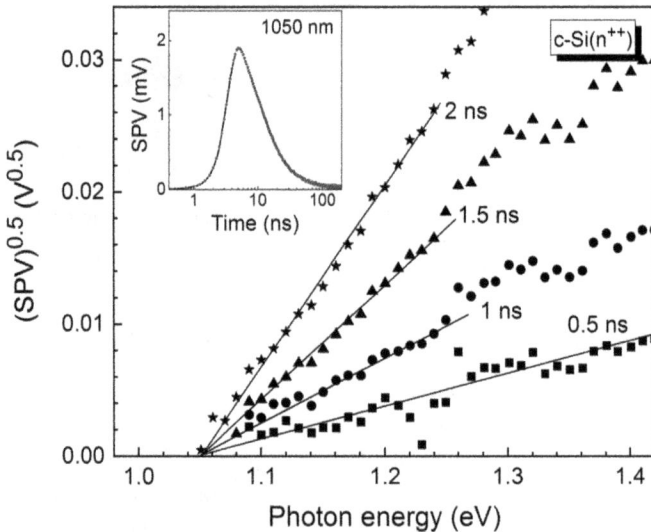

Figure 3.13. Square roots of the transient SPV spectra for highly n-type doped c-Si measured at 0.5, 1, 1.5 and 2 ns after switching on the laser pulses as an example for the determination of E_g from transient SPV spectroscopy (inset: single SPV transient (duration time of the laser pulse about 3 ns, wavelength of the laser pulse 1050 nm).

SPV spectra for highly n-type doped c-Si measured at constant photon flux and different times for the analysis of E_{ig}. The value of the indirect band gap of the highly n-type doped c-Si was obtained from the plots of the square roots of the transient SPV signals and amounted to 1.05 eV for all measurement times. The band gap of highly n-type doped c-Si was red-shifted due to the high density of defect states near the band gap (Overstraten and Mertens, 1978).

For the development of dye-sensitized solar cells and systems for water splitting, TiO_2 plays an important role as an electron conductor and a photocatalyst, respectively. TiO_2 can be prepared by very different techniques such as sol–gel and reactive sputtering. As an example, Figure 3.14(a) shows a ΔCPD spectrum of a reactively sputtered TiO_2 (anatase) layer in the range of the band gap (see Mwabora *et al.*, 2008 for more details). The drift of the baseline was very weak so that the corresponding continuous SPV spectra could be obtained with a high reliability for the given measurement.

At photon energies below E_g, the SPV signals increased exponentially with increasing photon energy (Figure 3.14(b)). The corresponding value of the exponential tails (E_t) was 70 meV.

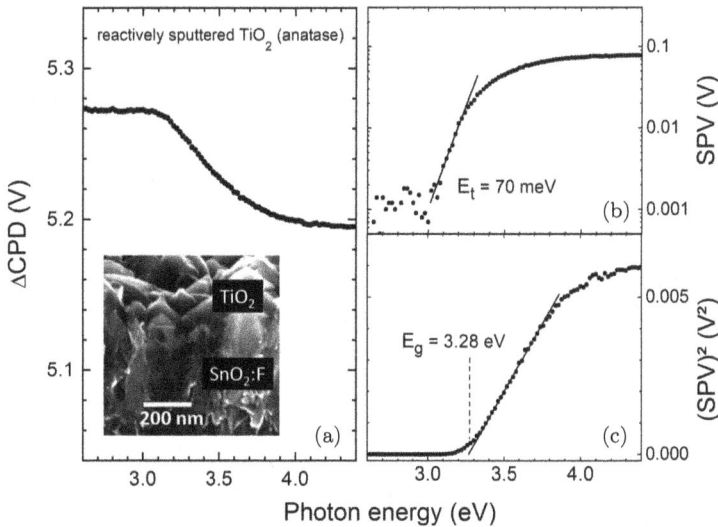

Figure 3.14. Spectrum of the ΔCPD of a TiO_2 (anatase) layer reactively sputtered onto a SnO_2:F substrate (a), corresponding SPV spectrum in a logarithmic scale (b) and squared SPV spectrum (c) as an example for the determination of E_g and E_t from continuous SPV spectroscopy with a Kelvin probe. The inset shows the cross section of the layer (see Mwabora *et al.*, 2008 for more details).

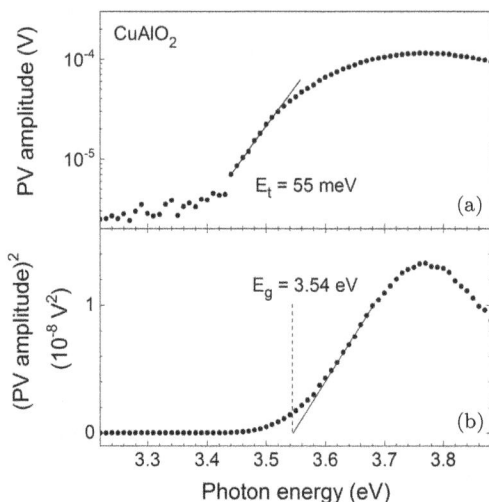

Figure 3.15. Modulated SPV spectrum in (a) a logarithmic scale and (b) squared modulated SPV spectrum of $CuAlO_2$ crystallites prepared from $LiAlO_2$ by ion exchange (see Dloczik *et al.*, 2004; Dittrich *et al.*, 2004 for more details) as an example for the determination of E_g and E_t from modulated SPV spectroscopy.

The anatase phase of TiO_2 has a direct band gap. Therefore, the value of E_g can be obtained from the plot of the squared absorption spectrum. Figure 3.14(c) shows the spectrum of the squared SPV signals. The value of E_g was 3.28 eV for the given layer.

$CuAlO_2$ is a promising p-type wide-gap semiconductor with a direct band gap. SPV signals of $CuAlO_2$ are negative (Dittrich *et al.*, 2004), which is typical for p-type doped semiconductors with a depletion region at the surface. Figure 3.15 shows the amplitude of modulated SPV spectra of $CuAlO_2$ crystallites prepared by ion exchange (see Dloczik *et al.*, 2004 for more details). For the given $CuAlO_2$ powder, the values of E_g and E_t amounted to 3.54 eV and 50 meV, respectively.

As an example for the study of molecular transitions in sub-monolayers of organic molecules, Figure 3.16 shows the spectra of modulated in-phase SPV signals of solitary Pd-centered porphyrin molecules deposited by soaking into an extremely thin nanoporous TiO_2 layer (see Zabel *et al.*, 2009a for more details).

Furthermore, transitions of charge transfer complexes and offsets between HOMO and LUMO bands of evaporated conjugated acceptor and donor molecules such as C_{60} and SubPc, respectively, were obtained from modulated SPV spectroscopy (Fengler *et al.*, 2015).

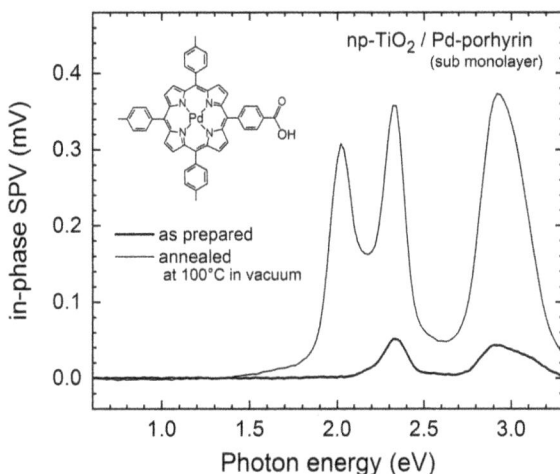

Figure 3.16. Spectra of the modulated in-phase SPV signals of as-prepared solitary Pd-centered porphyrin molecules on TiO$_2$ nanoparticles (thick solid line) and after annealing in vacuum at 100°C as an example for the characterization of molecular transitions with modulated SPV spectroscopy. Details can be found in Zabel *et al.* (2009a) and Funes *et al.* (2010).

The positive sign of the in-phase signals gives evidence for injection of electrons from excited porphyrin molecules into TiO$_2$. For the modulated SPV spectrum of the as-prepared sample, the typical Q and S bands with maxima at 2.33 and 2.92 eV, respectively, were well distinguished. After annealing in vacuum at 100°C, an additional peak with a maximum at 2.03 eV band appeared (see Funes *et al.*, 2010 for more details). The additional peak can be ascribed to the non-reversible formation of a charge transfer complex between the Pd central atom of the porphyrin molecule and a reduced Ti^{3+} atom at the surface of TiO$_2$ nanoparticles. Incidentally, the formation of a reversibly inducible charge transfer complex has been also observed between Ru-based dye molecules (N3) and TiO$_2$ in oxygen- and water-deficient atmospheres by using modulated SPV spectroscopy (Dittrich *et al.*, 2007).

3.3. Transient SPV Measurements at Short Times

3.3.1. *Realization of a high-impedance buffer*

The authors got experience with numerous high-impedance buffers which were developed, improved and fabricated by different specialists over two decades. A very reliable and robust high-impedance buffer for SPV

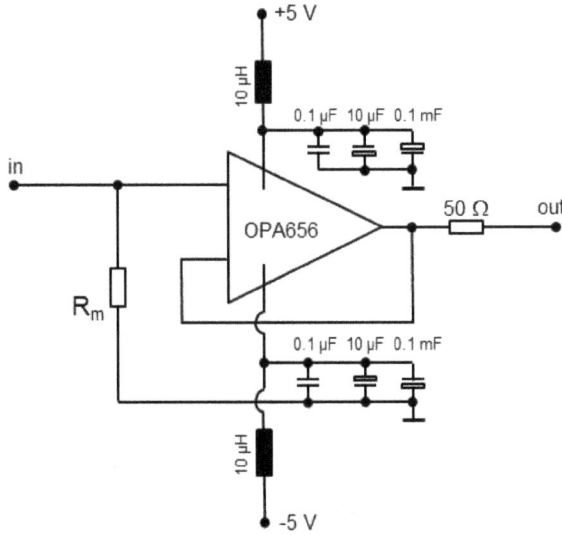

Figure 3.17. Circuit of a high-impedance buffer based on an operational amplifier (here OPA656), a very high measurement resistance, an output resistance and series inductances and parallel capacitors damping fluctuations from the power supply.

measurements from the nanosecond to the 100 ms range was based on the operational amplifier OPA656 (Bönisch *et al.*, 2004). Aside the OPA656, the circuit of the high-impedance buffer consisted of the very high measurement resistance, the output resistance and series inductances and parallel capacitors for damping of electrical fluctuations from the power supply (Figure 3.17).

Efficient damping and shunting of fluctuations from the power supply over a wide range of frequencies is very important for sensitive SPV measurements with low noise over many orders of magnitude in time. Therefore, three shunting capacitors with capacitances of 100 nF, 10 μF and 0.1 mF were connected with the inputs for the positive and negative potentials of the operational amplifier and combined with series inductances with an inductivity of 10 μH (buffer design shown in Figure 3.17). It is recommended to use batteries as power supplies.

The circuit of the buffer design used by the authors over many years was made on a double-sided copper-clad circuit board which was soldered into a little zinc-coated iron housing for high-frequency electronics. The layout of the buffer was made in a way that connecting lines were close to each other and to the shielding in order to reduce parasitic inductances. This was important for transient SPV measurements at very short times close to the limit of the operational amplifier. The BNC connectors for the input and output signals and the connectors to the power supply were directly incorporated into the housing of the buffer. Incidentally, BNC connectors are not suitable for measurements at times longer than about 0.1–1 s due to the shunt resistance of BNC connectors. After soldering and finishing the fabrication of the buffer, a careful cleaning of the whole buffer including especially the BNC connector for the input signal was important.

Very high resistances are sensitive to humidity, mechanical stress, contaminations like fats and other factors. Therefore, the careful handling of a very high resistance is crucial for successful operation of a high-impedance buffer. For robust SPV measurements at short times, it can be useful to reduce the value of R_m to a needed range of time. For example, a value of R_m of 1 $G\Omega$ is sufficient for measurements of up to about 10 ms.

The buffer shall be directly connected with the reference electrode without an intermediate BNC connector for SPV measurements from the nanosecond range to very long times of up to 1,000 s. The connection of high-impedance buffers for the measurement of extremely long SPV transients to high and ultrahigh vacuum chambers is challenging since feedthroughs must not shunt the extremely high measurement resistance. The production of a separate, extremely well insulating, feedthrough for a high-impedance buffer with an extremely high measurement resistance will be necessary for the realization of a related setup.

3.3.2. *Damping of parasitic LC oscillations*

Resistances and parasitic capacitances and inductances are part of configurations for transient SPV measurements (see also Chapter 2). Damped oscillations can be excited by SPV signals and disturb therefore measured SPV transients at short times. The values of C_{par} and L are given by the measurement configuration and can be varied only in a relatively narrow range for a given setup. For SPV measurements at very short times, typical values of L and C_{par} are of the order of 20 nH and 5 pF, respectively. These values result in a frequency of about 500 MHz, which is equal, for example, to the bandwidth of the OPA656. Incidentally, the resolution time (τ_{res}) of an electronic device is given by the reciprocal bandwidth (f_0) divided by 2π:

$$\tau_{res} = \frac{1}{2\pi \cdot f_0} \tag{3.8}$$

A bandwidth of 500 MHz corresponds to a resolution time of about 0.3 ns. However, this resolution time is increased due to the excited damped oscillations in the measurement configuration.

Oscillations excited in the measurement configuration are damped by the series resistance. A suitable way to reduce the influence of damped oscillations on measured SPV transients is the adjustment of the series resistance in the LCR circuit toward overdamping. The critical resistance (R_{crit}) for overdamping is reached when the squared natural frequency (Equation (2.24)) and the squared decay constant (Equation (2.25)) are equal:

$$R_{\text{crit}} = \sqrt{\frac{4 \cdot L}{C}} \tag{3.9}$$

The series resistance in the measurement configuration arises mainly at the reference electrode and its connection to the metal wire. The reference electrode is usually made from a transparent conductive oxide (TCO), such as SnO_2:F (see also Chapter 2). The sheet resistance of a TCO layer is of the order of 10–50 Ω. The resistance of a TCO layer is proportional to the ratio between the length and the width of the conducting strip. Typically, the length and width of the reference electrode and its connection to the wire are of the orders of 3–10 and 6–20 mm, respectively. Therefore, the value of the series resistance in a typical measurement configuration for transient SPV measurements can vary between about 5 and 100 Ω.

Figure 3.18 shows, as an example, the damping of oscillations for LCR circuits with values for L, C_{par} and R of 20 nH, 5 pF and 5, 50 and 100 Ω, respectively. Incidentally, R_{crit} is equal to 126 Ω for the given values of L and C_{par}.

For R equal to 5 and 50 Ω, the oscillations are damped within 30–40 and 3–4 ns, respectively. Therefore, depending on the measurement configuration, transient SPV measurements are usually not critical at times longer than about 5–50 ns. For R equal to 100 Ω, only one strongly damped peak remained at times shorter than about 1–2 ns. Therefore, transient SPV measurements can be performed without additional measures for analysis for times longer than 1–2 ns for an optimized measurement configuration.

The frequency of the damped oscillations (f_d) is reduced in comparison to the natural frequency due to damping:

$$f_d = \frac{\sqrt{\frac{1}{L \cdot C} - \frac{R^2}{4 \cdot L^2}}}{2\pi} \tag{3.10}$$

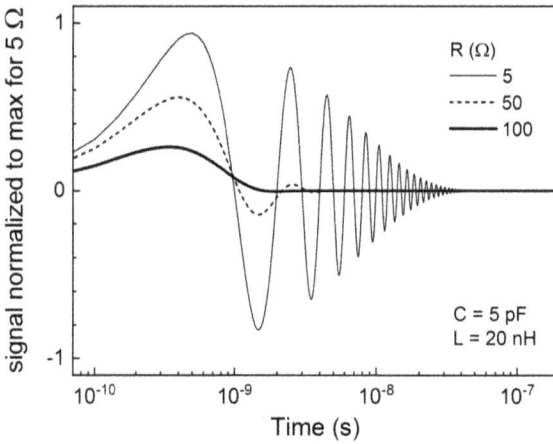

Figure 3.18. Example of damped oscillations for an inductance of 20 nH, a capacitance of 5 pF and resistances of 5, 50 and 100 Ω (thin solid, dashed and thick solid lines, respectively). The values were normalized to the maximum in the oscillation for 5 Ω.

The damping time is equal to the reciprocal damping constant. It is realistic to damp oscillations for times longer than about half of the oscillation period, i.e. the damping time is between the half and the fourth of the oscillation period. Therefore, an optimum series resistance for damping the influence of LCR in transient SPV measurements (R_d) can be defined as

$$R_d \in \sqrt{\frac{3 \cdot L}{C}} \cdots \sqrt{\frac{2 \cdot L}{C}} \qquad (3.11)$$

For example, the value of R_d is between about 90 and 110 Ω for L and C_{par} equal to 20 nH and 5 pF, respectively.

Electronic devices as operation amplifiers and capacitors are designed for operation under normal pressure in a limited range of temperatures. Therefore, high-impedance buffers are mounted outside cryostats and/or high and ultrahigh vacuum chambers. This means that an increase of the length of the wire connecting the reference electrode with the high-impedance buffer has to be taken into account. The natural frequency of the LCR circuit decreases with increasing L. As a consequence, the resolution time of the measurement configuration increases with increasing L and damping. By adjusting R_d with respect to Equation (3.11), SPV transients undisturbed by the measurement configuration can be well studied, for example, at times longer than about 3 ns when the reference electrode is separated from the high-impedance buffer over a distance of about 20 cm.

A consequence of damping of oscillations caused by the measurement configuration is that SPV transients can be measured with a resolution which is longer than the resolution time given in Equation (3.8) by about a factor of 3. Therefore, damping is an elegant practical way for the measurement of SPV transients undisturbed by the measurement configuration, but it does not allow for transient measurements at the highest resolution in time.

3.3.3. *Self-consistent extraction of very short source transients*

For reaching the shortest resolution time in transient SPV measurements, damped oscillations caused by the measurement configuration shall be considered in the analysis. A measured SPV transient consists of three parts:

(i) the undisturbed source transient ($U_{\text{source}}(t)$),
(ii) the damped oscillation ($U_{\text{osc}}(t)$), and
(iii) the onset function ($F_{\text{on}}(t)$).

The onset function describes the minimum resolution or onset time of the whole setup (τ_{on}). It is reasonable to suppose a logistic growth function for the onset function:

$$F_{\text{on}}(t) = \frac{1}{1 + \exp\left(-\frac{t-t_0}{\tau_{\text{on}}}\right)} \qquad (3.12)$$

The onset function can be shifted in time (t_0) with respect to the trigger. The function of the damped oscillations is given by Equation (2.27) and includes L, C, R, the amplitude (A_{LCR}) and a phase angle (ϕ_{LCR}).

A measured SPV transient is a superposition of U_{source}, U_{osc} and F_{on}:

$$\text{SPV}(t) = F_{\text{on}}(t) \cdot [U_{\text{osc}}(t) + U_{\text{source}}(t)] \qquad (3.13)$$

The source transient is unknown. For fitting, it is useful to approximate $U_{\text{source}}(t)$ by a function which shall be as simple as possible and as precise as needed. For example, source transients of highly n-type doped c-Si can be well described by an exponential combined with a stretched exponential:

$$U_{source}^{c-Si(n++)}(t) = A_1 \cdot \exp\left(-\frac{t}{\tau_1}\right) + A_2 \cdot \exp\left(-\left(\frac{t}{\tau_2}\right)^{\beta}\right) \qquad (3.14)$$

The parameters A_1, A_2, τ_1, τ_2 and β in Equation (3.14) denote the amplitudes and time constants of the exponential and stretched exponential and the stretching parameter, respectively.

Multi-parameter fits are challenging. Even in one of the simplest cases for transient SPV measurements on highly n-type doped c-Si, a fit of 12 (!) parameters is needed in order to extract $U_{source}(t)$ for one single measurement.

Multi-parameter fits may result in local minima. The probability to stack with a multi-parameter fit in a local minimum can be drastically reduced if the order after which the separate parameters are varied for fitting is varied randomly. Multi-parameter fits demand reasonable starting values for the parameters, well-defined criteria and rather complex software. The development of user-friendly and reliable software for general multi-parameter fits will also be beneficial for the analysis of other kinds of transient measurements, for example, transient photoconductivity, transient photoluminescence and transient absorption.

The precision of a multi-parameter fit can be increased by increasing the number of measured transients, whereas only one significant parameter is varied in the measurement. This means that the number of parameters for fitting increases only by several percent whereas the number of fits increases by several times. In transient SPV measurements, for example, the inductance in the measurement configuration can easily be varied by changing the length of the wire connecting the reference electrode with the high-impedance buffer. The length of the wire has only little influence on C_{par} and R. Therefore, such an extended procedure results, for example, in a fit of 15 parameters for four measurements with four different values of L on highly n-type doped c-Si.

A general scheme for the self-consistent extraction of $U_{source}(t)$ is depicted in Figure 3.19. For the demonstration of self-consistency, the source transient shall be fed into a device simulation including the equivalent circuit of the whole setup.

SPV transients can be simulated by applying professional simulation software such as, for example, MultiSim$^{\circledR}$ or Spice$^{\circledR}$. The software

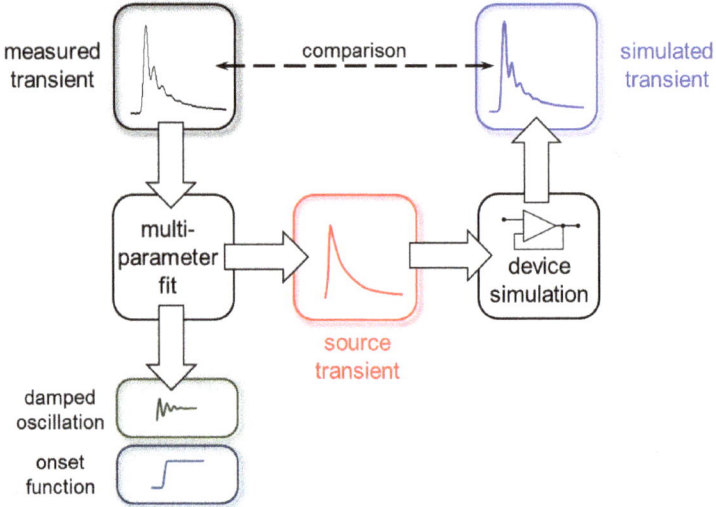

Figure 3.19. Scheme of the self-consistent extraction of the source transient from a measured SPV transient by a multi-parameter fit and simulation of a SPV transient from the source function and the electronic devices of the setup including the measurement configuration and the high-impedance buffer.

MultiSim, for example, includes libraries with the parameters of numerous electronic devices such as OPA656.

A typical equivalent circuit for simulations includes the high-impedance buffer (see, for example, Figure 3.17) and the elements of the measurement configuration. An ideal measurement configuration consists only of the potential source (U_{source}) and C_m (Figure 3.20(a)). An ideal measurement configuration is usually sufficient for the simulation of SPV transients at times longer than 20–30 ns. The measurement configuration applied for the simulation of SPV transients at very short times contains U_{source} and the elements C_m, C_{par}, R_{ser} and L (Figure 3.20(b), see also Figure 2.14(d)). Furthermore, U_{source} can be replaced by the equivalent circuit of a solar cell (Figure 3.20(c)).

Figure 3.21 shows a measured SPV transient of highly n-type doped c-Si for a relatively long wire connecting the reference electrode with the buffer. The damped oscillations are very pronounced for this example. The extracted source transient was the difference between the measured SPV transient and the damped oscillation corresponding to effective values of L, C and R of the measurement configuration of 254 nH, 3.8 pF and 55 Ω, respectively. With regard to Equation (3.14), the values of τ_{on},

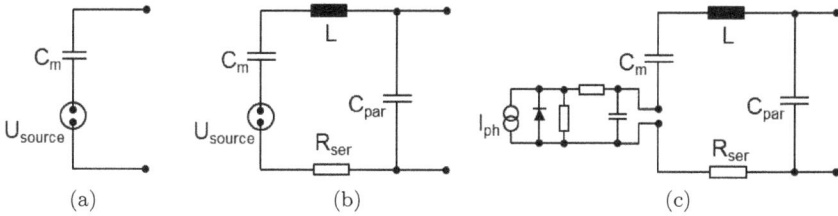

Figure 3.20. Equivalent circuits of SPV measurement configurations: (a) ideal, (b) with an LCR circuit, (c) with an LCR circuit and a solar cell as a potential source.

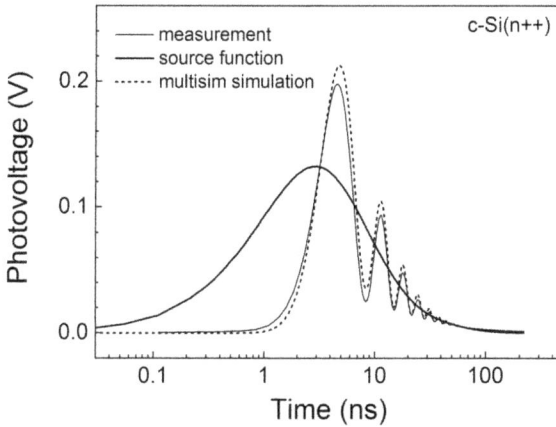

Figure 3.21. Example for a measured SPV transient of highly n-type doped c-Si (thin solid line), extracted source function (thick solid line, the effective values for L, C and R of the measurement configuration were 254 nH, 3.8 pF and 55 Ω, respectively) and SPV transient simulated for the source function with MultiSim (dashed line). The duration time and the wavelength of the laser pulse were 3 ns (FWHM) and 540 nm, respectively (see for details Dittrich *et al.*, 2019).

A_1, A_2, τ_1, τ_2 and β amounted to 0.5 ns, 260 mV, 31.5 mV, 7 ns, 25 ns and 0.85, respectively, for the source transient. The deviations were very low between the measured SPV transient and the transient simulated from the source transient and the corresponding equivalent circuit of the measurement configuration with the high-impedance buffer. Therefore, the source transient obtained from the multi-parameter fit corresponded to the real SPV response of the sample. Incidentally, the onset time of 0.5 ns gave evidence for a reduced resolution time of the setup below 1 ns (see also Dittrich *et al.*, 2019).

The example shown in Figure 3.21 demonstrates that a SPV measurement configuration can be described by a relatively simple equivalent circuit. This opens the opportunity for the self-consistent analysis of SPV transients at very short times even for setups in which the wire connecting the reference electrode with the high-impedance buffer has a length of the order of 20 cm and longer.

The situation is much more complicated for the analysis of SPV transients at very short times when an additional LCR circuit is required for the equivalent circuit. This can be the case, for example, if oscillations can significantly be excited in the circuit of the high-impedance buffer. A second LCR circuit and additional beating oscillations make a multi-parameter fit and therefore a self-consistent analysis of SPV transients at very short times practically impossible.

For a further reduction of the resolution time in transient SPV measurements, the properties of a given sample geometry, the measurement configuration and elements in the high-impedance buffer have to be analyzed as one electronic system. The following steps are required for the development of a very fast high-impedance buffer. First, an OPA with a very high bandwidth is chosen and incorporated into a design and layout of a high-impedance buffer. Second, the response of the high-impedance buffer to a very short source transient is optimized by varying the design and the layout. Third, the new high-impedance buffer is fabricated. Fourth, the new high-impedance buffer is tested with a model system, such as highly n-type doped c-Si, and the SPV transients are self-consistently analyzed following the scheme in Figure 4.19.

3.4. Variation of Parameters during SPV Measurements

3.4.1. *Temperature-dependent SPV measurements*

Temperature-dependent SPV measurements are important for the characterization of photoactive materials, for example, of the temperature dependence of band gaps, of activation energies in transport processes, of transport and recombination mechanisms.

The controlled variation of the temperature demands a cold reservoir at temperature T_{cold}, a heater, a power supply, a sensor for the measurement of the temperature and a temperature controller (Figure 3.22). Electrical heating with a resistance is recommended for SPV measurements. The controlled variation and the measurement of the temperature must not influence the SPV measurements. Therefore, the power of the electrical heater shall be changed as smooth as possible with a related controller. Incidentally, heating with variable current pulses by using low-cost temperature

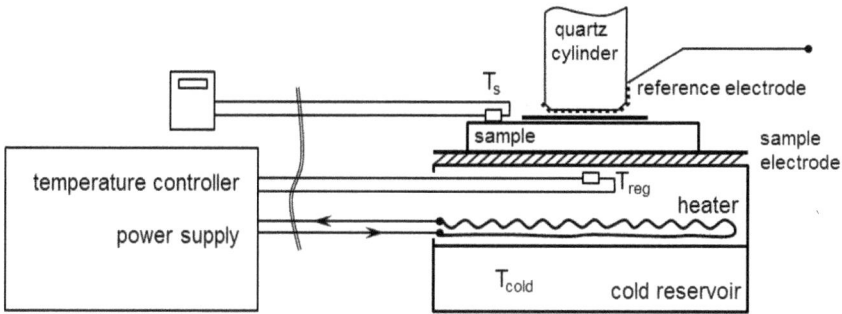

Figure 3.22. Schematic configuration for the temperature control in SPV measurements with a fixed capacitor. The temperature controller with the power supply is outside the cryostat.

controllers is not suitable for temperature-dependent SPV measurements at high sensitivity.

Passive sensors without a need for amplification of signals are recommended for the measurement of the temperature. Calibrated platinum resistances, such as PT100, are highly reliable for accurate temperature measurements over wide temperature ranges. SPV signals are practically not influenced by temperature measurements with resistances which can be easily measured with a well-stabilized constant current source at a low current.

The temperature T_{reg} is measured by the temperature controller and kept constant by varying the power. In SPV measurements with a fixed capacitor, the sample and the reference electrode with the quartz cylinder are in thermal contact via the mica sheet. An additional measurement directly at the sample (T_s) can be helpful. For very precise temperature-dependent SPV measurements, the fixed electrode and the sample shall be placed inside a massive oven.

Temperature-dependent SPV measurements with fixed capacitors are usually quite time consuming since a relatively long time is needed for equilibrating the temperature at each value. Temperature-dependent SPV measurements were carried out over a temperature range between about $-180°C$ and $300°C$.

As an example, Figure 3.23 shows the spectra of ΔCPD for a graphitic C_3N_4 powder at different temperatures (Dittrich *et al.*, 2011). For these spectra, the baselines were not well defined and the band gap was estimated from the point at which the light-induced change of ΔCPD set on (E_{on}).

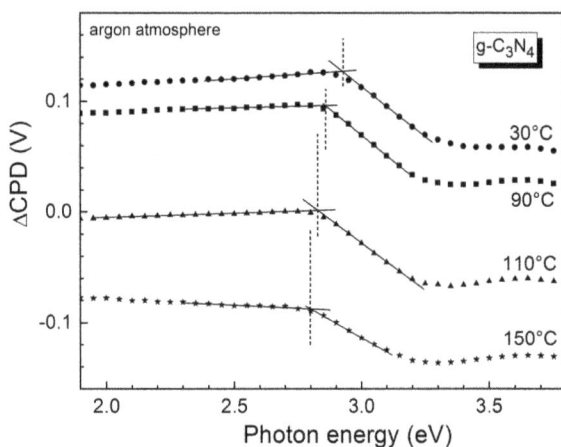

Figure 3.23. Spectra of ΔCPD of (graphitic) g-C$_3$N$_4$ measured at 30°C, 90°C, 110°C and 150°C (circles, squares, triangles and stars, respectively; see, for details, Dittrich *et al.*, 2011).

These onset energies are close to the band gap. The onset energies shifted toward lower photon energies with increasing temperature.

3.4.2. *Ambient gas-dependent SPV measurements*

For photocatalysis, the investigation of reactive surface species under illumination is of great interest. Reactive surface sites can be influenced by gas atmospheres and electrolytes. Therefore, model experiments in controlled gas atmospheres are desired. The influence of ambient gases on reactive surface sites under illumination can be tested by SPV measurements since the surface potential is varied by charge transfer, changes of trapping at donor and acceptor like defects and variation of surface recombination.

The variation and the measurement of the gas pressure shall not influence the SPV signals. One has to keep in mind that the gas molecules are also in contact with the reference electrode. Therefore, the reference electrode should be inert against chemical and photochemical reactions with the gas molecules introduced into the measurement chamber. For example, SnO$_2$:F, which is often used for reference electrodes in SPV measurements with fixed capacitors, is robust against numerous of reactive gases. In contrast, TiO$_2$ surfaces are very sensitive to defect formation in oxygen- and/or water-deficient atmospheres (see, for example, Duzhko *et al.*, 2001b).

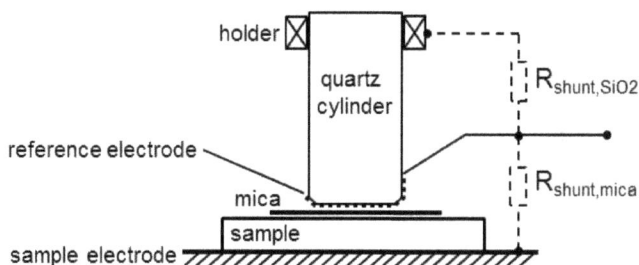

Figure 3.24. Schematic diagram of an electrode configuration with a fixed capacitor and partially shunted.

In fixed capacitor arrangements, molecules of water or other liquids can condensate at the surface of the insulating layer (mica sheet) and/or the insulating quartz cylinder holding the reference electrode (see also Chapter 2). Condensed molecules can partially shunt the very high measurement resistance (Figure 3.24). This may cause overshoots in transient SPV signals at longer times after switching off illumination and can also influence the response of modulated SPV signals.

In order to reduce possible influence of partial shunting, it is recommended to increase the area of the insulating sheet and the length of the quartz cylinder and to avoid pressure above the saturation pressure of water or other polar liquids.

As an example, Figure 3.25 shows modulated SPV spectra of nanoporous TiO_2 measured in air and measured at reduced pressure of N_2 atmosphere (Dittrich and Neumann, 2005). The in-phase SPV signals were positive for measurements in air but negative for measurements at low pressure, i.e. photogenerated electrons were preferentially separated toward the internal or external surface for measurements in air or at reduced pressure, respectively. Incidentally, the signs of the SPV signals were similar for transient measurements under the same conditions. For measurements in air, the SPV signals set on at photon energies between 3.1 and 3.2 eV and reached the maximum at about 3.5 eV. For measurements at low pressures, SPV signals set on at photon energies between 1.5 and 1.6 eV. Therefore, surface defects with a broad distribution were generated in the band gap of nanoporous TiO_2 at low pressures. The surface states were mainly charged positively so that electrons were predominantly separated toward the external surface.

As an example for the change of the slope of SPV transients in gas atmospheres, the influence of the pressure of H_2O or N_2 on SPV transients

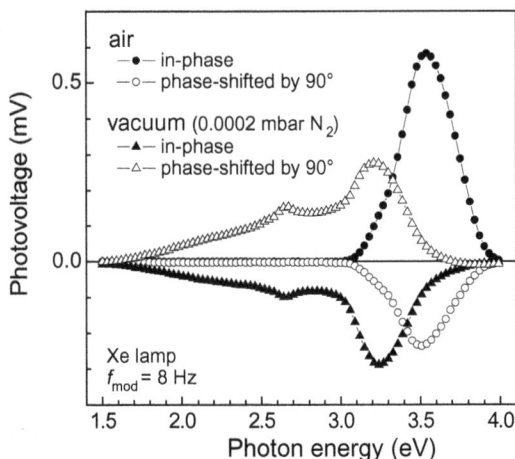

Figure 3.25. Modulated in-phase (solid symbols) and phase-shifted by 90° (open symbols) SPV signals of nanoporous TiO_2 in air (circles) and in N_2 atmosphere at 0.0002 mbar vacuum (triangles) (Dittrich and Neumann, 2005).

is shown for a sintered layer of nanoporous TiO_2 (P25, mixture of anatase and rutile) in Figure 3.26(a) or 3.26(b), respectively. The most characteristic feature of these SPV transients is a maximum shifting in time over several orders of magnitude as a function of the pressure of H_2O or N_2 molecules. Such a maximum is caused by independent diffusion and related to the dielectric relaxation time of the nanoporous TiO_2, i.e. to its conductivity (see also Chapter 1).

In vacuum, the SPV signals were negative and very low after switching on the laser pulse and increased with increasing time to about $-1.8\,mV$ at about 150 μs. After introducing H_2O or N_2 molecules at different pressures, the shapes of the SPV transient changed strongly. The SPV signals were positive at about 100 ns and increased with increasing pressures. The times at which the maximum SPV signals were reached amounted to 30, 1.5, 500, 70 and 2 μs for H_2O at 0.02, 0.1, 0.25, 2.2 and 19.5 mbar, respectively, and 20, 3, 0.8, 1 and 5 ms for N_2 at 0.19, 0.55, 2.7, 14 and 1014 mbar, respectively. For a more detailed analysis of related results with respect to the role of reactive surface species, a combination of SPV measurements with the characterization of chemical bonds at the surface will be very useful.

3.4.3.   *SPV measurements under bias potentials*

SPV measurements under bias potentials are important for the investigation of surface states. The band bending at semiconductor surfaces (φ_0) can be

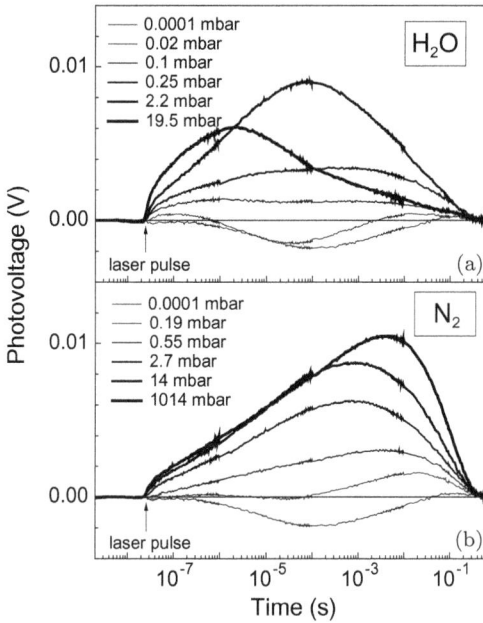

Figure 3.26. SPV transients of nanoporous TiO_2 at different pressures of (a) H_2O and (b) N_2 molecules. The duration time and the wavelength of the laser pulses were 0.15 ns and 355 nm, respectively. Incidentally, this is also an example for building of one transient by stacking of four transients measured at different time scales without a logarithmic read-out (Dittrich and Neumann, 2005).

measured by transient SPV measurements in the large signal case with a fixed capacitor by considering Equations (1.16), (1.17) and (1.26) in the analysis (Heilig, 1968, 1974). The surface band bending is caused by charge neutrality between the densities of charge in the surface space charge region (Q_{SC}) and in surface states (Q_{SS}). Q_{SC} is a function of band bending and doping. The charge in surface states and in the space charge region can be modified by influencing additional charge (Q_i) under a bias potential (U_{bias}). The variations of the potential drops across the insulating layer (ΔU_i) and the space charge region ($\Delta \varphi_0$) are given by:

$$\Delta U_{bias} - \Delta U_i - \Delta \varphi_0 = 0 \tag{3.15}$$

If hysteresis effects can be neglected, one can write for the charge neutrality (Lam, 1971):

$$\Delta Q_{SC} + \Delta Q_{SS} + \Delta Q_i = 0 \tag{3.16}$$

The value of ΔQ_i is given by the capacitance of the measurement capacitor (C_m) and ΔU_i:

$$\Delta Q_i = C_m \cdot \Delta U_i \tag{3.17}$$

The density of surface states is defined as

$$D_{it} \equiv -\frac{1}{q} \cdot \frac{\Delta Q_{SS}}{\Delta \varphi_0} \tag{3.18}$$

An expression of the density of surface states is obtained from Equations (3.15)–(3.18).

$$D_{it} = \frac{C_m}{q} \cdot \left(\frac{\Delta U_{bias}}{\Delta \varphi_0} - 1 \right) + \frac{\Delta Q_{SC}}{\Delta \varphi_0} \tag{3.19}$$

In Equation (3.19), the values of C_m and ΔU_{bias} are given by the experiment, $\Delta \varphi_0$ is obtained from the SPV measurements at different values of U_{bias} and ΔQ_{SC} is calculated from φ_0 and doping.

If there is no Fermi-level pinning at an investigated semiconductor surface, the analysis of the dependence of the amplitude of SPV transients on the applied bias potential allows for the measurement of D_{it} (Lam, 1971; Heilig *et al.*, 1979).

Depending on the thickness of the insulating mica sheet in a fixed capacitor, C_m is of the order of $100\,\mathrm{pF/cm^2}$. This means that bias potentials of 10–1000 V are required for recharging about $10^{10} - 10^{12}\,\mathrm{q/cm^2}$. The desired accuracy of the measurement of the amplitudes of SPV transients is 1 mV or less. Therefore, a well-stabilized high-voltage source is required for the investigation of the dependence of amplitudes of SPV transients.

The bias potential is applied at the backside of the sample and the sample electrode shall be well insulated against the metal of the shielding housing (Figure 3.27).

Furthermore, the abrupt change of the bias potential can damage the high-impedance buffer, i.e. the input of the high-impedance buffer shall be protected during switching the bias potential. The protection of the high-impedance buffer is realized by shunting the buffer input during switching of U_{bias}, for example, with a reed relay as a switch with low capacitance (Figure 3.27). The reed relay is always closed (off-state) and only opened during the measurement of an SPV transient (on-state).

During the application of U_{bias}, the sample with the mica can be polarized. After switching off U_{bias}, it can take some time for the relaxation

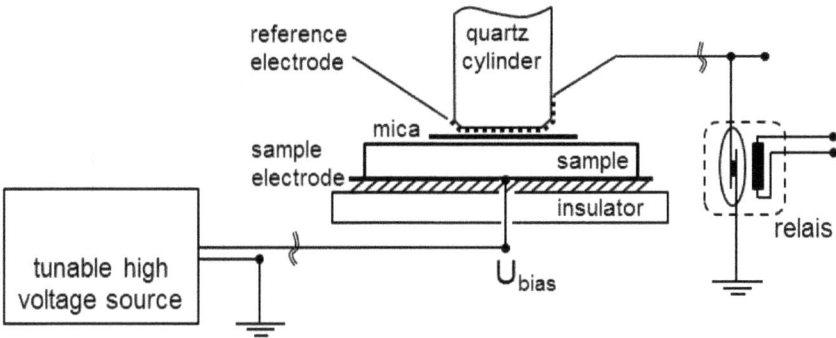

Figure 3.27. Schematic configuration for transient SPV measurements under bias potential.

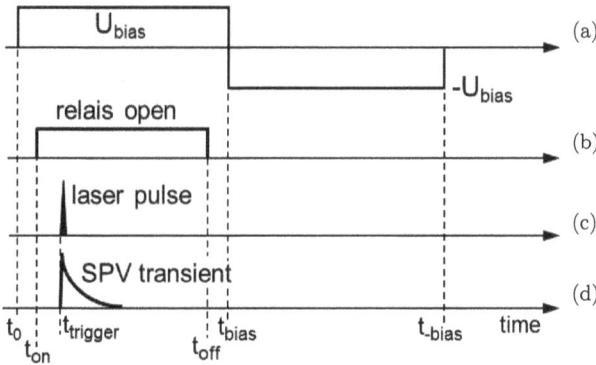

Figure 3.28. Regime for the measurement of the amplitude of an SPV transient at a bias potential (sequence of the application of U_{bias} between the times t_0 and t_{bias} and of $-U_{bias}$ between the times t_{bias} and t_{-bias}, signal for opening the relay at t_{on} and closing the relay at t_{off}, laser pulse triggered at $t_{trigger}$ and measured SPV transient ((a)–(d), respectively).

of the polarization charge. The relaxation time of the polarization charge can be reduced by erasing the polarization charge by applying a high potential of the opposite sign (U_{-bias}) after switching off U_{bias}. The regime for the measurement of the amplitude of an SPV transient at a bias potential is shown in Figure 3.28.

The bias potential is applied at the time t_0. After some time for stabilization (for example, of the order of ms), the relay is opened at the time t_{on}, the pulse laser is triggered at the time $t_{trigger}$ and the SPV transient is measured. After the SPV transient decayed, the relay is closed

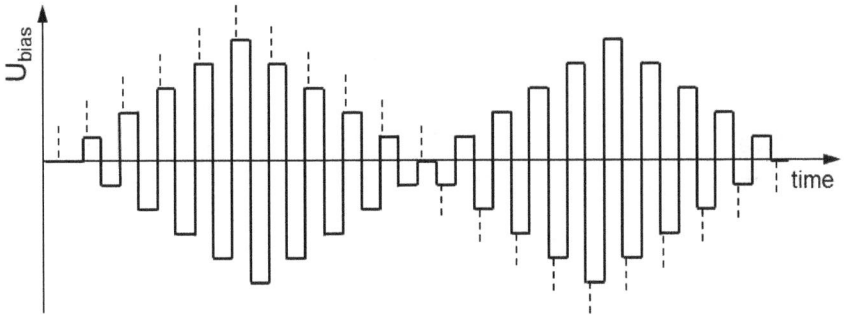

Figure 3.29. Example for the variation of the bias potential in potential-dependent transient SPV measurements. The dashed lines mark the trigger pulses for the laser.

at t_{off}. The bias potential is changed to the opposite sign at the time t_{bias} and the bias potential of the opposite sign is switched off at the $t_{-\text{bias}}$. The time intervals are equal for the application of U_{bias} and $U_{-\text{bias}}$.

In order to measure the dependence of the amplitude of SPV transients (SPV$_{\text{max}}$) as a function of U_{bias}, the measurement regime shown in Figure 3.28 is repeated for a series of tens of values of U_{bias}. In a whole cycle, the values of U_{bias} are first stepwise increased from a starting value to a positive maximum, then decreased to a negative maximum and again increased to the starting value (Figure 3.29).

The measurement regime shown in Figure 3.29 allows for extracting information about (i) the density of fast surface states from the slope of the dependence of SPV$_{\text{max}}$ on U_{bias} with regard to Equation (3.19), (ii) asymmetric charge transfer under polarization from the hysteresis at increasing and decreasing values of U_{bias} and (iii) fixed surface charge from the shift of the bias potential at the maximum slope of the dependence of SPV$_{\text{max}}$ on U_{bias}.

Figure 3.30 shows a schematic dependence of SPV$_{\text{max}}$ on U_{bias} for a sample with high hysteresis and fixed charge. Such a behavior, for example, is typical for hydrogenated silicon surfaces and depends on the regime of application of U_{bias} and $U_{-\text{bias}}$ (Dittrich *et al.*, 1999). The values of the fixed and polarization charge can be obtained from the corresponding U_{bias} and C_m. The hysteresis can increase, for example, if the Fermi energy at the surface shifts during the application of U_{bias} closer to the valence or conduction band edges which causes preferential charging of donor or acceptor states or molecules adsorbed at the semiconductor surface, respectively.

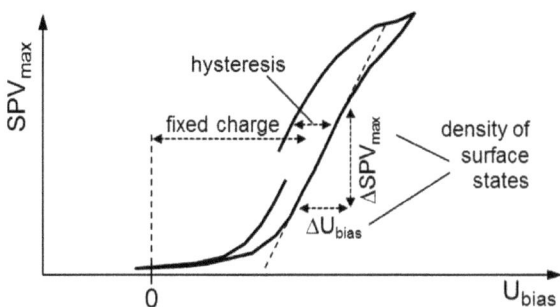

Figure 3.30. Schematic diagram of a dependence of the amplitude of a SPV transient on the bias potential. Information about the density of surface states, polarization charge and fixed charge is obtained from the slope, the difference between U_{bias} at increasing and decreasing parts for the same SPV_{max} and the shift of U_{bias}, respectively.

3.4.4. *Measurements at constant SPV signals*

At constant SPV signals, the diffusion length of photogenerated charge carriers (L_{diff}) can be obtained from the dependence of the photon flux on the absorption length (Goodman, 1961). The idea is based on the fact that the densities of photogenerated charge carriers (Δp) are equal for equal SPV signals excited with monochromatic light of different wavelengths (λ). For (i) planar geometry with a layer thickness larger than L_{diff} and larger than the absorption length (α^{-1}), (ii) steady state, (iii) low signal case and, (iv) a surface space charge region much thinner than L_{diff} and α^{-1}, Δp is given by (see reference 9 in Goodman, 1961):

$$\Delta p = \frac{QE(\lambda) \cdot (1 - R(\lambda)) \cdot L_{diff}}{D/L_{diff} + s} \cdot \frac{Int(\lambda)}{\alpha(\lambda)^{-1} + L_{diff}} \qquad (3.20)$$

where QE, Int, R, D and s in Equation (3.20) denote the quantum efficiency, intensity, reflectivity, diffusion coefficient and surface recombination velocity, respectively. QE, Int, R and α are functions of λ. In the case that QE and R are nearly independent of λ in the considered range,

$$\frac{Int(\lambda_1)}{\alpha(\lambda_1)^{-1} + L_{diff}} = \frac{Int(\lambda_2)}{\alpha(\lambda_2)^{-1} + L_{diff}} = \frac{Int(\lambda)}{\alpha(\lambda_1)^{-1} + L_{diff}} = const \qquad (3.21)$$

Therefore, L_{diff} is determine by the intersection point of dependence of the photon flux on α^{-1}.

Figure 3.31 shows a scheme for the spectral-dependent measurement of the light intensity at constant modulated SPV signals.

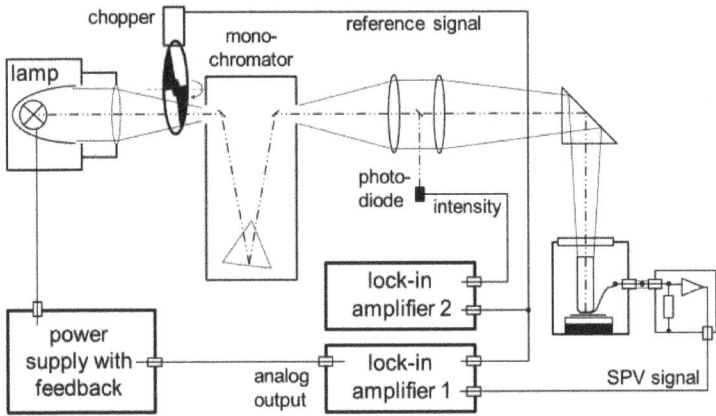

Figure 3.31. Simplified schematic setup for the spectral-dependent measurement of the light intensity at constant modulated SPV signals.

The SPV signals are measured with one of the two lock-in amplifiers in the setup. Lock-in amplifiers have analog outputs for measured signals. The range of analog outputs of lock-in amplifiers is usually between -10 and $+10\,\text{V}$. The SPV signal is given from the analog output of the first lock-in amplifier to the input of the power supply with a feedback unit for the halogen lamp. The feedback compares permanently the signal from the analog output of the lock-in amplifier with an internal reference potential and tunes the power of the lamp in order to keep the difference constant with regard to the desired SPV signal. The intensity of a halogen lamp can be changed relatively fast and smooth with the analog power supply and feedback unit. The photon flux is calibrated in relation to the light intensity. The modulation period of the chopper and the time constant of the feedback control shall be significantly shorter and longer, respectively, than the integration time constant of the lock-in amplifier in order to avoid oscillations in the feedback loop. The setup given in Figure 3.31 allows to keep constant SPV signals of the order of tens of microvolts.

The spectrum of the light intensity at constant SPV is measured with a calibrated photodiode and the second lock-in amplifier. For analyzing the diffusion length, the absorption spectrum of the photoactive material should be known. The diffusion length is equal to the negative value of the intersection point at the photon flux approximated to 0. As an example, Figure 3.32(a) shows the dependencies of the photon flux on the absorption length at different constant values of the SPV for a c-Si wafer.

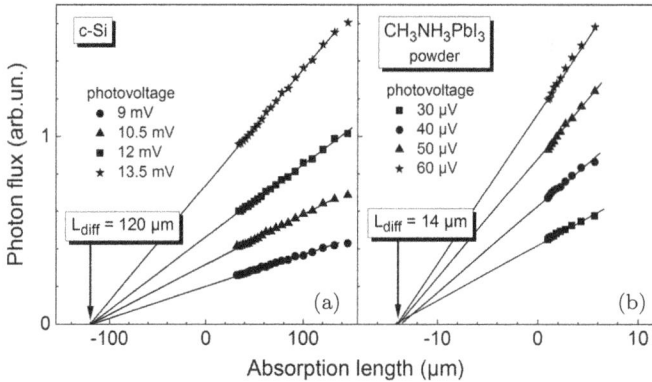

Figure 3.32. Dependence of the photon flux on the absorption length at different values of the constant SPV (a) for a c-Si wafer with a diffusion length of $120\,\mu$m and (b) for $CH_3NH_3PbI_3$ powder with a diffusion length of $14\,\mu$m.

All dependencies have one common intersection point at $-120\,\mu$m, i.e. the diffusion length is $120\,\mu$m for the given c-Si wafer.

The setup shown in Figure 3.31 allows also for the characterization of L_{diff} in powder particles such as $CH_3NH_3PbI_3$ (see the example in Figure 3.32(b)) and for quasi *in situ* measurements of L_{diff} under degradation as demonstrated for $CH_3NH_3PbI_3$ (Dittrich *et al.*, 2016a, 2016b).

3.5. Summary

A more detailed understanding about practical aspects of SPV techniques is important for their broader applications in research and development of photoactive materials and combinations of photoactive materials with other materials as well as for monitoring in production processes. Practical aspects of SPV techniques concern, for example, light sources, measurement regimes, data analysis and the design of experiments.

Basic setups of SPV spectroscopy contain a monochromatic light source, an optical alignment system, the measurement configuration, a measurement device and a computer control with software.

Monochromators with lamps are used as monochromatic light sources for continuous and modulated illumination, which are applied for spectral-dependent SPV measurements with Kelvin probes or fixed capacitors, respectively. Halogen and xenon arc lamps are preferentially applied for measurements in the near-infrared to violet and visible to UV ranges, respectively.

Monochromators are characterized by their spectral range, spectral resolution and stray light. Quartz prism monochromators are useful for continuous measurements of overview spectra from the near-infrared to the UV range. Stray light has a broad spectral distribution and can cause artifacts in SPV spectra in ranges with low absorption.

SPV signals are usually high for conventional semiconductors but can be low, for example, for photocatalytic nanoparticles or sub-monolayers of organic molecules. Therefore, a compromise between spectral resolution, light intensity and intensity of stray light shall be found depending on the SPV signal in relation to acceptable noise.

The intensity of monochromatic light increases with increasing opening ratio of a monochromator and increasing widths of the entrance and exit slits. The increase of the opening ratio causes a strong increase of the intensity of stray light. The increase of d_{slit} results in a decrease of the spectral resolution and an increase of the intensity of stray light. The intensity of stray light can be reduced with additional optical long-pass filters. An opening ratio of about 7 and a spectral resolution between 5 and 50 meV are well suited for spectral-dependent SPV measurements on most photoactive materials.

Pulsed tunable dye or Nd:YAG lasers with duration times of the order of a nanoseconds are available for transient SPV measurements in spectral ranges from near-infrared to UV. Tunable Nd:YAG lasers with a spectral cleaning unit have sharp laser lines at the desired wavelengths and strong suppression of light with other wavelengths. A homogeneous beam profile, high intensity stability and a relatively low repetition rate are important for transient SPV spectroscopy.

Spectral-dependent SPV measurements in the low signal case are very useful for the characterization of band gaps and transition energies. In the low signal case, SPV signals are proportional to the generation rate. If, in addition to the low signal case, the charge separation length is much shorter than the absorption length, the shapes of SPV spectra can be handled as absorption spectra. However, in contrast in the case of absorption spectra, one has to keep in mind that only photogenerated mobile charge carriers contribute to SPV signals.

A well-reliable and robust high-impedance buffer based on the OPA656 (bandwidth 500 MHz) was developed and the design was optimized with increasing experience and opportunities in transient measurements. Aside the sample and the reference electrodes, the measurement resistance and the high-impedance buffer, measurement configurations include also a wire connecting the reference electrode with the high-impedance buffer and connectors. Damped oscillations of the resulting equivalent LCR circuit can be excited by SPV signals. Depending mainly on the length of the connecting wire(s) and the measurement and parasitic capacitances, the period of the oscillations is in the nanoseconds range. The series resistance of the reference electrode causes damping of the oscillations within tens of nanoseconds.

Damped oscillations can strongly disturb SPV transients at times shorter than about 10–20 ns. By adjusting the series resistance, the oscillations can be damped close to critical damping so that resolution times of about half of the oscillation period can be reached. Therefore, transient SPV measurements can be well performed without additional measures at times longer than about 1–2 ns.

The source function of a measured SPV transient corresponds to the SPV transient undisturbed by the measurement configuration and resolution time. At very short times, the source function, the onset function and the parameters of the equivalent LCR circuit can be extracted from measured SPV transients by multi-parameter fits. An onset time, which corresponds to resolution time of the system, of about 0.5 ns was reached for transient measurements on highly n-type doped silicon. By further optimization of measurement setups and high-impedance buffers, a reduction of the resolution time up to about 100 ps seems realistic.

For self-consistency, SPV transients are simulated from the source function and the equivalent circuit including all elements of the measurement configuration and compared with the measured SPV transients. The development of reliable and robust algorithms for self-consistent multi-parameter fits will be a decisive step toward broad applications of transient SPV spectroscopy at very short times.

The control of external parameters, such as temperature, pressure of gases or bias potentials, is crucial for the design of dedicated SPV measurements. The control and the variation of external parameters shall not influence or disturb the measurement of SPV signals. From the electronic side, this can be realized with analog comparators and feedback loops connected with well-stabilized dc power supplies and passive sensors near samples. Furthermore, one has to take into account that gas molecules can also influence the reference electrode (variation of the work function, photoinduced reactions) and may lead to partial shunting between the samples and reference electrodes, for example, due to condensation of polar molecules.

The density of surface states can be found from the dependence of the band bending on an applied bias potential. Band bending is measured with SPV transient at large signals. The application of a high and well-stabilized bias potential demands the protection of the input of the high-impedance buffer during switching. The influence of phenomena related to polarization on SPV signals can be reduced by adjusting the measurement regime, i.e. pulsed application of U_{bias} and of $U_{-\text{bias}}$ with opposite polarity. The characterization of surface states by bias potential-dependent transient SPV measurements is possible for a semiconductor surface at which the Fermi energy is not pinned.

The diffusion length of photogenerated charge carriers in a semi-conductor can be obtained from measurements at constant SPV signals whereas the light intensity is plotted as a function of the absorption length (Goodman, 1961). Under ideal conditions, the absorption length is equal to the extrapolated negative value at zero intensity. The combination of a halogen lamp, modulation of light and a power supply with an analog feedback unit allows to keep SPV signals constant within tens of microvolts.

Chapter 4

Random Walk Simulation
of Surface Photovoltage Signals
in Small Systems

4.1. Introduction

Small systems are introduced in order to analyze the influence of localized
states on surface photovoltage (SPV) signals. A small charge-selective
system consists of two small volumes with electronic states which are
available differently for a photogenerated electron and/or hole. The number
of localized states in a small system is limited. This chapter starts with a
short description of localized and delocalized states, charge transfer between
localized states by tunneling and by trap limited transport and distributions
of localized states. Small systems can be combined to more complex systems
by introducing, for example, periodic boundary conditions and/or soft
lattices. It follows an explanation of principles for random walk simulations
of transient and modulated SPV signals in small systems by applying
a matrix of charge transfer rates. The influence of specific distributions
on the shape of simulated SPV transients is shown. The role of multiple
trapping on simulated SPV transients is demonstrated. The simulation of
the coefficients S_1–S_4 from modulated photogeneration and the calculation
of the corresponding in-phase and phase-shifted by 90° SPV signals are
explained. A striking advantage of small systems is that SPV signals
can be directly fitted by applying random walk simulations. The random
assignment of random rank numbers to the simulation parameters of a small
model system reduces the probability for sticking in local minima of the

error square during fitting. It is self-consistently shown that random walk simulations can be applied to multi-parameter fitting of SPV transients. The empirical fitting of SPV transients with analytical functions and their relation to distributions of localized states is briefly discussed.

4.2. Charge Transfer between Localized States in Small Systems

4.2.1. *Localized and delocalized states in small systems*

Localized states (Figure 4.1) are extended over very small ranges that are of the order of the length of a chemical bond. The distance between two localized states i and j (r_{ij}) is defined as the distance between the centers of both states. The diameter of a localized state (δ) is smaller or much smaller than the distance between two localized states. In the following, the diameters of localized states are neglected and only the distances between localized states are considered.

In a small system, the number of localized states is limited and the localized states are distributed over a small volume (V). The volume can be defined by any shape such as cubes or spheres. Localized states are often distributed randomly in space (disorder in photoactive materials).

A small system can consist of localized and delocalized states. Charge carriers are not mobile in localized states (also called traps for this reason). Delocalized states are extended over the whole volume of a small system

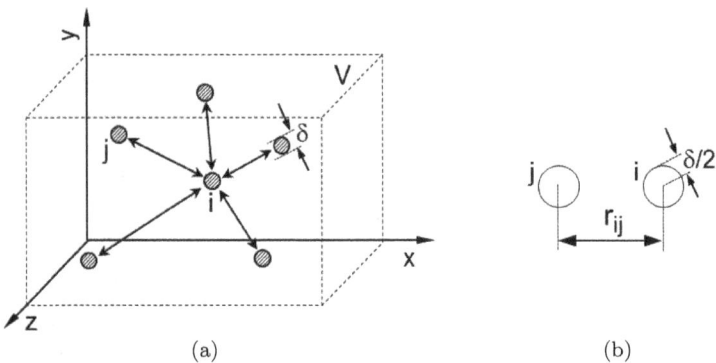

(a)   (b)

Figure 4.1. Localized states distributed over a small volume V (a) and definition of the distance between two localized states i and j (b). For a delocalized state, the expanse is considered. The extension of a localized state and the distance between two localized states are denoted by δ and r_{ij}, respectively.

or a part of the volume of a small system. Charge carriers are mobile in delocalized states.

Charge carriers can stay for a relatively long time in a localized state depending on distances between localized states and/or on the energy of a trapped charge carrier in relation to other localized states or to an extended state and on temperature. Localized states can limit the relaxation of a disturbance of a space charge or of SPV signals in photoactive materials. The theoretical analysis of the influence of localized states on charge separation across interfaces is important for a better understanding of SPV signals and opens ways for the interpretation of related experimental data. For this purpose, simulation algorithms have to be developed and applied for different interface configurations.

The incorporation of reliable and fast simulation algorithms into broader analysis is realistic for systems with a relatively small number of localized states. Such systems are called small systems in the following.

In small systems, charge separation is realized by combining two distinguished volumes with localized and/or delocalized states which are differently available for positive (holes) and negative (electrons) charge carriers (V_A and V_B in Figure 4.2). The asymmetry of the distributions of localized and/or delocalized states defines a direction of charge separation (x) and a charge-selective plane (yz).

Recombination in a small system via localized states is defined as the transfer of an electron or a hole into a localized state occupied by a hole or an electron, respectively. In analogy, recombination via delocalized states

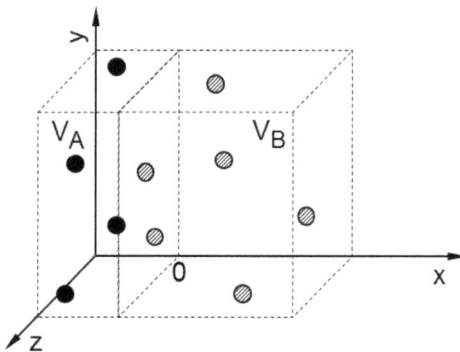

Figure 4.2. Schematic diagram of a small charge-selective system with different distributions of localized states in the volumes V_A and V_B. The charge-selective contact is at $x = 0$ in the yz-plane.

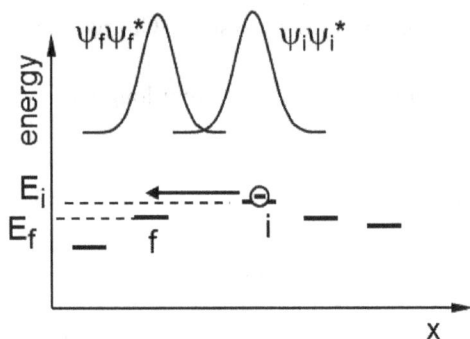

Figure 4.3. Schematic diagram of charge transfer between localized states by tunneling of an electron from the initial state (i) to the final state (f) due to a partial overlap of the probability density of the wave function of the initial state ($\psi_i \psi_i^*$) and final state ($\psi_f \psi_f^*$).

is defined as the transfer of an electron or a hole into an unoccupied below or occupied state above the Fermi energy, respectively.

4.2.2. *Charge transfer between localized states by tunneling*

Charge carriers can be transferred between localized states by tunneling (Figure 4.3) from an initial state (i) to a final state (f). Tunneling is possible due to a partial overlap of the wave function of a charge carrier (ψ) trapped at the initial state with the wave function of the final state (see, for example, Griffith, 2004).

The probability of a tunneling step decreases exponentially with increasing distance between the initial and finals states (r_{if}). The extension of wave functions in space is merged to one controlling parameter which is called the inverse tunneling length (α_{tunn}). The tunneling rate between the initial and final states (ν_{if}) depends also on a bouncing frequency ($\nu_{0,\text{tunn}}$) at which a trapped charge carrier probes the tunneling barrier.

$$\nu_{if} = \nu_{0,\text{tunn}} \cdot \exp(-2 \cdot \alpha_{\text{tunn}} \cdot r_{if}) \qquad (4.1)$$

Typical values of α_{tunn} are on the order of $2\,\text{nm}^{-1}$. A typical value of $\nu_{0,\text{tunn}}$, for example, can be of the order of a phonon frequency. The normalized tunneling rate is equivalent to the tunneling probability.

The energies of the initial and final states (E_i and E_f, respectively) can be different. This has an influence on the tunneling rate. If E_i is higher than or equal to E_f, Equation (4.1) remains unchanged. If E_f is higher than E_i, ν_{if} is thermally activated by the difference between E_i and E_f. The influence of thermal activation on tunneling processes can be expressed as (Miller and Abrahams, 1960)

$$\nu_{if} = \nu_{0,\text{tunn}} \cdot \exp\left(-2 \cdot \alpha \cdot r_{if} - \frac{E_f - E_i + |E_f - E_i|}{2 \cdot k_B \cdot T}\right) \qquad (4.2)$$

In an electric field, the energy difference between two sites is changed by the drop of the potential energy between both sites $(q \cdot U_{ij})$. This shall also be considered in the tunneling rate.

$$\nu_{ij} = \nu_0 \cdot \exp\left(-2 \cdot \alpha \cdot r_{ij} - \frac{E_j - E_i + |E_j - E_i|}{2 \cdot k_B \cdot T} - \frac{q \cdot U_{ij}}{k_B \cdot T}\right) \qquad (4.3)$$

4.2.3. *Charge transfer between localized states by trap limited transport*

Trap limited transport consists of the steps of thermal excitation of a charge carrier from a localized (i) into an delocalized state (dl), movement of the charge carrier in the delocalized state and trapping of the charge carrier at another localized state (j) (Figures 4.4(a)–4.4(c), respectively).

The excitation of a charge carrier from a localized into a delocalized state is also called detrapping or thermal emission. The trapping of a charge carrier from a delocalized state into a localized state is also called retrapping of the charge carrier.

The probability for thermal excitation or activation of a trapped charge carrier from a localized into a delocalized state depends exponentially on the difference between the energy of the delocalized state (E_{dl}) and the energy of the localized state (E_i) and on the reciprocal temperature (T). The maximum detrapping rate is denoted by $\nu_{0,\text{th}}$. Typical values of $\nu_{0,\text{th}}$ are of the order of a phonon frequency or a multiple of a photon frequency. The thermal activation or detrapping rate ($\nu_{i-\text{th}}$) is defined as the product of $\nu_{0,\text{th}}$ and the exponential of the negative ratio of $E_{\text{dl}} - E_i$ and $k_B \cdot T$.

$$\nu_{i-\text{th}} = \nu_{0,\text{th}} \cdot \exp\left(-\frac{E_{\text{dl}} - E_i}{k_B \cdot T}\right) \qquad (4.4)$$

Retrapping rates are usually much larger than detrapping rates. It is reasonable to assume that retrapping rates are equal for all localized states in a small system and on the order of $\nu_{0,\text{th}}$.

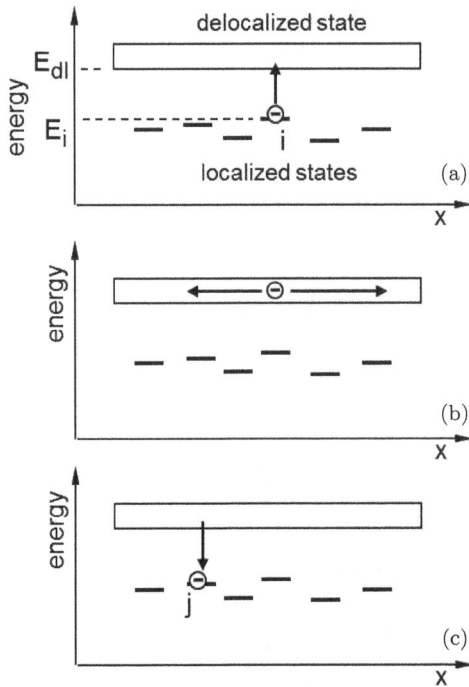

Figure 4.4. Schematic diagram of charge transfer between localized states by trap limited transport consisting of the steps of thermal excitation of a trapped charge carrier into a delocalized state (a), charge transport in the delocalized states (b) and trapping of the charge carrier at another localized state (c).

4.2.4. *Distributions of localized states*

Charge transfer from localized states into delocalized states or other localized states depends on their distributions in space and energy. Distribution functions of localized states of real samples depend very much, for example, on the nature of localized states and on preparation conditions of bulks and interfaces. In an ideal layer, localized states are distributed homogeneously in space (g_{hom}). Figure 4.5(a) shows g_{hom} for a homogeneous distribution over a layer thickness d_{H}.

$$g_{\text{hom}}(x)|_{x \in (0, d_{\text{H}})} = 1 \tag{4.5}$$

Localized states with energies close to the conduction or valence band edges of a semiconductor are often distributed exponentially. These states are also called tail states since they cause the exponential absorption tails

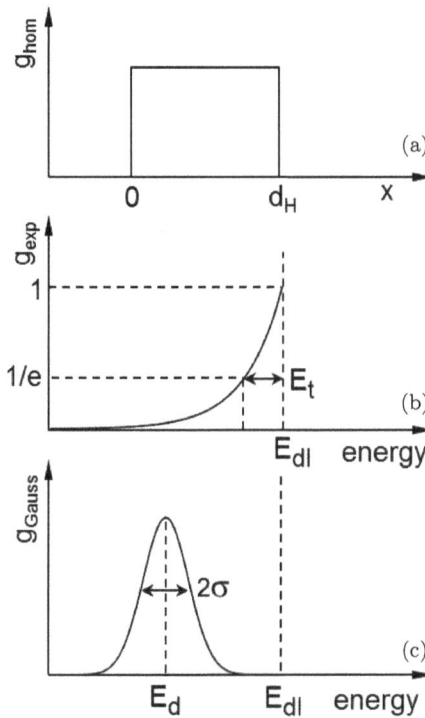

Figure 4.5. Schematic diagram of a homogeneous distribution over a thickness d_H (a) and exponential (b) and Gaussian (c) distribution functions in energy. The parameters E_t, E_{dl}, E_d and σ denote the energy parameter of the exponential, the energy of the delocalized transport state and the peak energy and width of the Gaussian distribution, respectively.

in absorption spectra below the band gap. The characteristic parameter of the exponential distribution is given by E_t, the so-called tail energy. The normalized exponential distribution (g_{exp}, Figure 4.5(b)) is reduced to $1/e$ times ($e \approx 2.72$) when the difference between the delocalized state and the energy of the localized state ($E_{dl} - E$) is equal to E_t.

$$g_{exp}(E) = \exp\left(-\frac{E_{dl} - E}{E_t}\right) \qquad (4.6)$$

Defect states deep in the band gap of a semiconductor can be often approximated by a Gaussian distribution function (g_{Gauss}, Figure 4.5(c)). The characteristic parameters of the Gaussian distribution are the energy

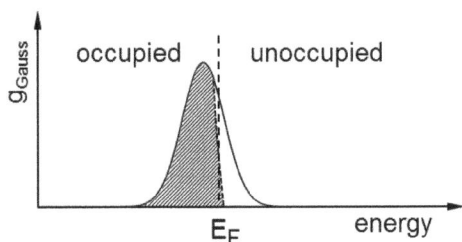

Figure 4.6. Gaussian distribution with occupied and unoccupied states.

at the maximum (E_d) and the standard deviation or width (σ).

$$g_{\text{Gauss}}(E) = \frac{1}{\sqrt{2\pi \cdot \sigma^2}} \cdot \exp\left(-\left(\frac{E_d - E}{4 \cdot \sigma}\right)^2\right) \qquad (4.7)$$

In charge transfer, only unoccupied (occupied) states are available for electrons (holes). The occupation of electronic states is described by the Fermi energy (E_F) and follows the Fermi-function. Occupied localized states are not available for an electron transfer (see, for example, Figure 4.6). Therefore, the number and energies of available states in a small system with a defined distribution of localized states can be additionally modified by implementing E_F.

The introduction of (quasi) Fermi energies into a small system allows for the consideration of the motion of single charge carriers in random walk simulations (Anta *et al.*, 2008).

4.3.   Principles of Random Walk Simulations of SPV Signals

4.3.1.   *Random walk of a single charge carrier in a small system*

A small system with localized states is defined by a limited, relatively small, number of localized states (N). Therefore, a charge carrier trapped at state i of a small system has transfer rates to $N - 1$ localized states. The states available for charge transfer from state i are denoted by the index j. The transfer rates of a charge carrier from state i to all localized states in a small system are written as a vector $m_i = (\nu_{i1} \cdots \nu_{iN})$ whereas the value of ν_{ii} is set to 0. All possible transfer rates in a small system are summarized in a matrix M.

$$M = \begin{pmatrix} \nu_{11} & \cdots & \nu_{1N} \\ \vdots & \ddots & \vdots \\ \nu_{N1} & \cdots & \nu_{NN} \end{pmatrix}, \quad \text{with } \nu_{ii} = 0 \tag{4.8}$$

In systems with localized and delocalized states, the charge transfer rate from state i to state j is the product of the tunneling and thermal activation rates to a delocalized state. The tunneling rates to the different states are different with respect to r_{ij} and E_i–E_j. Thermal activation rates depend on E_j–E_i (Miller and Abrahams, 1961) and on E_{dl}–E_i (trap limited transport).

Charge transfer rates correspond to probabilities. For the simulation of one charge transfer step in a small system, the final state of a charge transfer shall be found by a procedure taking into account the charge transfer rates to all localized states. In this procedure, first, the charge transfer rates from state i to all available states in the small system are summed ($\Sigma\nu_i$). Second, a random number (R_{rdm}) between 0 and 1 is generated. Third, a number Z_i is formed as the product of R_{rdm} and $\Sigma\nu_i$.

$$Z_i = R_{rdm} \cdot \sum_{j=1}^{N} \nu_{ij} = R_{rdm} \cdot \sum_{\nu_i} \tag{4.9}$$

Fourth, the possible charge transfer rates for transfers from state i are added (Σ_{run}) one after the other and the sum is compared with Z_i. The index j at which this sum gets equal or larger than Z_i corresponds to the final state (f) for the charge transfer step.

$$\Sigma_{run} = \sum_{j=1}^{f-1} \nu_{ij} < Z_i \tag{4.10$'$}$$

$$\Sigma_{run} = \sum_{j=1}^{f} \nu_{ij} \geq Z_i \tag{4.10$''$}$$

The time interval for the charge transfer step from the initial state i to the final state $f(\Delta t_{if})$ corresponds to an event with a certain probability, i.e. the available time intervals are also distributed. Non-repeating events are distributed exponentially, like prime numbers. For an exponential

distribution of time intervals, the natural logarithm of a random number
between 0 and 1 is considered (Nelson, 1999; Anta *et al.*, 2002).

$$\Delta t_{if} = -\frac{\ln(R_{\mathrm{rdm}})}{\nu_{if}} \tag{4.11}$$

After a charge transfer step, the time (t) is increased by Δt_{if} and the
sum of t and Δt_{if} is set as the new value of t. For continuing the motion
of a given charge carrier with a new charge transfer step, the value of f is
set to the new index i and a new number Z_i is defined. Incidentally, the
generator of random numbers is started at the value of the actual time in
order to avoid that random numbers repeat when starting the program.

The simulation of subsequent charge transfers between localized states
is called a random walk. A charge carrier disappears after transfer to
a recombination site which can be localized or delocalized. A complete
random walk of a single charge carrier from its photogeneration until its
recombination is called a random walk sample (RWS).

A RWS starts at $t = 0$ with an initial distribution of photogenerated
positive and negative charge carriers in space. As an example, Figure 4.7(a)
shows an initial distribution for a hole in a delocalized state in volume V_A
and an electron in one of the localized states in volume V_B. Volume V_A acts
as a delocalized recombination site and V_A can be replaced by the yz-plane
for the given example. Incidentally, Figure 4.7(a) is a suitable model for

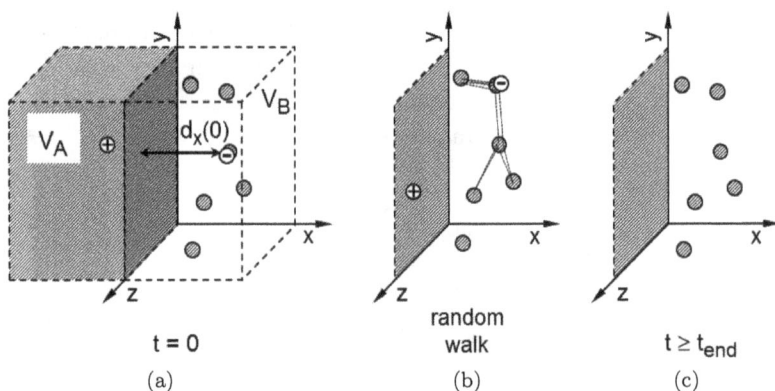

Figure 4.7. Example for an initial distribution of a photogenerated hole and electron
into a delocalized state in volume V_A (recombination site) and into one of the localized
states in volume V_B, respectively (a), a random walk of the electron over localized states
(b) and finish of the random walk after recombination (c).

internal photoemission of an electron from a metal layer such as platinum into a disordered semiconductor such as amorphous titania.

After photogeneration of charge carriers, a random walk follows (Figure 4.7(b)). The random walk of a single charge carrier or a RWS is finished after reaching a recombination site (Figure 4.7(c)).

The time of a charge transfer from site i to site j is also called the waiting time. In a random walk of a single charge carrier, the system remains unchanged during the waiting time, i.e. no simulation time is required for the characterization of a small system during a waiting time. This allows for the simulation of transient SPV signals over many orders of magnitude in time.

Random walks of electrons and holes have to be considered in simulations when localized states are present in both volumes V_A and V_B. The random walks of the electron and of the hole can start in the different volumes or in the same volume depending on initial photogeneration and charge separation. Internal photoemission, for example, can lead to the motion of the electron and of the hole in separate volumes if states for transport of a given charge carrier are not available in the other material (Figure 4.8(a)). Photogeneration by, for example, fundamental absorption causes random walks of an electron and of a hole in the same volume. Charge separation becomes possible if localized states are preferentially available for only one of the charge carriers (Figure 4.8(b)). In the second case, an SPV signal can arise due to transfer of the electron or of the hole into available states into the material where photogeneration was absent.

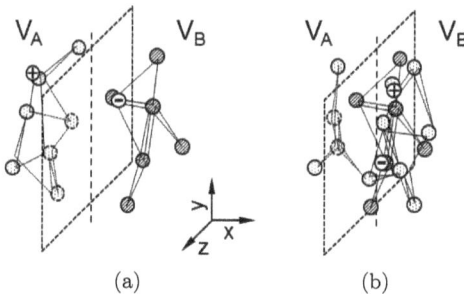

Figure 4.8. Examples for random walks of an electron and of a hole in separate volumes V_A and V_B, respectively (a) and in the same volume V_B whereas the hole can escape into available states in volume V_A (b).

The random walks of a photogenerated hole and electron are not treated independently. The values of Δt_{if} are calculated for the hole and for the electron and the charge carrier with the shorter Δt_{if} will be transferred.

4.3.2. *Extension of random walks to more complex systems*

Small systems can be combined to more complex systems. For this purpose, small units for random walk simulations and rules for their combination(s) shall be defined. A small unit is defined as a volume with a given number of localized states. Furthermore, a delocalized state can be distributed within a small unit. As an example, Figures 4.9(a) and 4.9(b) show small units with three and six localized states with a constant density, respectively. As a rule for keeping the density of localized states constant, small units can be expanded or repeated in one dimension proportionally to the increasing number of localized states.

A layer with a homogeneous distribution of localized states can be built by combining small units with the shape of rectangular prisms. A nomenclature including the number of small units in the x, y and z directions (i_x, i_y, i_z) can be introduced for such systems. As examples, Figures 4.9(c) and 4.9(d) show combinations of small units with 3 (Figure 4.9(c)) and 6 (Figure 4.9(d)) localized states with the nomenclatures (1,2,2) and (1,3,3), respectively.

The number of small units in small systems with localized states shall be kept as small as possible with regard to simulation time suitable for reliable data analysis. Periodic boundary conditions can be applied in order

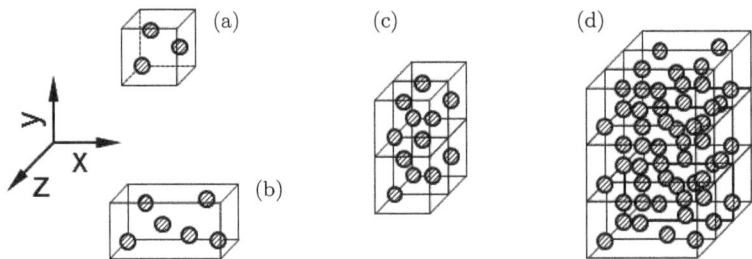

Figure 4.9. Examples for small units with three (a) and six (b) localized states with constant density and lateral arrangements of four small units with three localized states (c) and of nine small units with six localized states (d). For the given examples, the area of the unit cells in the y–z plane (parallel to a charge-selective contact) was constant.

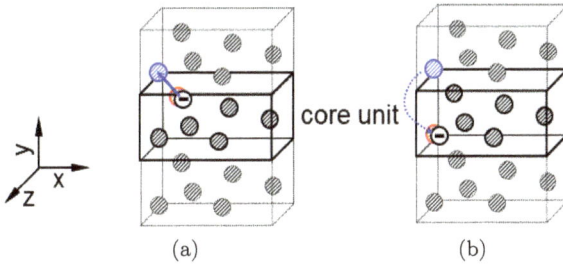

Figure 4.10. Principle of the periodic boundary condition for an electron transfer from a localized state in a small core unit into a localized state in a neighbored unit (a). Before starting the next charge transfer step, the electron is set into the equivalent position of the core unit (b).

to reduce the number of small units to nearest neighbor units (Figure 4.10). For periodic boundary conditions, the neighbored cells do not exist in reality.

Under periodic boundary conditions, a random walk is simulated within one smaller core unit. Charge transfer from a localized state in the core unit is possible into localized states within the core unit as well as into localized states of neighbored units. Before the simulation of a next charge transfer step is started, the position of the transferred charge carrier is proven and set into the equivalent position of the core unit if the charge has been transferred into a neighbored unit.

The definition of homogeneously distributed localized states can result in extremely short distances between localized states which are not reasonable from physical point of view, for example, if considering the lengths of a chemical bond. Therefore, additional criteria are required for the definition of randomly distributed localized states.

An inhomogeneous distribution of localized states in space can be simulated, for example, by a random distribution of localized states at the surface of a sphere (Figure 4.11(a)). In this case, the distance between a localized state i and the center of the sphere is fixed and the angles φ_i and θ_i are found randomly. Incidentally, instead of a sphere, other geometric bodies such as an ellipsoid can also be used for the definition of an inhomogeneous distribution of localized states in space.

The variation of distances between nearest neighbors is the dominating parameter for a random walk in a disordered system. A definition of a large number of localized states distributed randomly in space is not needed for

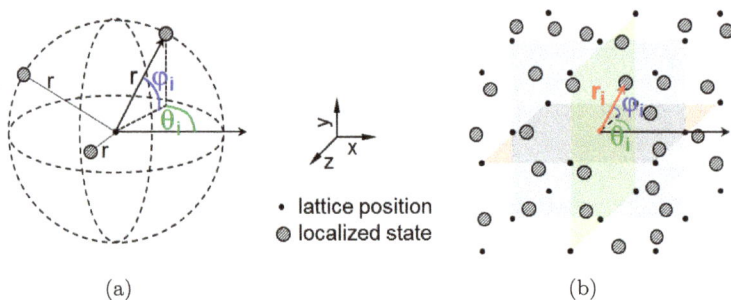

Figure 4.11. Schematic diagrams of three localized states distributed randomly at the surface of a sphere (a) and of a cubic soft lattice for the definition of randomly distributed distances between nearest neighbors of localized states (b).

describing a random distribution of distances between nearest neighbors. In the model of the so-called soft lattice (Fengler and Dittrich, 2016), localized states are assigned to positions in a well-defined lattice whereas the position and the angles of a localized state vary randomly with regard to the corresponding lattice position (Figure 4.11(b)). For example, the position of a localized state i in a soft lattice is given by the position of the lattice site i, the distance between the lattice site and the localized state (r_i) and its angles φ_i and θ_i. The minimum distance between two localized states can be defined by the range of r_i. The parameters r_i, φ_i and θ_i are found randomly.

The volume of a sphere can define the distribution of a delocalized state (Figure 4.12(a)) or of localized states (Figure 4.12(b)). The combination of spheres can be well applied for the definition of localized and delocalized states in more complex systems with a hierarchy in morphology such as porous semiconductor electrodes (Figure 4.12(c)). Hard spheres are defined as spheres which cannot geometrically interpenetrate each other.

The charge transfer between two neighbored hard spheres with delocalized states will be limited by tunneling if introducing a distance between both spheres. This is the case, for example, in systems with colloidal semiconductor quantum dots. A barrier between two connected hard spheres with delocalized states can also arise due to dielectric screening if the dielectric constant of the sphere is larger than the dielectric constant of the surrounding ambient (Keldysh, 1979). This can be the case, for example, in systems of sintered metal oxide nanoparticles such as nanoporous TiO_2. Incidentally, hard spheres can also be arranged in soft lattices.

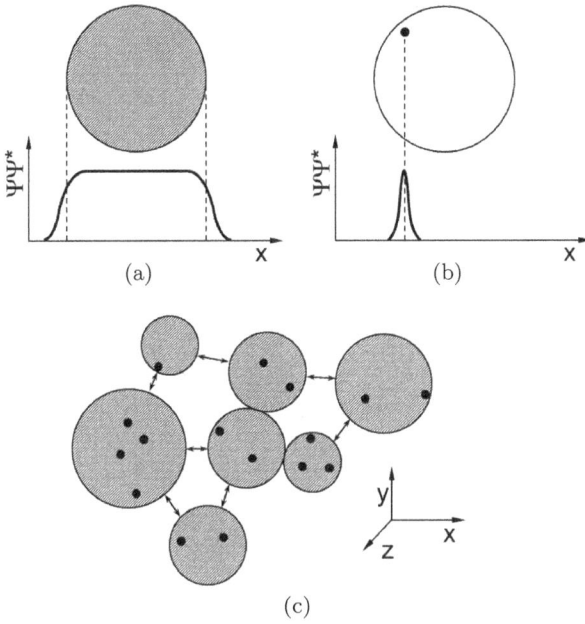

Figure 4.12. Schematic diagrams of a hard sphere with a delocalized state and its wave function (a) and of a localized state in a sphere with its wave function (b) and a combination of hard spheres with delocalized and localized states and varying distances between the spheres (c).

4.3.3. *Random walk simulation of SPV transients*

SPV signals are formed as the superposition of numerous random walk samples (RWSs) as illustrated in Figure 4.13. The natural numbers K and k denote the total number of RWSs and the index of the running RWS, respectively. The center of charge approach is taken into account. For calculating the center of charge, the projections of the positions of the charge carrier(s) parallel to the surface normal of the sample electrode are taken (d_x). The SPV signal at a certain time is obtained by normalizing the sum of the projected positions of all RWSs at that time ($d_{xk}(t)$) to the areal density of charge carriers ($q/(K \cdot A)$) and by considering the dielectric constant ($\varepsilon_r \cdot \varepsilon_0$).

$$\text{SPV}(t) = \frac{q}{K \cdot A} \cdot \frac{1}{\varepsilon_r \cdot \varepsilon_0} \cdot \sum_{k=1}^{k=K} d_{xk}(t) \tag{4.12}$$

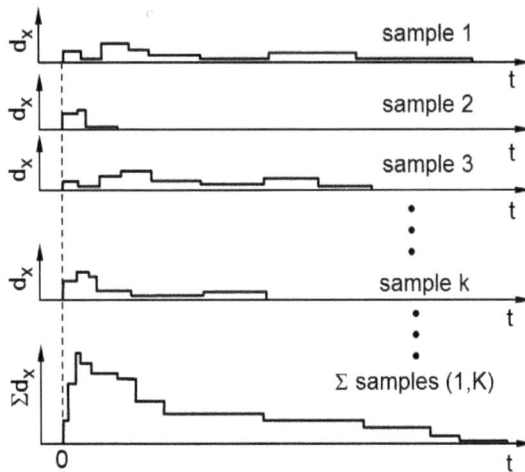

Figure 4.13. Principle of the simulation of the time-dependent charge separation length by superposition of random walk samples.

The number of RWSs simulated for one SPV transient is usually of the order of 10^4–10^6.

In order to reduce the number of data points in SPV transients simulated over many orders of magnitude in time, the superposition of numerous RWSs is combined with sorting the times of charge transfer steps into a histogram with time intervals increasing exponentially in time. This is also useful with regard to the exponential distributions of charge transfer rates and with regard to real measurements of SPV transients.

The time intervals of a simulated SPV transient are numbered by the index ℓ and t_ℓ corresponds to the total time at the number ℓ. The total number of time intervals is denoted by L. The values t_ℓ of the time vector $(t_1 \ldots t_L)$ are determined by a calculation rule for the time interval between two times (Δt_ℓ), i.e. Δt_ℓ is a function of t_ℓ and $t_{\ell+1}$ is equal to the sum of t_ℓ and Δt_ℓ. L is of the order of 10^3 and t_ℓ varies from the (sub) nanosecond to second range.

Simulated SPV transients are obtained from data pairs of time (t_ℓ) and d_x at that time $(d_{xk\ell})$. The index ℓ denotes the number of a data pair in a transient. The value of t denotes the running time. At the beginning of the simulation of a transient, all values of the d_x vector $(d_{x1} \ldots d_{xL})$ are set to 0. Furthermore, the value of t is set to a time with respect to the light source at the beginning of the simulation of each new RWS. After each

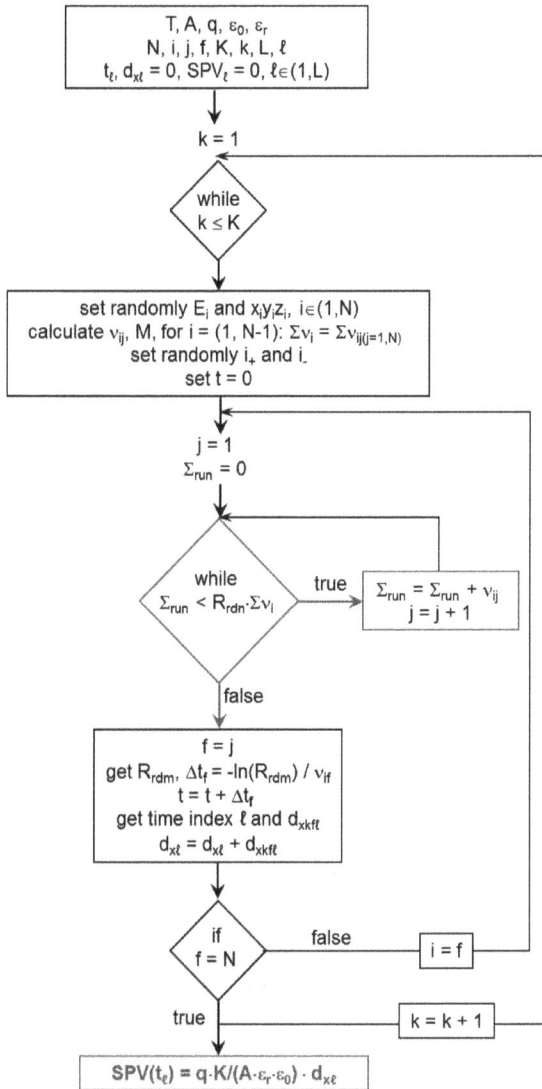

Figure 4.14. Principle of random walk simulations for SPV transients in small systems with localized states and one moving charge carrier.

charge transfer step from state i to state f of the RWS k, the corresponding time t_ℓ is found and the value of $d_{xkf\ell}$ is added to the corresponding $d_{x\ell}$.

Figure 4.14 summarizes the simulation procedure of an SPV transient in a system with localized states for one moving charge carrier.

At the beginning of the simulation, the starting conditions are defined. The starting conditions contain the definitions of the constants $(A, q, \varepsilon_0, \varepsilon_r)$, temperature (T), the integer numbers N, K and L for the total numbers of localized states, RWS and time values, respectively, and the integer numbers i, j, f, k and ℓ for the indices of the initial, available and final states and for the running RWS and time step, respectively. Furthermore, the distribution functions of localized states in energy $(g_e(E_i))$ and space $(g_s(x_i y_i z_i))$ and of the initial charge separation $(g_i(+, -))$ as well as the time intervals for the documentation of the transient and the initial vectors for $d_{x\ell}$, t_ℓ and SPV_ℓ are defined. Incidentally, additional integer numbers and indices are introduced for the simulation of SPV transients in more complex systems with more moving charge carriers.

The calculation of each RWS starts with setting randomly the values for the energy and the position of each localized state and with setting the initial positions for the separated charge carriers $(i_+$ and $i_-)$ in accordance to the distribution functions. Furthermore, the charge transfer rates between all available states, i.e. the matrix M, and all values of $\Sigma\nu_i$ are calculated and the running time is set to 0.

The cycle for the simulation of a charge transfer step starts with getting a new R_{rdm} and calculating Z_i by using Equation (4.9). After that, Σ_{run} is set to 0. It follows the cycle for determining the final state for the corresponding charge transfer step by applying the condition given by Equations (4.10') and (4.10''). After finding the final state, the new values of the running time within the given RWS and of d_{xf} are calculated. The charge transfer step is finished by assigning the running time to a time index k and adding d_{xf} to the value of d_{xk}. Before starting with a new charge transfer step, it is verified whether the positions of the positive and negative charges coincide (recombination). If not, the simulation of a new charge transfer step is started. After a recombination event, a RWS is finished and it is checked whether the total number of RWSs is reached. If not, the simulation of a new RWS is started. After finishing the simulations of all RWSs, the simulation of the transient is finished and the vectors of $d_{x\ell}$ and t_ℓ are transformed to the transient $\text{SPV}_\ell(t_\ell)$, i.e. SPV (t).

The procedure given in Figure 4.14 can be applied to the simulation of SPV transients in small systems with any distributions of localized states and any initial charge separation, i.e. the influence of dedicated parameters of a small system with localized states on the behavior of SPV transients can be studied in detail. Furthermore, the procedure given in Figure 4.14 can be extended to independent random walks of positive and negative charge carriers in a small system. In such a case, localized states available for positive and negative charge carriers are defined and charge transfer steps and $d_x(+)$ and $d_x(-)$ are calculated separately for the positive and negative charge carriers, respectively. The SPV is calculated by considering the difference between $d_x(+)$ and $d_x(-)$.

4.3.4. *Random walk simulation of modulated SPV signals*

In modulated SPV measurements, the photon flux changes periodically (modulation period, T). The periodic modulation of the photon flux results in a periodic distribution of starting times for random walks of single charge carriers (Figure 4.15). The starting time for the random walk of a single charge carrier is denoted by t_{0k} for the sample number k. The RWSs are

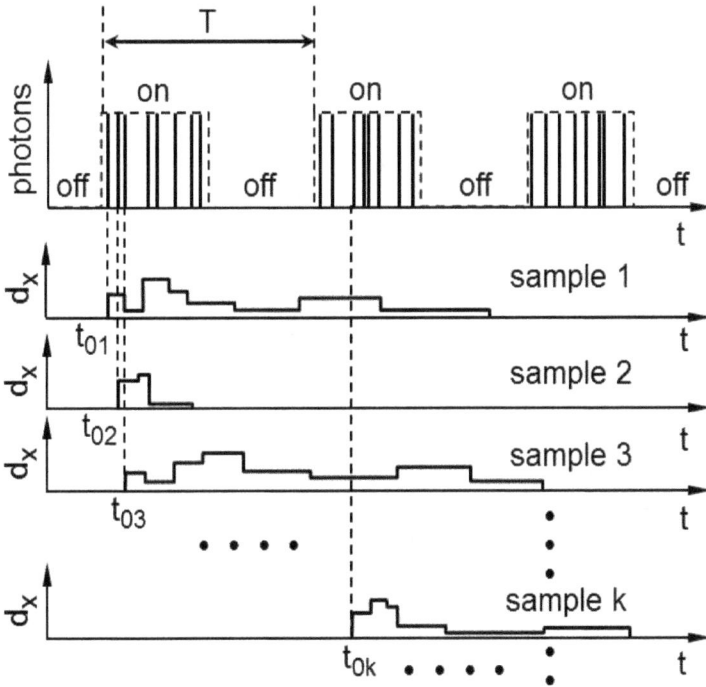

Figure 4.15. Schematic diagram of a modulated light intensity with photons (top) which lead to photogeneration of charge carriers separated in space and projected distances perpendicular to the charge-selective contact for some random walks of a single charge carrier (samples). T and t_{0k} denote the modulation period and the starting time of sample k, respectively.

simulated in the similar way as shown in Figure 4.14 by taking into account the different values of t_{0k}.

The total time of the random walk of a single charge carrier can last significantly longer than one modulation period. Therefore, the simulation of modulated SPV signals shall consider time ranges significantly longer than one modulation period. Furthermore, since RWSs can last over times longer than T, modulated charge separation can depend on time. Therefore, the simulation of modulated SPV signals demands the subsequent simulation of individual RWSs. The total number of RWSs for the simulation of modulated SPV signals is denoted by K and on the order of 10^5–10^6.

The density of starting times of samples depends on the photogeneration rate, i.e. on the photon flux of the modulated light source, on the spectrum of the photon flux and on the absorption spectrum of the photoactive material (Figure 4.16(a)).

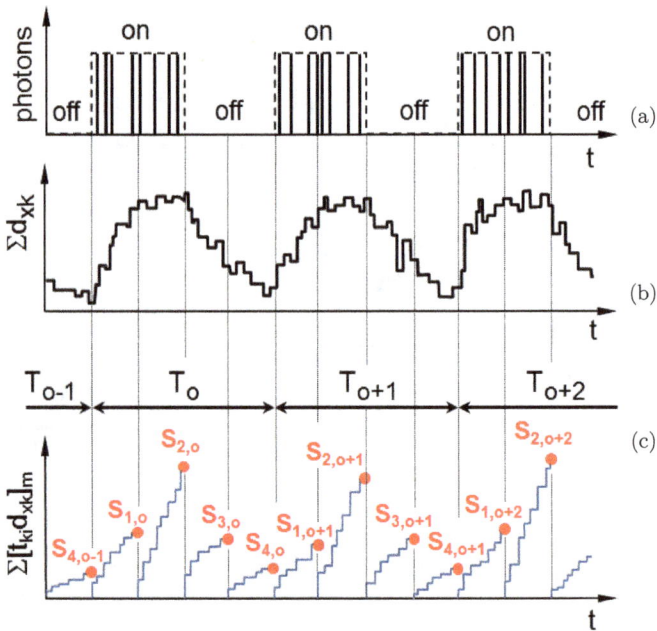

Figure 4.16. Schematic diagram of photons leading to modulated photogeneration of charge carriers separated in space (a), the sum of projected distances perpendicular to the charge-selective contact (b) and for the integrated over quarter periods sums (S_{1i}–S_{4i}, $i = $ o-1, o, o$+1$ and o$+2$) of projected distances d_{xk} (c). T_{o-1}, T_o, T_{o+1} and T_{o+2} mark the periods with the numbers o-1, o, o$+1$ and o$+2$, respectively.

The simulated RWSs are added whereas the time increases continuously (Figure 4.16(b)). The extraction of modulated SPV signals from the sum of simulated RWSs requires the same periodic time base and a minimum of data points allowing for the calculation of in-phase (X) and phase-shifted by 90° (Y) signals as well as for averaging over the number of periods.

Modulated SPV signals can be characterized by a minimum of four data points obtained by integration within one modulation period (S_1–S_4). The values of S_1–S_4 are calculated for each period by adding up the integrals of d_{xk} within the first, second, third and fourth quarters of a given modulation period, respectively. Therefore, a corresponding time base shall be generated for calculating and averaging S_1–S_4. The number of a given modulation period within a train of modulation periods (total number O) is denoted by the index o (see also Figure 4.16(c)).

$$S_j = \frac{1}{O} \sum_{o=1}^{O} \left\{ \int_{t=(\frac{j-1}{4})\cdot T + o\cdot T}^{t=(\frac{j}{4})\cdot T + o\cdot T} \left[\sum_{k=1}^{k=K} d_{xk}(t) \right] dt \right\}, \quad j = 1,\ldots,4 \quad (4.13)$$

The X- and Y-signals are calculated by using the parameters S_1–S_4 simulated by applying Equation (4.13), the definitions of the in-phase and phase-shifted by 90° signals and a normalization function with respect to the areal density of photogenerated charge carriers (f(K/A)).

$$X = f\left(\frac{K}{A}\right) \cdot \frac{q}{\varepsilon_r \cdot \varepsilon_0} \cdot (S_1 + S_2 - S_3 - S_4) \quad (4.14')$$

$$Y = f\left(\frac{K}{A}\right) \cdot \frac{q}{\varepsilon_r \cdot \varepsilon_0} \cdot (S_1 - S_2 - S_3 + S_4) \quad (4.14'')$$

The function $f(K/A)$ depends on the photon flux and absorption and is varied in additional time steps for the simulation of SPV spectra.

The amplitude and the phase angle of modulated SPV signals can be calculated by using the calculated values of the X- and Y-signals and the definitions given by Equations (2.31) and (2.32), respectively.

4.4. Examples for Simulated SPV Signals in Small Systems with Localized States

4.4.1. *SPV transients in extremely thin layers*

For extremely thin layers, the layer thickness (d_H) is much shorter than the average distance between localized states (d_{avg}). Therefore, a random walk of charge separated across an interface with an extremely thin layer is given by single steps of charge transfers from sites at which charge carriers are trapped after photogeneration to recombination sites at the interface. The single steps of charge transfer are limited only by tunneling recombination ($R_{tunn-rec}$) or/and by thermally activated recombination via a delocalized state coupling directly to a recombination state ($R_{therm-rec}$). Therefore, the consideration of extremely thin layers with localized states is useful for the analysis of the influence distributions of localized states in space (tunneling recombination) or energy (thermally activated recombination) on the shape of SPV transients.

In the simplest case, recombination is not limited by the distribution of recombination states, which is the case for a homogeneous and dense distribution of recombination states in energy (Figure 4.17). This is realized, for example, at a metal surface.

It is reasonable to start the simulation at an elementary time (τ_0) corresponding to the reciprocal of the maximum charge transfer rates. Furthermore, it is assumed that the maximum charge transfer rates are identical for tunneling and thermally activated recombination via delocalized states.

$$\tau_0 = \frac{1}{\nu_{0,tunn}} = \frac{1}{\nu_{0,th}} \tag{4.15}$$

In the following, the values of τ_0 and of the reciprocal tunneling length (α_{tunn}) were fixed at 0.1 ns and 1 nm^{-1}, respectively, and the transients are represented in log-lin scales.

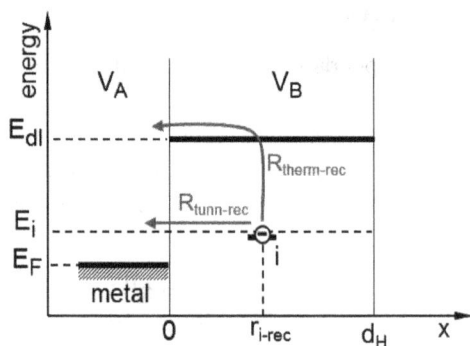

Figure 4.17. Energy diagram for the transfer of an electron from a localized state (i) in an extremely thin layer (in volume V_B) into homogeneously distributed recombination states (metal, volume V_A) by tunneling recombination ($R_{tunn-rec}$) or by thermally activated recombination via a delocalized state ($R_{therm-rec}$).

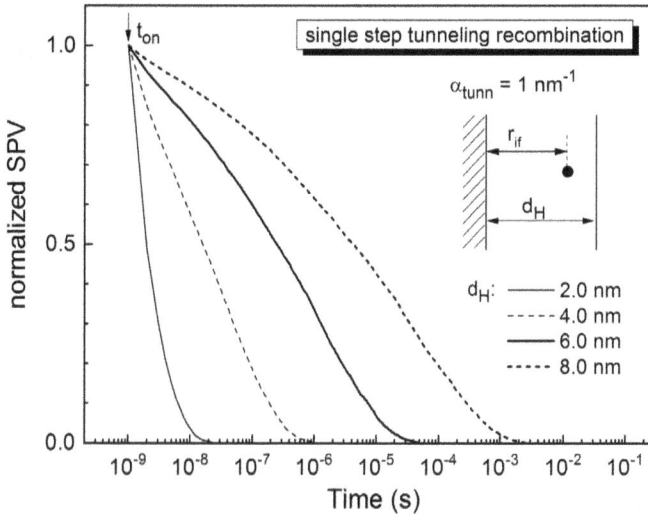

Figure 4.18. Normalized simulated SPV transients limited by tunneling recombination of charge carriers from localized states distributed homogeneously in an extremely thin layer into a delocalized recombination state on the substrate for layer thicknesses of 2, 4, 6 and 8 nm (thin solid, thin dashed, thick solid lines and thick dashed lines, respectively). The arrow marks the starting time (t_{on}). The reciprocal tunneling length was set to $1\,nm^{-1}$.

Figure 4.18 shows normalized simulated SPV transients for tunneling recombination of charge carriers from localized states with a homogeneous distribution in an extremely thin layer with different thicknesses into recombination states with a homogeneous distribution in energy. In this model, the distance between the initial and final states is given by the distance of the localized state to the yz-plane and thermal activation is not considered.

The decays of the SPV transients set on within the elementary time due to localized states being very close to the recombination states. The SPV transients decayed within 10–20 ns and 0.5–1 μs for d_H equal to 2 and 4 nm, respectively. The decay time increased up to the millisecond range for d_H equal to 8 nm. Furthermore, the decays got steeper on the logarithmic timescale with increasing time for thicker layers.

Localized states are usually not distributed homogeneously near interfaces due to different chemical interactions between atoms of a substrate and atoms of a layer and between atoms within a layer. For a passivated interface, for example, an inhomogeneous distribution of distances between

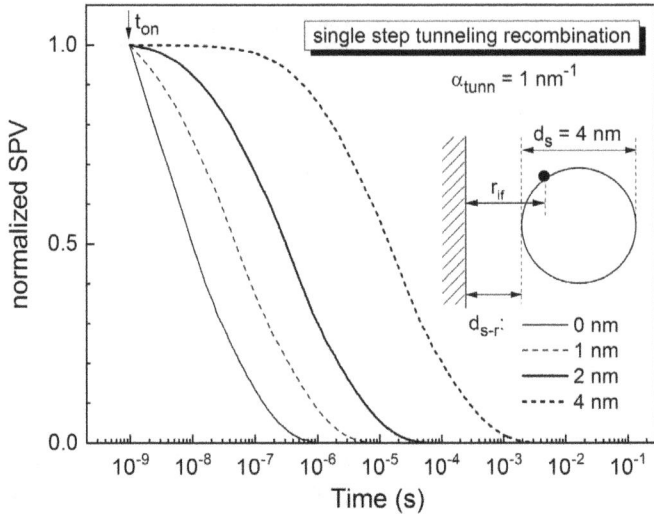

Figure 4.19. Normalized simulated SPV transients limited by tunneling recombination of charge carriers from localized states distributed homogeneously at the surface of a sphere into a delocalized recombination state on the substrate. The diameter of the sphere was set to 4 nm. The minimum distance between the surface of the sphere and the plain of recombination states was 0, 1, 2 and 4 nm (thin solid, thin dashed, thick solid lines and thick dashed lines, respectively). The arrow marks the starting time (t_{on}). The reciprocal tunneling length was set to $1\,\text{nm}^{-1}$.

localized states and recombination states homogeneously distributed in energy can be approximated by localized states homogeneously distributed at the surface of a sphere which has a certain minimum distance to the plain of recombination states (d_{s-r}). Such a sphere can be, for example, a colloidal semiconductor nanoparticle separated from a metal surface by a monolayer of organic molecules.

Figure 4.19 gives an impression about the behavior of normalized simulated SPV transients for tunneling recombination of charge carriers from localized states with a homogeneous distribution at a surface of a small sphere (diameter, $d_s = 4\,\text{nm}$) which has different minimum distances (between 0 and 4 nm) into recombination states with a homogeneous distribution in energy.

For d_{s-r} equal to 0 nm, the decay of the SPV transient set on within the elementary time due to localized states near the surface of recombination states. Furthermore, the shape of the SPV transient for d_{s-r} equal to 0 nm decayed within $0.5{-}1\,\mu s$ similar to the SPV transient for localized states distributed homogeneously over d_H equal to 4 nm. This is not surprising since both decays were limited by the same longest distances between localized and recombination states.

For d_{s-r} equal to 4 nm, the SPV transient did not remarkably decay within the elementary time due to the strongly reduced tunneling rate over 4 nm (Equation (4.1)). The SPV transient for $d_{s-r} = 4$ nm decayed to 90% of its initial value after about 600 ns. For comparison, the SPV transient for $d_{s-r} = 2$ nm decayed to 90% of its initial value after a time which was about 50 times shorter than for $d_{s-r} = 4$ nm, which is well comparable with a tunneling rate increased by e^4 times due to the reduction of d_{s-r} by 2 times ($2 \cdot \alpha_{tunn} = 2$ nm^{-1}). Furthermore, the shapes of the SPV transients for d_{s-r} equal to 2 and 4 nm were very similar for the further decay on the logarithmic timescale.

In amorphous semiconductors, the transport is often limited by trapping into exponentially distributed tail states close to the conduction or valence bands. Figure 4.20 shows normalized SPV transients for thermally activated recombination via a delocalized state coupling directly into a

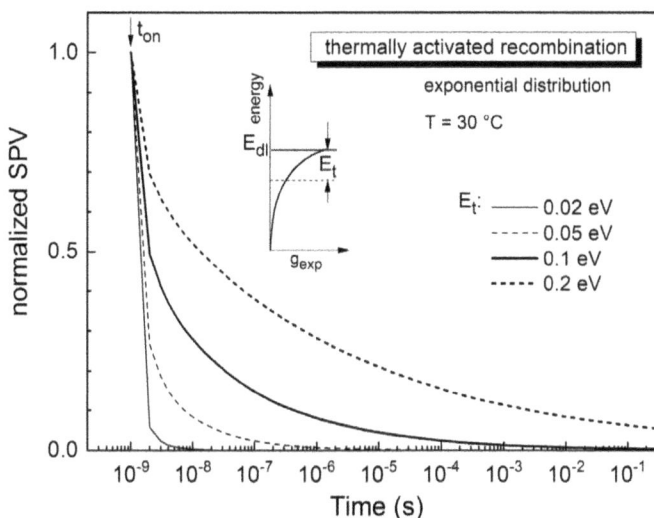

Figure 4.20. Normalized simulated SPV transients limited by thermally activated recombination from a localized state into a recombination state via a delocalized state coupling directly to the recombination state. The localized states are distributed exponentially with E_t equal to 0.02, 0.05, 0.1 and 0.2 eV (thin solid, thin dashed, thick solid and thick dashed lines, respectively). The temperature was kept at 30°C. The arrow marks the starting time (t_{on}).

recombination state, i.e. the charge transfer from the delocalized into the recombination state was much shorter than the elementary time. For the simulation of these SPV transients, the temperature was kept at 30°C and the parameter of the exponential tails was varied.

The values of E_t were set to 0.02, 0.05, 0.1 and 0.2 eV. For comparison, exponential tails are about 0.015 and 0.05–0.06 for $CH_3NH_3PbI_3$ and intrinsic amorphous silicon (De Wolf *et al.*, 2014) and 0.13–0.17 eV for In_2S_3 in a nanoporous TiO_2 matrix (Juma *et al.*, 2013). The temperature was kept at 30°C which corresponds to a value of $k_B \cdot T$ of 0.026 eV.

In general, all SPV transients limited by thermal activation from exponentially distributed localized states set on within τ_0 and decayed within a ns to 67, 48, 25 and 1% of its initial values for E_t equal to 0.02, 0.05, 0.1 and 0.2 eV, respectively. Therefore, all SPV transients contained very fast components. This is a general behavior of SPV transients limited by thermally activated recombination from exponentially distributed tail states.

The steepest decay of the practically complete SPV transient within 1 ns for $E_t = 0.02$ eV was not surprising since the energy of most of the localized states was less than $k_B \cdot T$. On the other side, the SPV transients decayed to 1% of the initial values within about 1 μs, 1 ms and even longer than 1 s for E_t equal to 0.05, 0.1 and 0.2 eV, respectively. Therefore, SPV transients contain components over many orders of magnitude in time if recombination is limited by thermal activation from exponentially distributed tail states with E_t much larger than $k_B \cdot T$.

The transients shown in Figure 4.20 are plotted in a log–log scale in Figure 4.21. The transients follow a power decay function (f_{power}) whereas the power coefficient is given by the ratio between $k_B \cdot T/E_t$.

$$\text{SPV}(t) \sim \left(\frac{1}{t - t_{\text{on}}} \right)^{\frac{k_B \cdot T}{E_t}} \tag{4.16}$$

A power decay function dependence of SPV transients is a signature for direct coupling of exponentially distributed localized states into charge transfer processes. The time dependence of SPV transients given by Equation (4.16) can be used for the measurement of E_t.

Deep defect states are often distributed in energy by Gaussians. The standard deviation of Gaussian defect distributions can be caused, for example, by a random variation of the nearest surrounding of a defect state. Figure 4.22 shows the influence of the standard deviation of localized

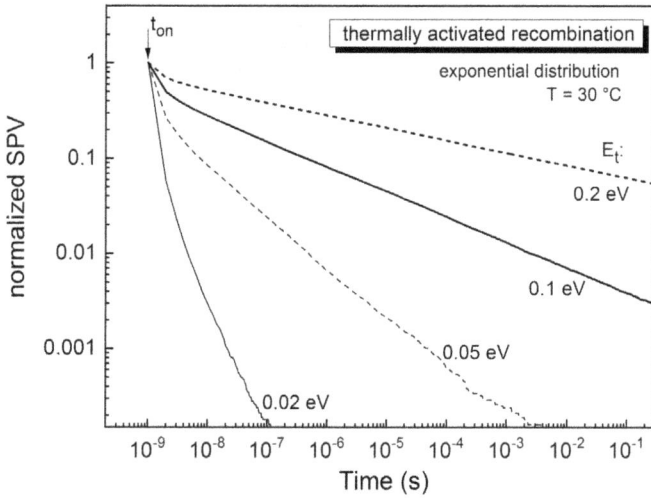

Figure 4.21. Normalized simulated SPV transients limited by thermally activated recombination from a localized state into a recombination state via a delocalized state coupling directly to the recombination state. The localized states are distributed exponentially with E_t equal to 0.02, 0.05, 0.1 and 0.2 eV (thin solid, thin dashed, thick solid and thick dashed lines, respectively). The temperature was kept at 30°C. The arrow marks the starting time (t_{on}).

states on normalized SPV transients limited by thermal activation from deep localized states.

In Figure 4.22, the value of $E_{dl}-E_d$ was set to 0.3 eV which is much larger than $k_B \cdot T$. The values of σ covered a wide range from narrow to very broad distributions and were set to 0.01, 0.05, 0.1 and 0.2 eV. The SPV transients limited by thermally activated recombination from deep defect states decayed to 99% or 95% of their initial values within about 80 or 500 ns, 15 or 100 ns and 0.3 or 6 ns for σ equal to 0.01, 0.05 and 0.1 eV, respectively. The shift of the onset of a remarkable decay of SPV transients to shorter times with increasing σ is caused by the increase of the amount of deep defects with energies closer to E_{dl}.

As a common feature, all SPV transients shown in Figure 4.22 decayed to 50% of their initial values ($\tau_{0.5}$) at nearly the same time of 6–7 μs. Therefore, the time at which SPV transients decay to 50% of their initial value can be used for the analysis $E_{dl}-E_d$, for example, in temperature-dependent experiments.

The SPV transients limited by thermally activated recombination from deep defect states decayed to 1% of the initial values within about 50 μs,

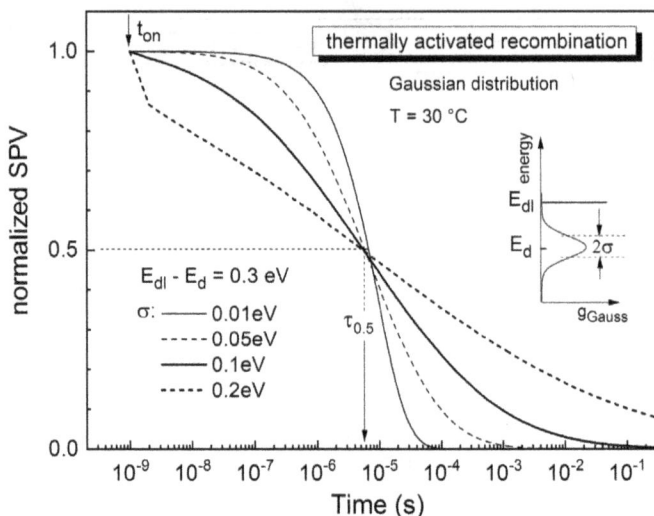

Figure 4.22. Normalized simulated SPV transients limited by thermally activated recombination from localized states distributed by a Gaussian with a peak energy at 0.3 eV below the energy of the delocalized state and standard deviations of 0.01, 0.05, 0.1 and 0.2 eV (thin solid, thin dashed, thick solid and thick dashed lines, respectively). The temperature was 30°C. The arrow marks the starting time (t_{on}).

0.8 ms, 40 ms and even longer than 1 s for σ equal to 0.01, 0.05, 0.1 and 0.2 eV, respectively. Therefore, SPV transients contain components over many orders of magnitude in time if recombination is limited by thermal activation from defect states with a very broad distribution.

4.4.2. SPV transients in a thin layer system

In thin layers with localized states, d_{avg} is shorter than d_H so that a random walk of a charge carrier can take numerous steps between initial photogeneration and recombination. For illustration, Figure 4.23 depicts an energy diagram with deep localized states and a delocalized state in volume V_B and an energetically sharp delocalized recombination state in volume V_A. The recombination state is located at a distance of d_{rec-dl} from the delocalized state in volume V_A. A charge carrier located in volume V_B can reach the recombination state in volume V_A by thermally activated tunneling from a localized state or by tunneling from the delocalized state into the recombination state. In volume V_B, a charge carrier can tunnel from one localized into another localized state. Furthermore, a charge carrier

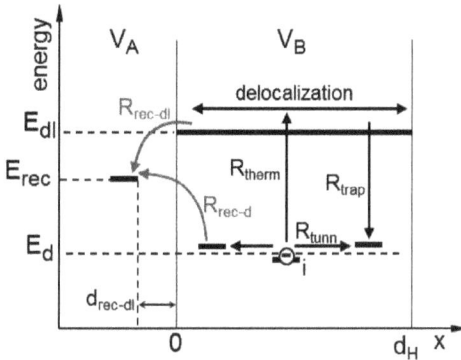

Figure 4.23. Energy diagram containing localized deep defect states and a delocalized state in volume V_B and an energetically sharp delocalized recombination state in volume V_A with a distance d_{rec-dl} to the delocalized state in volume V_B. Processes of tunneling between localized states and of thermally activated charge transfer via the delocalized state in volume V_B are denoted by R_{tunn} and R_{therm} and R_{trap}, respectively. The processes of recombination by thermally activated tunneling from a localized state or by tunneling from the delocalized state in volume V_B are denoted by R_{rec-d} and R_{rec-dl}, respectively.

can be thermally activated from a localized into a delocalized state and trapped afterwards at any localized state.

In Figure 4.24, examples are given for normalized simulated SPV transients limited by thermally activated tunneling from localized states in thin layers with densities of three localized states in volumes of $2 \times 2 \times 2$ (a) and $5 \times 5 \times 5$ (b) nm^3 (another example was applied in the next paragraph about self-consistent fitting). The localized states were distributed homogeneously within the cell and by a Gaussian in energy. The difference between the energy of the recombination state and the peak energy of the Gaussian distribution $(E_{rec}-E_d)$ and the standard deviation were set to 0.2 and 0.025 eV, respectively.

The shapes of the temperature-dependent SPV transients shown in Figure 4.24 were similar but not identical on the logarithmic timescale. With increasing temperature, the SPV transients shifted toward shorter times and became steeper. For example, the times at which the transients simulated for 200 and 400 K decayed to 90% or 10% of their initial values decreased by about 200 or 600 times, respectively (simulation for the density of three localized states in the volume of $2 \times 2 \times 2$ nm^3). Furthermore, the SPV transients were steeper and shifted toward shorter times for the layer with the higher density of localized states ($2 \times 2 \times 2$ nm^3).

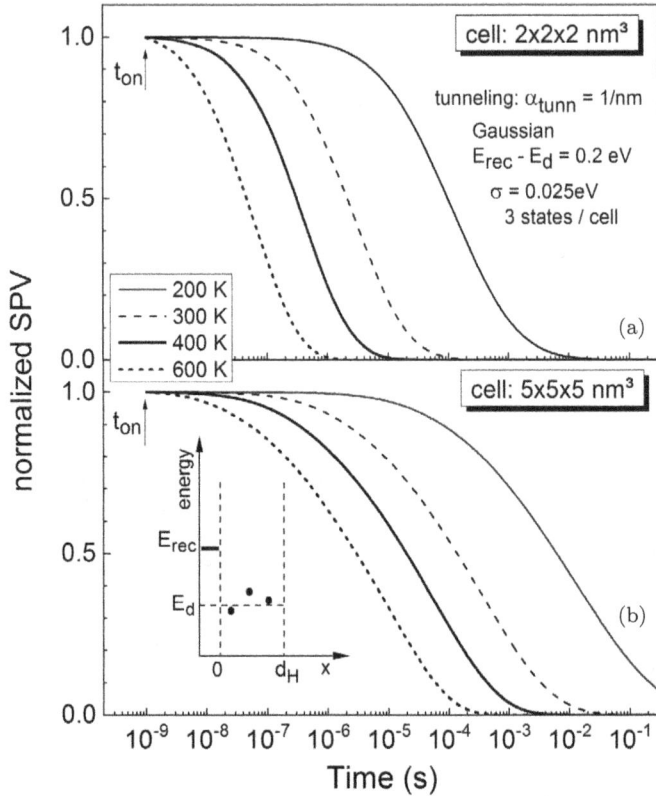

Figure 4.24. Normalized simulated SPV transients limited by thermally activated tunneling recombination from a layer with homogeneously distributed in space deep defect states into a recombination state for three localized states in a cell volume of $2 \times 2 \times 2\,\text{nm}^3$ (a) and $5 \times 5 \times 5\,\text{nm}^3$ (b) for temperatures of 200, 300, 400 and 600 K (thin solid, thin dashed, thick solid and thick dashed lines, respectively). The deep defect states are distributed by a Gaussian with $E_{\text{rec}}-E_d = 0.2\,\text{eV}$ and $\sigma = 0.025\,\text{eV}$. The arrow marks the starting time (t_{on}).

Figure 4.25 shows the Arrhenius plots of $1/\tau_{0.5}$ for thermally activated tunneling recombination under the conditions as in Figure 4.21. The activation energy (E_A) was equal to 0.2 eV for both densities of localized states. The value of $1/\tau_{0.5}$ approximated to $1/T = 0$ corresponds to the maximum recombination or charge transfer rate of the system. The value of $1/\tau_{0.5}$ approximated to $1/T = 0$ amounted to $10^9\,\text{s}^{-1}$ for three localized states in a volume of $2 \times 2 \times 2\,\text{nm}^3$ which was about 50 times larger than for three localized states in a volume of $5 \times 5 \times 5\,\text{nm}^3$. This relatively strong

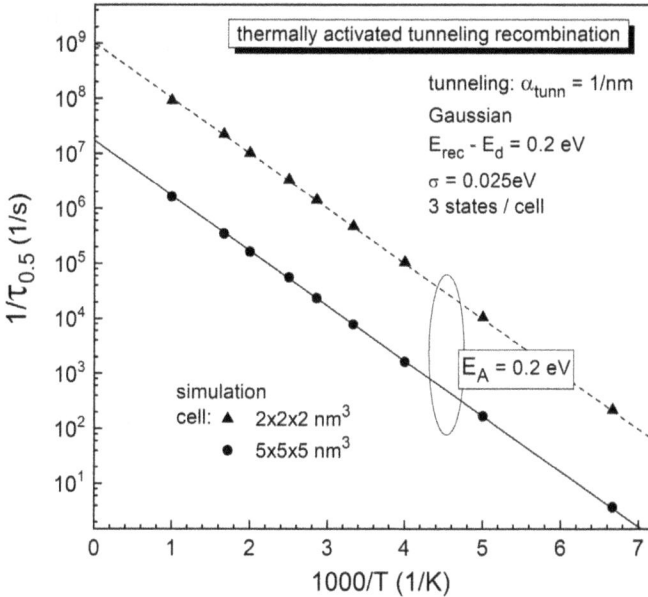

Figure 4.25. Arrhenius plots of the reciprocal decay time to 50% of the initial value of SPV transients for thermally activated tunneling recombination under conditions as in Figure 4.21 for three localized states in a cell volume of $2 \times 2 \times 2$ and $5 \times 5 \times 5 \, \text{nm}^3$ (triangles and circles, respectively).

change of the maximum recombination or charge transfer rate is caused by the different average distances for tunneling for both densities of localized states.

SPV transients dominated by tunneling or by thermally activated recombination can have, depending on the distributions of localized states and on parameters of the system, rather similar shapes. The simulation of temperature-dependent SPV transients is needed in order to distinguish the limiting recombination processes.

4.4.3. *Modulated SPV spectra in a thin layer system*

The modulated light intensity is kept constant at a fixed wavelength for several seconds and the wavelength is varied stepwise in modulated SPV spectroscopy. Therefore, in contrast to the simulation of transient SPV signals, RWSs start at different times whereas the density of starting times depends on modulation and photogeneration.

In-phase and phase-shifted by 90° SPV signals are measured in case of modulated photogeneration and charge separation (see also Chapter 2). As an example, real measurement conditions were adapted to the simulation of modulated SPV spectra for a thin layer system with inhomogeneous photogeneration in space.

Charge carriers photogenerated in thin layers are not necessarily distributed homogeneously in space. Reasons for inhomogeneous initial distributions of charge carriers can be, for example, a high absorption coefficient for exciting light, dissociation of excitons at an interface or an asymmetry in trapping at a charge-selective contact. In the following, initial charge separation by dissociation of excitons at a sharp charge-selective contact will be considered whereas the electron is trapped at an electron acceptor plane and the hole is trapped at a localized state close to the electron acceptor plane (Figure 4.26).

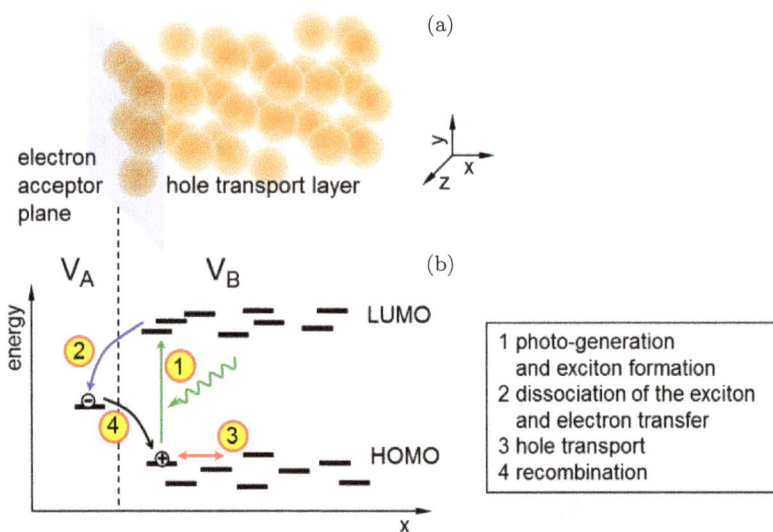

Figure 4.26. Model system consisting of an organic absorber and hole transport layer and an electron acceptor plane (a). The localized states in the hole transport layer are approximated by hard metal spheres arranged in an soft fcc lattice. The absorber layer consists of localized HOMO and LUMO states, the electron acceptor plane is given as an electron trap (b). An exciton is formed by photogeneration (1), dissociates by trapping of the electron at the electron acceptor plane (2), the hole can move from a localized state close to the electron acceptor plane into the layer by tunneling between localized HOMO states (3) and the electron can recombine with a hole by tunneling (4). (a) adapted with permission from Fengler and Dittrich, 2016; copyright 2016 American Chemical Society.

For the given model, localized states in the hole transport layer are approximated by hard metal spheres arranged in a soft fcc lattice (Figure 4.26(a)). The absorption spectrum is approximated by considering Gaussian distributions of highest occupied molecular orbital (HOMO) and lowest unoccupied molecular orbital (LUMO) states in the absorber layer (volume V_B in Figure 4.26(b)). The absorber layer acts also as the hole transport layer in which the hole can tunnel between localized states. The electron trapped at an acceptor state (volume V_A in Figure 4.26(b)) can recombine with a hole by tunneling.

A simulation of modulated SPV spectra includes numbers of sequences with a constant modulated generation rate for each sequence, each corresponding to one wavelength. A typical simulated sequence of the sum of projected charge separation lengths of numerous RWSs is shown in Figure 4.27(a).

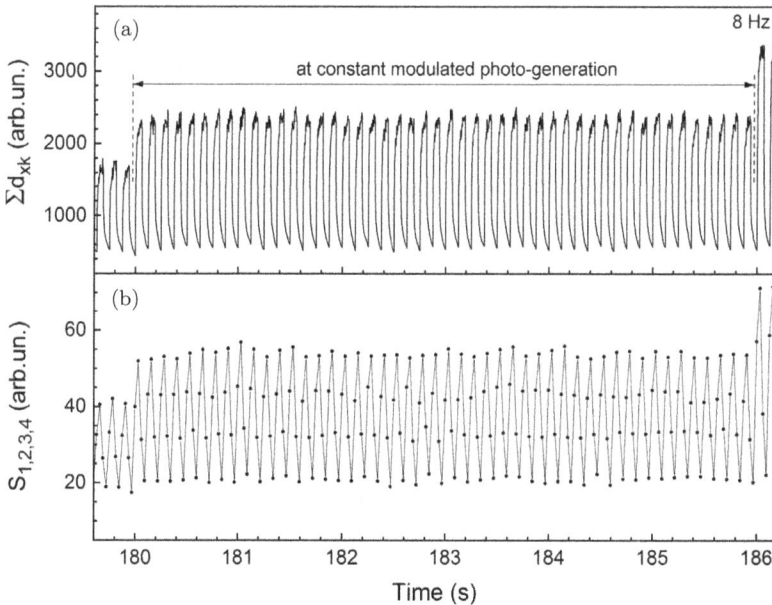

Figure 4.27. Example for a simulated sequence of the sum of numerous d_{xk} for modulated photogeneration (a) and corresponding sequence of the parameters S_1–S_4 which were used for the calculation of the in-phase and phase-shifted SPV signals (b). The modulated photogeneration was kept constant between 180 and 186 s. The photogeneration was lower at times shorter than 180 s and higher at times longer than 186 s. The modulation frequency was 8 Hz.

The sequence shown in Figure 4.27 was part of the simulation of a modulated SPV spectrum. The modulated photogeneration was kept constant at the fixed wavelength for 6 s for the given example. The parameters S_1–S_4 were calculated for each modulation period (Figure 4.27(b)) and averaged afterwards by using Equation (4.13). The periods within the first second were not considered for averaging since the first second of a sequence corresponded to the change of the wavelength and a waiting time in a real measurement for the given example. The in-phase and phase-shifted by 90° SPV signals were calculated from the S_1–S_4 parameters by using Equations (4.14′) and (4.14″), respectively.

The measurement conditions of a modulated SPV spectrum, which were adapted for the presented simulation, are summarized in Figure 4.28(a). The spectrum of the light intensity corresponds to that of a quartz prism monochromator (see also Figure 3.3). The absorption spectrum was hypothetic with respect to a Gaussian distribution. The modulation frequency was 8 Hz.

Figure 4.28(b) shows the simulated in-phase (X), phase-shifted by 90° (Y) and the amplitude SPV spectra corresponding to the conditions given in Figure 4.28(a) and the model explained in Figure 4.26. The SPV spectrum was simulated for a time interval of 500 s. The simulation was performed for a temperature of 30°C. For the given conditions, the X spectrum did roughly follow the photogeneration rate, i.e. the product of the absorption spectrum and the spectrum of the photon flux. The Y spectrum showed maxima in the regions with low absorption and a decrease in the region of high absorption.

The shape of the Y spectrum depended sensitively on the modulation frequency (the two maxima disappeared at higher frequencies, not shown, see for details (Fengler and Dittrich, 2016)). In addition, the onset energies (see Figure 2.22 for definition) were different for the X and Y spectra and shifted to lower photon energies in comparison to the absorption spectrum. Therefore, the extraction of rough empirical data from modulated SPV spectra does not necessarily provide parameters describing precisely a property of an absorption spectrum. The nature of absorption and charge transfer processes and the measurement conditions shall be considered for the interpretation of modulated SPV spectra on small systems with localized states.

The spectrum of the phase angles is shown in Figure 4.28(c) for the given simulation conditions. At photon energies around 1–1.15 and

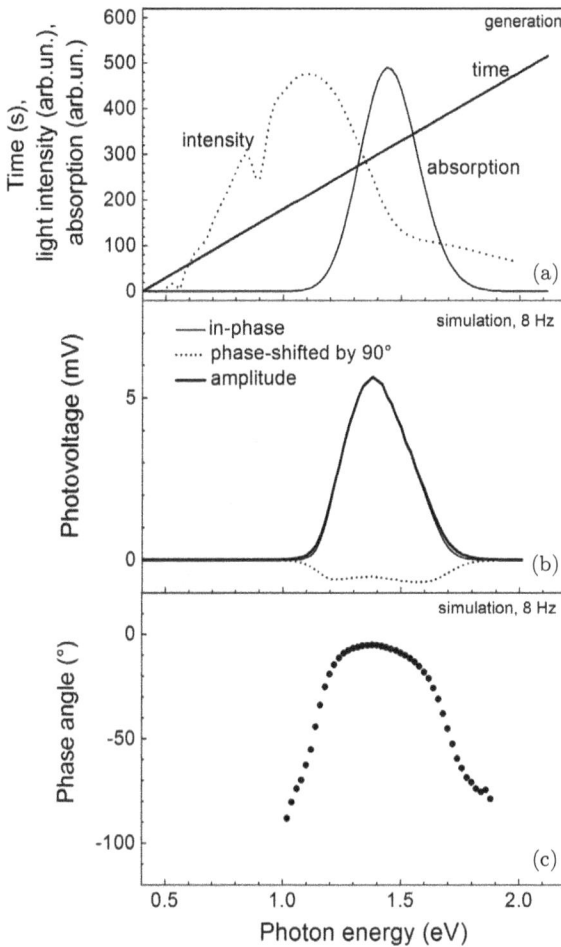

Figure 4.28. Conditions for the measurement of a modulated SPV spectrum as a function of photon energy (a) which were applied to the simulation of modulated in-phase, phase-shifted by 90° and amplitude (b) and phase angle (c) SPV spectra. The measurement conditions included the time regime, the spectrum of the light intensity and the absorption spectrum.

1.7–1.85 eV, the phase angles were about −80 to −50°, i.e. phase-shifted signals had a relatively strong influence. At photon energies between about 1.2 and 1.6 eV, the phase angles were between −20 and −5°, i.e. the in-phase signals dominated the SPV.

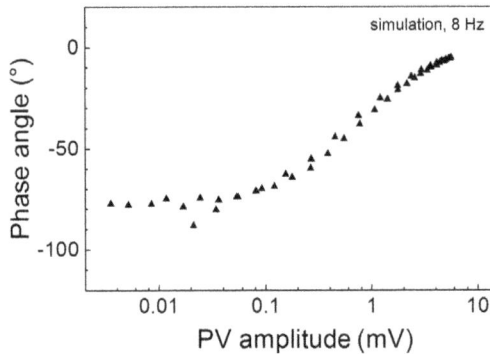

Figure 4.29. Example for the simulated dependence of the phase angle on the amplitude obtained from modulated SPV spectra.

The dependence of the phase angle on the amplitude of modulated SPV signals provides an impression about the occupation of localized states. For example, deeper defect states are occupied at higher photogeneration and more states with higher transfer rates are involved in modulated charge separation. As a result, the modulated response gets faster and the phase angles shift toward 0°. This is demonstrated in Figure 4.29 for the given simulation conditions. The phase angles were close to 80° at the lowest amplitudes and increased toward 5° at the highest amplitudes. Furthermore, a hysteresis in the dependence of the phase angle on the amplitude occurred depending whether the amplitude increased or decreased. This gives evidence for occupation of the deepest localized states for time longer than one sequence for the measurement at a fixed wavelength. Related phenomena are often observed in modulated SPV measurements.

4.5. Fitting of SPV Signals with Random Walk Simulations

4.5.1. *Multi-parameter fits by random walks in small systems*

SPV signals depend on numerous parameters related to photogeneration, localized states, charge transport and recombination. The extraction of complex parameters from SPV measurements is challenging for this reason. As shown before, transient and modulated SPV signals can be simulated in small systems by random walks. For random walk simulations of SPV signals, initial distributions of photogenerated charge carriers in time and space and of available localized and delocalized states in space and energy shall be defined. Initial distributions of photogenerated charge carriers in time and space depend on parameters of the light source and on optical parameters of the small system. Distributions of available localized and delocalized states depend on the structure, the nature of chemical bonds and the occupation of electronic states. The aim of multi-parameter fits is to obtain reliable information about parameters dominating SPV signals.

A measurement results in a measured function (f_{meas}) such as an SPV transient. Measurement parameters (p_{meas}) are varied and controlled in the experiment. It is possible to probe a sample with different measurement parameters, for example, time, light intensity, photon energy or temperature.

A simulated function (f_{sim}) is generated by using a model containing measurement parameters and simulation parameters (p_{sim}) such as parameters of distribution functions of localized states in space and energy. For simplicity, the measurement parameters are equal for f_{meas} and f_{sim} for multi-parameter fits. Furthermore, all values of p_{meas} are assumed as mathematically sharp in multi-parameter fits.

The deviation between f_{meas} and f_{sim} is defined as the error square (χ^2) which is the sum of the squared deviations between f_{meas} and f_{sim} for all values of p_{meas}. The running number and the total number of values of a certain measurement parameter are denoted by pi and mi, respectively.

$$\chi^2 = \sum_{pi=1}^{mi} \sum_{\ell=1}^{L} \left(f_{meas}(t_\ell, p_{meas,pi}) - f_{sim}(t_\ell, p_{meas,pi}; p_{sim,1} \cdots p_{sim,n}) \right)^2$$

(4.17)

The time is the most important measurement parameter for simulations of SPV signals in small systems. For transient SPV measurements and simulations, f_{meas}, f_{sim} and $p_{meas,p}$ correspond to SPV_{meas}, SPV_{sim} and t_ℓ, respectively, i.e. p_{meas}, pi and mi in Equation (4.17) are equal to t, ℓ and L, respectively (see also Figure 4.14).

In a multi-parameter fit, χ^2 is minimized by varying p_{sim} in a fit procedure. The run number of a simulation parameter and the total number of simulation parameters are denoted by r and R_{sim}, respectively.

Multi-parameter fits can result in finding rather different values of $p_{sim,r}$ due to local minima of χ^2. The results of a multi-parameter fit can strongly depend, for example, on starting values of parameters, procedures and algorithms. Therefore, a major issue of algorithms for multi-parameter fitting is the reduction of the probability of sticking in a local minimum of χ^2 during a fitting algorithm.

The probability of sticking in a local minimum of χ^2 in multi-parameter fits is drastically reduced by varying randomly the order in that parameters are optimized during fitting. For this purpose, a so-called random rank number (ra) is introduced. The practical elimination of sticking in a local minimum of χ^2 in a multi-parameter fit would demand huge numbers of simulations. The number of simulations shall be limited, for example, by the definition of a reasonable minimum value of χ^2 ($\chi^2_{\min}$) for all parameters or/and by defining a maximum number of simulations ($S_{\max}$) which is of the order

$$S_{\max} \approx \sqrt{10^{R_{\text{sim}}}} \qquad (4.18)$$

A multi-parameter fit consists of four loops (Figure 4.30) considering the maximum of transients simulated (N_{tra}), the order in which the R_{sim} parameters are optimized, the optimization of single parameters (N_{turns}) and the total number of fitted transients (N_{meas}). The values of N_{meas} and R_{sim} are given be the number of measurements and by the simulation model, respectively. At the beginning, the constants are set, the limits for the simulation parameters and the values of N_{tra}, N_{turns} and $\chi^2_{\min}$ are defined. The whole program runs until χ^2 is below $\chi^2_{\min}$ or the number of simulated transient (n_{tra}) is above N_{tra}. Until meeting this condition, the following procedures are performed.

At first, a ra is found for each p_{sim} ($p_{\text{sim,ra}}$). Afterwards, an initial value is generated for each $p_{\text{sim,ra}}$ and the corresponding increments $\Delta p_{\text{sim,ra}}$ are calculated. In the following, all simulation parameters are optimization in the order given by the values of ra. An individual optimization loop consists of a number of sets of simulations (n_{run}) and is finished when a maximum number of turns (N_{turns}) is reached. The simulation parameters are identical for all measured SPV transients to be fitted (one set). The simulations can be performed as described in Figure 4.14.

For each simulation set, the value of $p_{\text{sim,ra}}$ is varied by adding $\Delta p_{\text{sim,ra}}$. If $\chi^2_{n\text{run}}$ is larger than $\chi^2_{n\text{run}-1}$, the value of $\Delta p_{\text{sim,ra}}$ changes to $-\Delta p_{\text{sim,ra}}/3$ for the given fits. This defines one turn ($n_{\text{turns}} = n_{\text{turns}} + 1$).

For the analysis, data sets consisting of χ^2 and $p_{\text{sim,ra}}$ are saved after each set of simulations, each forming a vector in the matrix of results (M_{res}). Optionally, the corresponding values of ra can be saved as well in order to

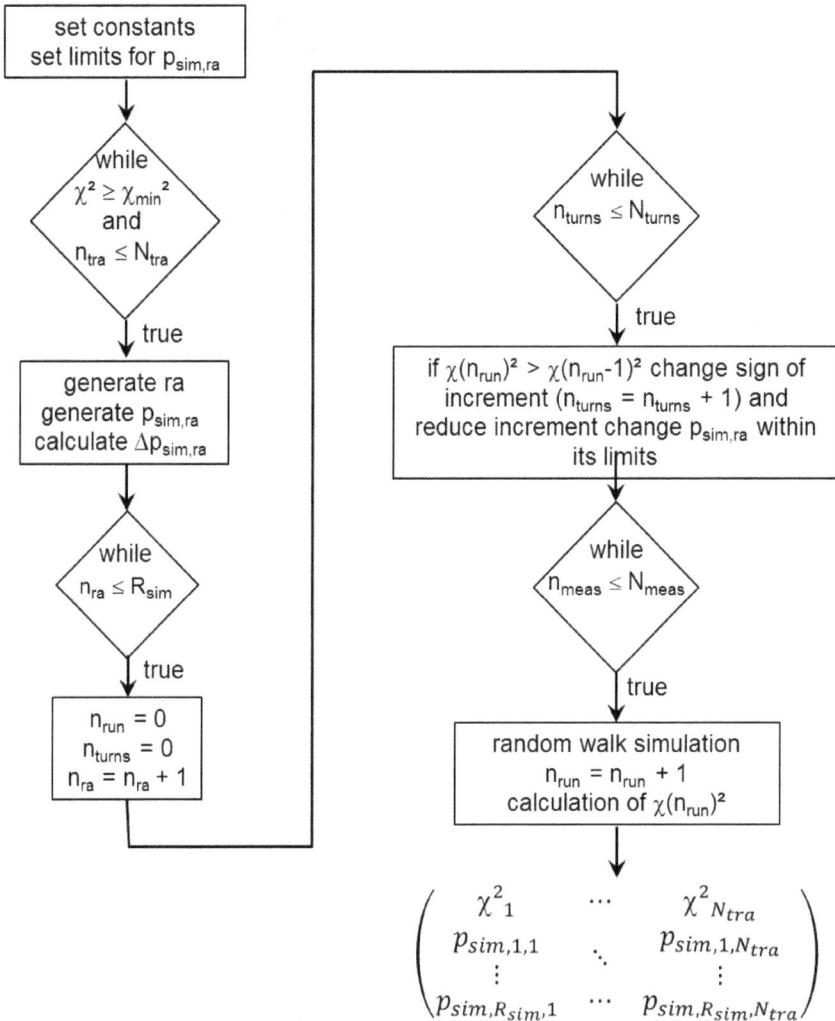

Figure 4.30. Principle scheme for multi-parameter fitting of SPV transients.

find out the influence of the order in that the parameters are optimized.

$$M_{\text{res}} = \begin{pmatrix} \chi_1^2 & \cdots & \chi_{N_{\text{tra}}}^2 \\ \vdots & \ddots & \vdots \\ p_{\text{sim},R_{\text{sim}},1} & \cdots & p_{\text{sim},R_{\text{sim}},N_{\text{tra}}} \end{pmatrix} \qquad (4.19)$$

The parameters corresponding to the lowest χ^2 are the set of the most probable parameters describing SPV_{meas} within the given model.

Multi-parameter fits can be repeated in order to reduce further χ^2. The order of the assignment of random rank numbers to r can be important for the result of a fitting procedure, some parameters have a stronger influence on SPV_{sim} than others. Therefore, there is a certain distribution of $r(ra)$ with respect to its influence on χ^2. The number of totally simulated transients can be drastically reduced for further fitting if considering the distribution of $r(ra)$ with respect to its influence on χ^2. Incidentally, lower random rank numbers can be assigned to simulation parameters with a stronger dependence of χ^2.

The accuracy of multi-parameter fits can be strongly increased if combining two (or more) measurement parameters which have a strong influence on the number of available states in a random walk simulation. The temperature has usually a strong influence on SPV_{meas}. Then, χ^2 is calculated by taking into account the variations in time and in temperature. The running parameter of the temperature and the maximum number of temperature variations are denoted by k and K, respectively.

$$\chi^2 = \sum_{k=1}^{K} \sum_{\ell=1}^{L} [SPV_{meas}(t_\ell, T_k) - SPV_{sim}(t_\ell, T_k, p_{sim,1} \cdots p_{sim,R_{sim}})]^2$$

(4.20)

However, when applying Equation (4.20), one shall keep in mind that certain simulation parameters may depend as well on temperature and that those dependencies are unknown.

A main advantage of small systems is that millions of RWSs can be simulated within short times. This opened the opportunity to fit SPV transients by random walk simulations for small systems.

Incidentally, different $p_{sim,r}$ can have rather different influence on the behavior of f_{sim}. In principle, there is no need for numerous variations of parameters which have only minor influence on multi-parameter fits. If the relative influence of each $p_{sim,r}$ on the behavior of f_{sim} would be known, the number of RWSs could be drastically reduced. For this purpose, the matrix of Equation (4.19) shall be analyzed many times. This is not really possible with the present algorithm. In future, it may be that self-learning algorithms will be able to separate between parameters with major and minor influence on f_{sim}. This will lead to a drastic reduction of time required for multi-parameter fits at one side and to a drastic increase of the accuracy of multi-parameter fits on the other side. Furthermore, self-learning algorithms for multi-parameter fits would allow for the implementation of more complex models containing, for example, detailed information about molecular orbitals.

4.5.2. Self-consistent fitting of simulated SPV transients

In the following, a multi-parameter fit will be self-consistently demonstrated for SPV transients simulated at different temperatures, i.e. the simulated transients correspond to $f_{\mathrm{meas}}(t, T)$. For this purpose, the model shown in Figures 4.31(a)– 4.31(c) was applied.

The model consisted of a cubic volume V_A (extension d_H) with homogeneously distributed localized states (peak energy of the Gaussian E_d, standard deviation σ, number N_d) and a delocalized state (energy E_{dl})

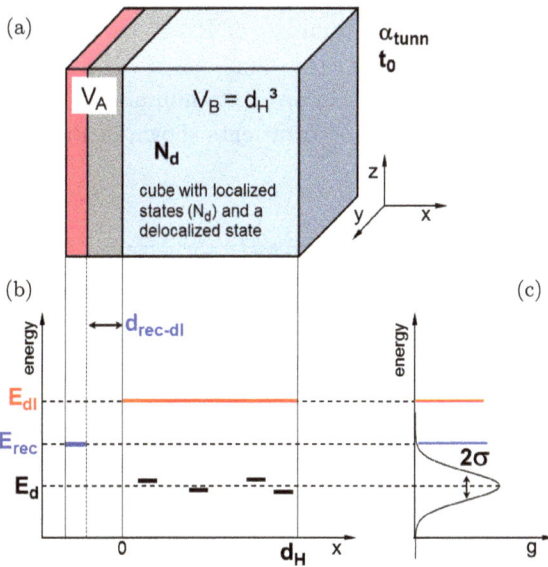

Figure 4.31. Spatial model for the simulation and self-consistent multi-parameter fit of SPV transients including a volume with a recombination state distributed homogeneously in the yz-plane in volume V_A and a delocalized state and localized states (number N_d) distributed homogeneously in the cubic volume V_B (a), band diagram including the energies of the recombination state (E_{rec}), the delocalized state (E_{dl}) and peak energy of localized states (E_d), the extension of the cubic volume V_B (d_H) and the distance between the recombination state and the delocalized state ($d_{\mathrm{rec-dl}}$) (b) and distribution functions (σ denotes the standard deviation of the localized states) of the electronic states in the model (c).

and a volume V_B with a recombination state (E_{rec}) delocalized over the yz-plane at a distance to the delocalized state in volume V_A (d_{rec-dl}). Therefore, the model contained eight simulation parameters (α_{tunn}, the minimum charge transfer time τ_0, N_d, d_H, d_{rec-dl}, E_d-E_{rec}, σ and $E_{dl}-E_{rec}$). Initially, the electron was randomly put in V_A and the hole was placed on the recombination state in V_B. Such an initial distribution is possible, for example, if electrons and holes have been photogenerated in V_A and the holes were separated into V_B by fast charge transfer.

For the simulation of SPV transients, the parameters were set as $\alpha_{tunn} = 1\ 1/\text{nm}$, $\tau_0 = 10^{-10}\,\text{s}$, $N_d = 10$, $d_H = 5\,\text{nm}$, $d_{rec-dl} = 0.3\,\text{nm}$, $E_d-E_{rec} = -0.1\,\text{eV}$, $\sigma = 0.1\,\text{eV}$ and $E_{dl}-E_{rec} = 0.1\,\text{eV}$. Figure 4.32 shows normalized SPV transients simulated for the temperatures of 30°C, 150°C and 300°C.

Fitting was performed in three approximation steps. For the given example, in the first approximation step, the ranges of the parameters were roughly found by limiting the number of RWSs to $4 \cdot 10^3$. In the second and third approximation steps, the numbers of RWSs were increased to $8 \cdot 10^3$ and $2 \cdot 10^4$, respectively, in order to minimize χ^2. Furthermore, the fits were performed for the three transients shown in Figure 4.32 with the same parameters excluding T.

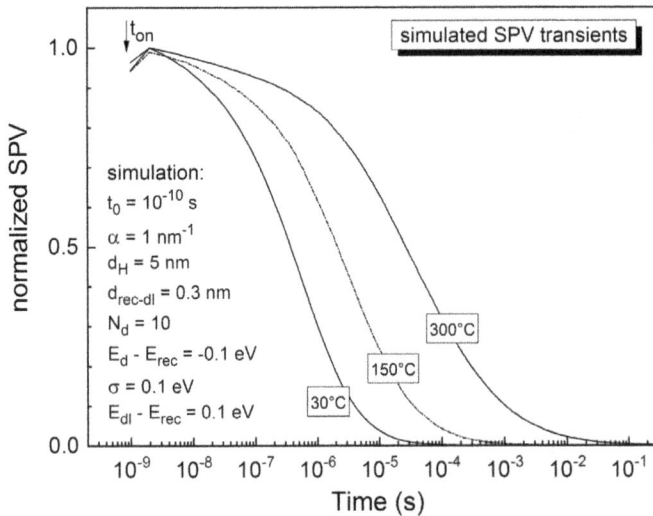

Figure 4.32. Normalized SPV transients simulated at 30°C, 150°C and 300°C following the model given in Figure 4.31. The arrow marks the starting time (t_{on}).

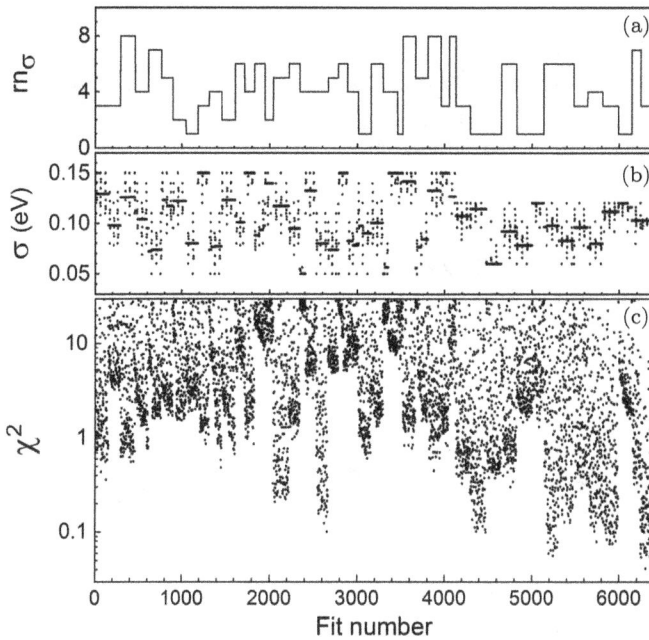

Figure 4.33. Example for the evolution of the random rank number assigned to σ (a), the values of σ (b) and χ^2 (c) as a function of the fitting cycles in a multi-parameter fit of the transients shown in Figure 4.32.

Figure 4.33 shows, as an example, the random rank number assigned to the simulation parameter σ (rn_σ, Figure 4.33(a)), the values of σ for each fitting cycle (Figure 4.33(b)) and the corresponding values of χ^2 (Figure 4.33(c)) as a function of the number of fitting cycles. The random rank number varied randomly between 1 and 8, the maximum number of simulations in each set of fitting cycles was set to 4. The values of χ^2 decreased within one set of fitting cycles. At certain combinations, the values of χ^2 reached minima.

The simulation parameters (α_{tunn}, $d_{\text{rec-dl}}$, d_{H}, N_d, τ_0, E_d-E_{rec}, σ and $E_{\text{dl}}-E_{\text{rec}}$ obtained from the multi-parameter fits of the transients shown in Figure 4.33 are plotted in Figures 4.34(a)–4.34(h), respectively, as a function of χ^2. In general, the distributions of the simulations parameters got sharper with decreasing χ^2. This can be nicely seen especially for E_d-E_{rec} (Figure 4.34(f)). The tendency of sharpening the distributions of the simulations parameters with decreasing χ^2 became stronger if setting one of the parameters, such as α_{tunn} or τ_0, to a fixed value. At the lowest

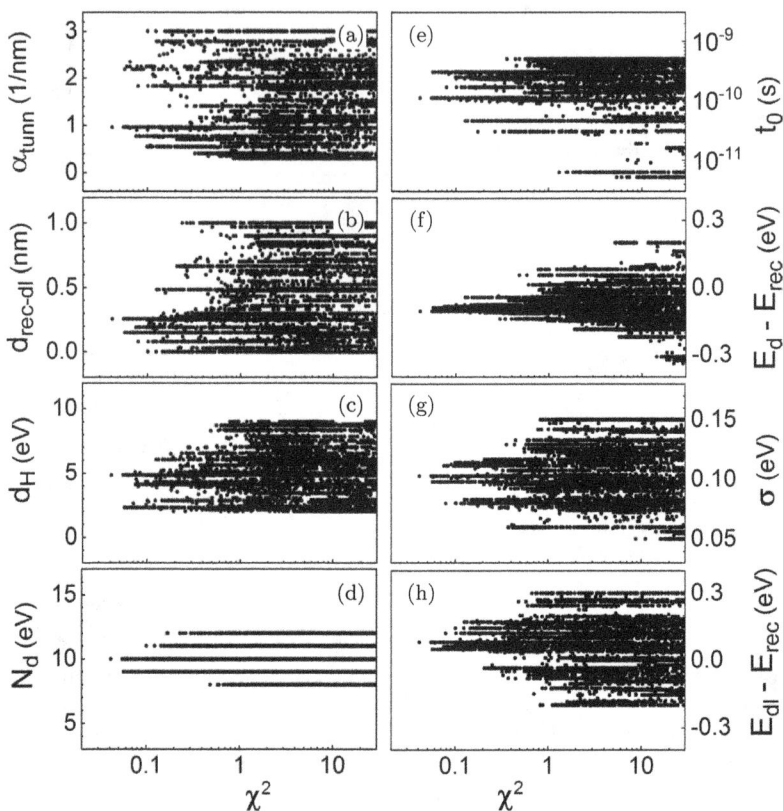

Figure 4.34. Plots of the fitted parameters α_{tunn}, $d_{\text{rec}-\text{dl}}$, d_{H}, N_d, t_0, E_d–E_{rec}, σ, E_{dl}–E_{rec} ((a)–(h), respectively) as a function of χ^2 of the simulated transients given in Figure 4.32. The distributions of the fitted parameters get narrower with decreasing χ^2.

value of χ^2, the fitted simulation parameters reached $\alpha_{\text{tunn}} = 0.96\,1/\text{nm}$, $d_{\text{rec}-\text{dl}} = 0.26\,\text{nm}$, $d_{\text{H}} = 4.8\,\text{nm}$, $N_d = 10$, $\tau_0 = 1.1 \cdot 10^{-10}$ s, E_d–$E_{\text{rec}} = -0.1\,\text{eV}$, $\sigma = 0.1\,\text{eV}$ and E_{dl}–$E_{\text{rec}} = 0.08\,\text{eV}$. These values were very close to those used for the simulation of the original transients.

The simulated SPV transients and the SPV transients resulting from the multi-parameter fit are compared in Figure 4.35. It can be seen that the simulated and fit transients correspond very well to each other. Therefore, a multi-parameter fit has been successfully applied in order to extract the distributions of localized, delocalized and recombination states in space and energy for the given simulated SPV transients.

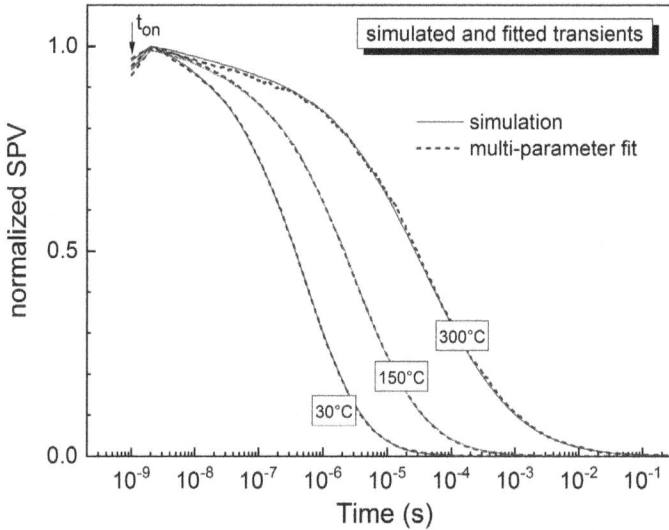

Figure 4.35. Normalized SPV transients simulated at 30°C, 150°C and 300°C following the model given in Figure 4.31 and shown in Figure 4.32 and corresponding multi-parameter fits with eight free parameters according to the model. The arrow marks the starting time (t_{on}).

However, the situation is more complicated for the application of multi-parameter fits to the interpretation of measured SPV transients since one appropriate model has to be found out from numerous models. For this purpose, the variation of several measurement parameters, such as time, temperature and wavelength, is useful.

4.5.3. *About fitting with analytical functions*

For practical reasons, measured SPV transients are often fitted with analytical functions. The advantage of fitting with analytical functions is that SPV transients can be described empirically with only a few parameters over wide ranges in time.

The most common analytical fit function is the stretched exponential. A stretched exponential decay function ($f_{stretched}$) is described by the time constant (τ), the stretching parameter (β, $0 \leq \beta \leq 1$) and the amplitude (A). A stretched exponential becomes an exponential for $\beta = 1$. The introduction of a starting time (t_{on}) is useful for the documentation of transients on a logarithmic time scale.

$$f_{\text{stretched}} = A \cdot \exp\left(-\left(\frac{t - t_{\text{on}}}{\tau}\right)^{\beta}\right) \qquad (4.21)$$

The decay of an exponential can be practically neglected at times shorter or longer than τ by 1–2 orders of magnitude depending on the experimental error. The introduction of a small stretching parameter can extend the decay toward times shorter or longer than τ by numerous orders of magnitude (see the example in Figure 4.36). Stretched exponentials have an S-shape on a logarithmic time scale.

A power decay function (f_{power}, equivalent to Equation (4.16)) describes a very fast decay on a logarithmic time scale (see also Figure 4.36). A power decay function is described by the power parameter (α_p) and the amplitude (A). For a given starting time, power decay functions have a L-shape on a logarithmic time scale.

$$f_{\text{power}} = A \cdot \left(\frac{1}{t - t_{\text{on}}}\right)^{\alpha_p} \qquad (4.22)$$

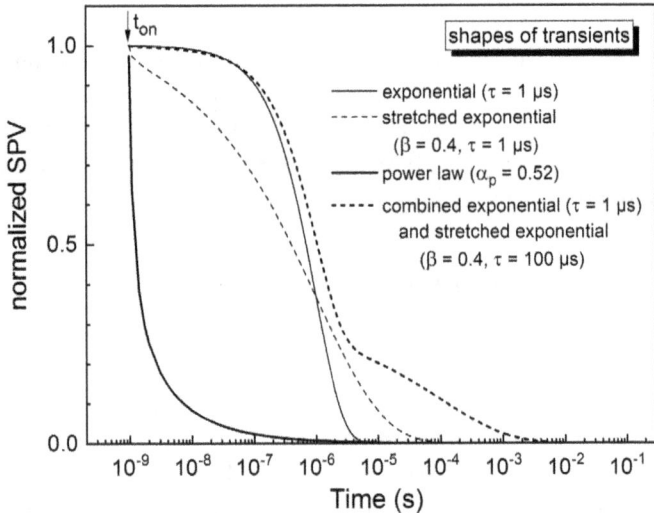

Figure 4.36. Examples of shapes of normalized transients following exponential, stretched exponential, power law and combined exponential and stretched exponential decays (thin solid, thin dashed, thick solid and thick dashed lines, respectively) for parameters shown in the legend. The arrow marks the starting time (t_{on}).

More complex decays can be described by a superposition of several decay functions. Figure 4.36 shows an example of a superposition of an exponential with a short τ (1 μs) and a stretched exponential with a longer τ (100 μs). Fits of SPV transients with a superposition of several decay functions can give evidence for the influence of several processes and/or distributions of electronic states in space and energy.

SPV transients are measured over about 2–3 orders of magnitude in the signal height. Therefore, it is useful to consider fits of SPV transients with analytical functions over about 3 orders of magnitude. As an example, Figure 4.37 shows simulated and fitted SPV transients limited by an exponential distribution (see also the parameters in Figure 4.20, $E_t = 0.05\,\mathrm{eV}$) and limited by tunneling from a localized state at the surface of a sphere over a short distance (see also the parameters in Figure 4.19, $d_{s-r} = 2\,\mathrm{nm}$, $d_s = 4\,\mathrm{nm}$). The transients are shown on a double logarithmic scale over about 3.5 orders of magnitude.

The SPV transient limited by an exponential distribution of localized states was excellently fitted over the whole range by Equation (4.22). The value of α_p was 0.52 which was equal to $k_B \cdot T / E_t$. As a general criterion,

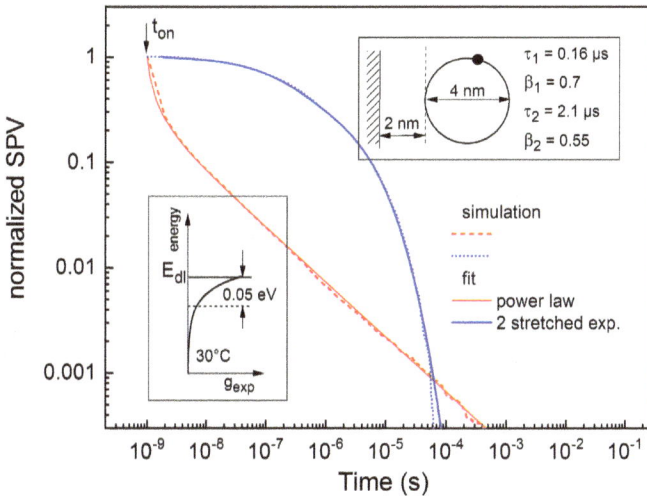

Figure 4.37. Simulated (dashed lines) and fitted (solid lines) of normalized SPV transients for recombination limited by an exponential distribution (red lines, $E_t = 0.05\,\mathrm{eV}$, fit with one power decay function) and limited by tunneling from a localized state at the surface of a sphere over a short distance (blue lines, $\alpha_{\mathrm{tunn}} = 1\,\mathrm{nm}^{-1}$, fit with two stretched exponentials). The arrow marks the starting time (t_{on}).

an SPV transient that can be well fitted by a power decay function gives evidence for limitation by an exponential distribution of localized states.

The SPV transient limited by tunneling from a localized state at the surface of a sphere over a short distance was very well fitted over the whole range by two stretched exponentials (Equation (4.21)). The values of τ_1, β_1, τ_2 and β_2 amounted to $0.16\,\mu s$, 0.7, $2.1\,\mu s$ and 0.55, respectively. As a general criterion, an SPV transient that can be well fitted by stretched exponentials gives evidence for limitation by a more homogeneous distribution of localized states.

Incidentally, more than two stretched exponentials are required for satisfactory fitting of SPV transients limited by a Gaussian over about 3 orders of magnitude in the signal height. Therefore, the number of stretched exponentials required for satisfactory fitting of SPV transients can give information about more homogeneous ranges of distributions and/or numbers of parameters of distributions of localized states within a small system.

The sign of SPV signals can change within a given SPV transient. The change of the sign gives evidence for charge separation into opposite directions. The fit of SPV transients with variation(s) in the sign requires a superposition of separate positive and negative decay functions.

4.6. Summary

The application of small systems to the simulation of SPV transients is very helpful for getting a deeper understanding about the influence of localized and delocalized states on processes of charge transfer in photoactive materials. Small systems can be part of larger complex systems including, for example, disordered surface regions of crystalline photoactive materials and semiconductors.

Two small volumes with electronic states, which are available differently for a photogenerated electron and/or hole, are combined in a small system in order to provide charge separation (charge-selective contact) and enabling therefore the simulation of SPV transients. The number of localized states in a small system is limited to a relatively small number in order to reduce the overall simulation time at limited resources.

Localized states are distributed in space and energy, for example, by exponential or Gaussian distributions. The occupation of localized states is defined by the introduction of a Fermi energy. Small systems are isolated in space and can be combined to more complex and even hierarchical systems.

The motion of a charge carrier in a small system is described by tunneling between localized and/or delocalized states and/or by charge transfer via thermal activation and trapping into delocalized and localized states, respectively.

The subsequent steps of charge transfers are called a random walk of a charge carrier. A random walk starts with the event of photogeneration and finishes with the event of recombination. A charge transfer step is obtained randomly from the distribution of transfer rates of all available states in a small system. The time between two charge transfer steps (waiting time) is obtained from the reciprocal rate of the given charge transfer and the natural logarithm of a random number between 0 and 1.

For the simplest simulation of SPV signals, the charge-selective contact of a small system is located in the yz-plane and the SPV signals are obtained from the charge separation distances projected on the x-direction.

Numerous RWSs with the same starting times are calculated for the simulation of one SPV transient. With respect to real measurement conditions, the events of charge transfers are sorted into a logarithmic distribution of time intervals. An SPV signal at a certain time is obtained from the sum of all projected charge separation distances in the given time interval.

For the simulation of modulated SPV signals, the starting times of RWSs are distributed in time with respect to modulation and photogeneration. The events of charge transfer are sorted into a linear distribution of time intervals. The coefficients S_1–S_4 are calculated by integrating the projected charge separation distances over the first to fourth quarters of modulation periods, respectively. After that, the coefficients S_1–S_4 are averaged over a sequence of periods with constant modulated photogeneration rate. The in-phase and phase-shifted SPV signals are calculated from the coefficients S_1–S_4. Modulated SPV spectra and the dependence of the phase angle on the amplitude, for example, can be simulated by random walks.

The general influence of distribution functions on SPV transients can be investigated by the simulation of small systems with only one localized state. Exponential distributions result in power decay functions whereas the power coefficient is given by the parameter of the exponential distribution (E_t) and the temperature. Gaussian distributions result in decay functions with a characteristic time at which the SPV signals decrease to the half of the initial signal ($\tau_{0.5}$). The thermal activation energy of $\tau_{0.5}$ is equal to the peak energy of the Gaussian distribution in relation to a delocalized or recombination state (E_{dl}–E_d).

A striking advantage of the application of small systems for the analysis of SPV signals is the fact that very large numbers of RWSs can be simulated within a short time. For this reason, random walk simulations can be applied directly to the analysis of SPV signals by fitting the parameters of a certain model.

SPV signals measured for small systems can be fitted by random walk simulations with multi-parameter fits. In a multi-parameter fit, random rank numbers are randomly assigned to the simulation parameters of a small model system. The random rank numbers define the order in which the parameters are optimized during the fitting procedure in order to minimize the error square (χ^2). The fitting procedure is repeated many times for numerous combinations of random rank numbers so that an overall minimum of χ^2 can be found from a statistical analysis. The introduction of random rank numbers and the large number of simulations reduces the probability for sticking in a local minimum of χ^2 during a fitting procedure. The measurement and fitting at several measurement parameters such as time, temperature and/or photogeneration rate, is recommended in order to increase the probability for finding an overall minimum of χ^2. The set of simulation parameters at the overall minimum of χ^2 represents the properties of a small system describing the SPV measurement with the highest probability. Incidentally, it will be very useful to develop complex and reliable algorithms for multi-parameter fits for broader applications in the future.

From an empirical point of view, it is often useful to fit measured SPV transients with an analytical function or a combination of several analytical functions over a range of 2–3 orders of magnitude in signal height. The advantage of such fitting procedures is that a set of only a few parameters describes the SPV transient over numerous orders of magnitude in time and that the distributions contain information about distribution functions of localized states. For example, power decay or stretched exponential decay functions are characteristic for exponential distributions or homogeneous distributions of localized states, respectively.

The combination of the simulation of SPV signals in small systems with the simulation of SPV signals in macroscopic systems is still challenging. For example, to date it is even not possible yet to simulate satisfying SPV transients over wide ranges for Si/SiO_2 interfaces. A main reason for this problem is the absence of an adequate description of surface states. Localized states distributed over a certain range close to the undisturbed bulk of a semiconductor crystal are the origin of surface

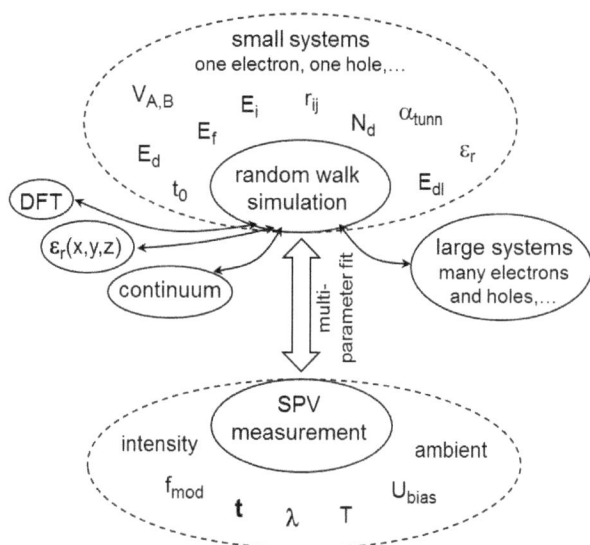

Figure 4.38. Scheme for summarizing parameters varied in SPV measurements and considered in random walk simulations in small systems. Simulations and measurements are connected via multi-parameter fits. Models can be extended by implementing density functional theory for describing localized states, dielectric functions, continuum models and large systems.

states, i.e. surfaces shall be locally treated as small systems with localized states. The empirical description of surface states by their capture cross sections in macroscopic models is not sufficient from that point of view. The behavior of small systems shall be considered in the further development of simulation tools for charge transfer in systems with photoactive materials.

Figure 4.38 summarizes parameters which are controlled during SPV measurements and have to be considered in random walk simulations. Furthermore, some of the typical parameters of small systems for random walk simulations are shown.

Simulations and measurements are connected via multi-parameter fits allowing for a quantitative analysis of parameters limiting the charge carrier dynamics. However, small systems are limited in their capability to describe more detailed and/or complex phenomena within samples. For example, the local chemical environment has strong influence on the nature of localized states. There is the hope that the implementation of DFT allows for more sophisticated defect analysis in the future. Furthermore, the local variation of the dielectric function and the consequent combination of random walk simulations with continuum theory is needed for SPV analysis of more complex photoactive materials. In this sense, large systems include the transfer, recombination and motion of many charge carriers in samples with complex morphology.

Chapter 5

Selected Applications of Surface Photovoltage Techniques

5.1. Introduction

Opportunities for the application of surface photovoltage (SPV) techniques are illustrated on the basis of selected examples: characterization of silicon (Si) surfaces, electronic transitions in gallium arsenide (GaAs) and charge separation in layers of colloidal quantum dots (QDs) and in metal oxide layers such as titanium oxide (TiO_2) and bismuth vanadate ($BiVO_4$).

Hydrogenated and oxidized Si surfaces were investigated by varying the occupation of surface states with influenced charge. Information about electronic states closer to the valence and conduction bands was obtained by transient SPV spectroscopy. Electrochemical grafting on Si surfaces was studied *in situ* by SPV.

Specific sensitivities of SPV spectroscopy techniques were shown for a system consisting of a native gallium oxide layer, an epitaxial GaAs layer and a compensated GaAs substrate. Numerous peculiarities in phase angle–PV amplitude plots gave information about the number of processes involved in modulated charge separation.

Electronic properties of cadmium selenide (CdSe) QDs were studied for different contact systems. Parameters of electronic states in monolayers of CdSe QDs were obtained by fitting of measured SPV transients with random walk simulations within the single QD approximation model.

Injected electrons were applied as probes for studying electron transport in doped metal oxides. Electronic states were distinguished in oxidized and reduced TiO_2.

The specific role of facet-selectively deposited dual cocatalysts for local charge separation in $BiVO_4$ crystallites was demonstrated by spatially resolved SPV spectroscopy with a Kelvin probe force microscope.

5.2. Characterization of Silicon Surfaces

5.2.1. *"Slow" and "fast" surface states at silicon surfaces*

Across surfaces, electrons and/or holes can be transferred from delocalized states in a bulk semiconductor, such as a silicon crystal (c-Si), into localized electronic states outside the bulk. The energies of delocalized electrons and holes in the bulk are given by the energies of the conduction and valence bands, respectively. The charge transfer depends on the density of electrons and/or holes in the bulk and at (or near) the surface of the bulk, the densities and energies of localized states at and near the surface and their distances to the surface. Historically, surface states were separated into fast and slow surface states ("slow" and "fast" surface states) with respect to the time response of charge transfer rates (Many *et al.*, 1965).

At silicon surfaces, dangling bonds at Si atoms with backbonds to three other Si atoms ($\bullet Si \equiv Si_3$) can cause donor and acceptor states near the middle of the bandgap of c-Si. At Si/SiO_2 interfaces, the $\bullet Si \equiv Si_3$ dangling bonds are called P_b centers which can also be generated under bias stress (see, for example, Poindexter, 1995 and references therein). The transfer of electrons and holes to P_b centers is fast and P_b centers are highly recombination active for this reason. Incidentally, not all $\bullet Si \equiv Si_3$ dangling bonds are recombination active, for example, dangling bonds at silicon dimers at reconstructed Si(100)-2 × 1 surfaces (Dittrich *et al.*, 2002).

At $c-Si/SiO_2$ interfaces, dangling bonds can also have backbonds to two Si atoms and one oxygen atom ($\bullet Si \equiv Si_2O$), one Si atom and two oxygen atoms ($\bullet Si \equiv SiO_2$) or to three oxygen atoms ($-Si \equiv O_3$). Obviously, depending on the local configuration, localized states related to $\bullet Si \equiv Si_2O$, $\bullet Si \equiv SiO_2$ and $\bullet Si \equiv O_3$ dangling bonds can have a certain distance to the c-Si surface and therefore rather different tunneling or charge transfer rates. Furthermore, stress at $c-Si/SiO_2$ interfaces causes distortions of chemical Si–Si bonds near the c-Si surface being responsible, in analogy to disorder in amorphous silicon, for electronic surface states near the conduction and valence band edges of c-Si (see, for example, Füssel *et al.*, 1996 and references therein).

Interactions between dangling bonds and additional chemical species such as hydrogen, oxygen (electron acceptor) and water (electron donor) molecules cause additional pathways for charge transfer across c-Si surfaces. Such interactions can take place, for example, at c-Si/SiO$_2$ interfaces (Pointdexter, 1995) or at wet chemically treated c-Si surfaces (Dittrich *et al.*, 1999).

Incidentally, the occupation of localized states near but not very close to c-Si surfaces by electrons or by holes is applied for surface passivation in c-Si solar cells, for example, with Al$_2$O$_3$ (Hoex *et al.*, 2008) or SiN$_x$:H (Lauinger *et al.*, 1996) layers, respectively. The distance between localized states used for surface passivation and a c-Si surface is long in relation to the tunneling length so that the charge is fixed at the passivating localized states (Q_{fix}, fixed charge).

Acceptor or donor surface states are negatively or positively charged when occupied or unoccupied, respectively. Figures 5.1(a)–5.1(d) show idealized configurations of an ideal c-Si surface and of c-Si surfaces with a P_b center, neutral and charged chemisorbed acceptor molecules (A^0 and A^-), respectively) and a neutral and charged (D^0 and D^+, respectively) chemisorbed donor molecules, respectively.

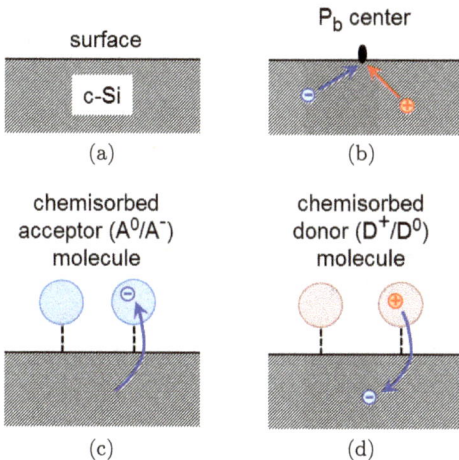

Figure 5.1. Idealized configurations of an ideal c-Si surface (a) and of c-Si surfaces with a P_b center (b), neutral and negatively charged chemisorbed acceptor molecules (c) and neutral and positively charged chemisorbed donor molecules (d).

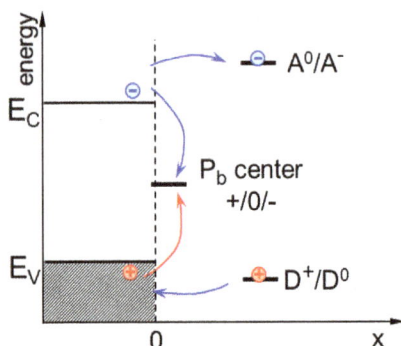

Figure 5.2. Schematic diagram of c-Si surface with localized states of a P_b center and of donor (D^+/D^0) and acceptor (A^0/A^-) states at a certain distance from the surface.

Acceptor or donor surface states are charged by transfer of charge carriers from the bulk. Therefore, acceptor or donor surface states are usually closer to the conduction or valence band edges of the bulk semiconductor, respectively. Figure 5.2 shows a schematic band diagram near a c-Si surface with localized states of a P_b center and of a donor and an acceptor molecule. The occupation of donor and acceptor surface states depends on the densities of holes or electrons in the bulk semiconductor. Usually, acceptor states are occupied and donor states are unoccupied at the surface of n-type or p-type doped semiconductors, respectively. This fact has been used, for example, for the study of donor and acceptor states at reconstructed c-Si surfaces in ultra-high vacuum (UHV) by SPV spectroscopy with an electron beam (Clabes and Henzler, 1980).

The application of a bias potential between the sample and reference electrodes in the fixed capacitor configuration (U_{bias}) allows for the study of surface states at c-Si surfaces (see also Chapter 4). By measuring the maximum SPV signal at high light intensity and at short times (SPV_{max}), the surface band bending and therefore the charge in the space charge region can be obtained. Furthermore, the band bending is related to the Fermi energy at the surface and therefore to the charge in surface states. As an example, two dependencies of SPV_{max} on U_{bias} (SPV_{max}–U_{bias} cycles) are given for a hydrogenated n-type doped c-Si(111) (n-Si(111):H) surface in Figure 5.3. The corresponding (approximated) scale of the Fermi energy at the surface ($E_{F,s}$) and the Fermi energy in the bulk ($E_{F,\text{bulk}}$) are given as well. An SPV_{max}–U_{bias} cycle started in the middle between the minimum

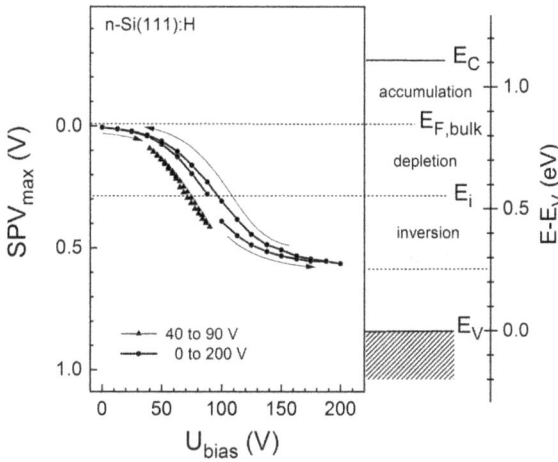

Figure 5.3. SPV_{max}–U_{bias} cycles for an electrochemically hydrogenated n-Si(111):H surface measured in ranges between 40 and 90 V (virgin cycle, triangles) and between 0 and 200 V (after repeated cycles, circles). The thin lines with the arrows mark the direction of variation of the bias potential. At the right side: approximate energies of the valence and conduction band edges, the Fermi energy and the energy at mid gap for the bulk. Data from (Dittrich *et al.*, 1999).

and maximum values of U_{bias}, i.e. at 65 and 100 V for ranges between 40 and 90 V and between 0 and 200 V, respectively.

For the SPV_{max}–U_{bias} cycle between 40 and 90 V, SPV_{max} amounted to about 0.25 V at the beginning and to 0.42 and 0.09 V at 90 and 40 V, respectively. Furthermore, for the SPV_{max}–U_{bias} cycle between 0 and 200 V, SPV_{max} amounted to about 0.38 V at the beginning and to 0.56 and 0.01 V at 200 and 0 V, respectively.

The maximum slopes of the SPV_{max}–U_{bias} cycles were reached near midgap (E_i) and did not depend on the scan direction of U_{bias}. Therefore, the maximum slope was independent of "slow" surface states and corresponded, with respect to analysis after Equation (4.19), to a $D_{it,min}$ of about $1 \cdot 10^{10}\,eV^{-1}cm^{-2}$ (Dittrich *et al.*, 1999). This value of $D_{it,min}$ near E_i is given by the density of P_b centers at the c-Si surface.

A hysteresis between forward and backward directions in SPV_{max}–U_{bias} cycles gives evidence for the influence of "slow" states. For the SPV_{max}–U_{bias} cycles between 40 and 90 V and between 0 and 200 V, a hysteresis of U_{bias} of about 3 and 12 V at SPV_{max} equal to 0.3 and 0.4 V, respectively, was observed. Interestingly, the hysteresis of the SPV_{max}–U_{bias} cycle between 0 and 200 V decreased to about 6 V at E_i for the second part of the $\mathrm{SPV_{max}}$–U_{bias} cycle ($100 \rightarrow 0 \rightarrow 87.25$ V). Therefore, the hysteresis increased for switching $E_{F,s}$ into inversion but decreased when switching $E_{F,s}$ into depletion. This means that the hysteresis was caused by trapping of holes at "slow" surface states, i.e. the "slow" surface states at a hydrogenated c-Si surface are localized acceptor molecules.

For hydrogenated c-Si surfaces, the slopes of the $\mathrm{SPV_{max}}$–U_{bias} cycles depended on the scan direction for $E_{F,s}$ shifting toward accumulation (neutralization of "slow" acceptor states) or strong inversion (charging of "slow" acceptor states). Furthermore, the measurement regime, i.e. the time intervals of switching U_{bias}, had a strong influence on the slopes of the $\mathrm{SPV_{max}}$–U_{bias} cycles for $E_{F,s}$ shifting toward accumulation or strong inversion. Therefore, the value of D_{it} is a characteristic of "fast" and "slow" surface states for $E_{F,s}$ shifting toward accumulation or strong inversion depending on the measurement conditions.

The values of U_{bias} at which $\mathrm{SPV_{max}}$ was 0.3 V were about 72 and 92 V for the $\mathrm{SPV_{max}}$–U_{bias} cycle between 40 and 90 V and between 0 and 200 V, respectively. This shift of U_{bias} gave evidence for an increase of Q_{fix} due to repetitive cycling toward strong inversion (Q_{fix} in the meaning of the time interval of measurements). An increase of Q_{fix} is possible for subsequent transfers of positive charge from acceptor states closer to the c-Si surface toward acceptor states at longer distances to the c-Si surface. A related mechanism of charge transfer can be, for example, a Grotthuss mechanism of proton transfer between adsorbed water molecules (see, for example, Agmon, 1995). Proton transfer between adsorbed water molecules can be suppressed by intense drying.

Incidentally, no hysteresis has been observed on thermally oxidized, i.e. dry, c-Si/SiO$_2$ interfaces. In contrast, for injection of holes into wet anodic oxides, a correlation between the density of "fixed" positive charge and the density of recombination active defects was observed (Dittrich *et al.*, 2001). Furthermore, the "fixed" positive charge decreased within hundred(s) of seconds. Therefore, the transfer of positive charge into water molecules at c-Si surfaces causes a depassivation of P_b centers.

As an example, a schematic model for the evolution of "fast" and "slow" surface states at a c-Si surface is presented in Figure 5.4. Injection of a hole is equivalent to the protonation of an H atom passivating a P_b center. As a result, the P_b center is depassivated and the positive charge is transferred to a protonated water molecule. The protonation of the passivating H atom and the P_b center belong to the action of "fast" surface states, whereas

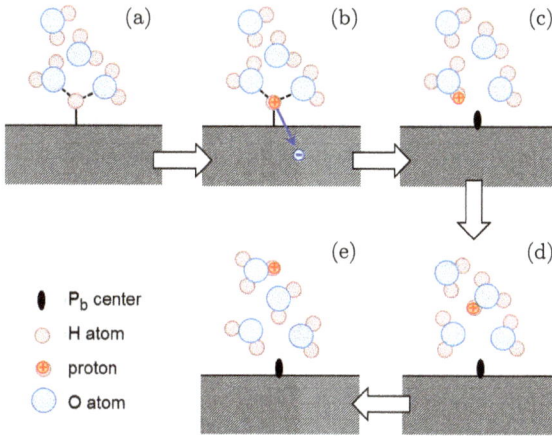

Figure 5.4. Example model for the evolution of "fast" and "slow" surface states at a c-Si surface: hydrogen-passivated H-Si $\equiv$ Si$_3$ and adsorbed water molecules (a), protonation of the passivating H atom by charge transfer (b), formation of a P_b center and of a protonated water molecule (c), proton transfer between adsorbed water molecules ((d) and (e)). The protonation of the passivating H atom and the P_b center belong to the action of "fast" surface states. The proton transfer between adsorbed water molecules is responsible for the evolution of "slow" surface states and of "fixed" positive charge.

the transfer of protons causes "slow" surface states and "fixed" positive charge.

Incidentally, hydrogenated c-Si surfaces are hydrophobic, i.e. water molecules are repelled from the c-Si surface. However, adsorption of water molecules is possible at reactive surface sites remaining at c-Si surfaces after finishing a hydrogen passivation by an etching process. Protonation of water molecules adsorbed at reactive surface sites can cause a depassivation of P_b centers also at hydrogenated c-Si surfaces.

5.2.2. On the role of roughness of hydrogenated silicon surfaces

Hydrogen-passivated c-Si surfaces are a result of etching processes in fluoride solutions. The (microscopic) surface roughness is related to the density of reactive surface sites at hydrogenated c-Si surfaces. Reactive surface sites arise due to breaking of chemical bonds and can cause the formation of defects and adsorbed molecules. The minimum density of surface states ($D_{it,min}$) can be obtained from SPV$_{max}$–U_{bias} cycles.

The etch rate of silicon oxide is much larger than the etch rate of c-Si in acidic fluoride solutions. Therefore, the roughness of hydrogen-passivated c-Si surfaces is influenced by the oxidation process. Furthermore, the etch rate of c-Si in fluoride solutions depends on the components, their concentrations and the pH value of the solution. Therefore, the roughness of hydrogen-passivated c-Si surfaces can also be influenced by adjusting the fluoride solution and the etch time.

Atomically flat facets or terraces terminated with H–Si≡Si$_3$ bonds are obtained at c-Si(111) crystals after etching in ammonium fluoride (NH$_4$F) solution (Higashi *et al.*, 1991; Dumas *et al.*, 1992). The lowest value of $D_{it,\text{min}}$ of $1.5 \cdot 10^{10}\,\text{eV}^{-1}\text{cm}^{-2}$ has been achieved on hydrogenated c-Si(111) crystals after etching in NH$_4$F solution (see, Angermann, 2014 and references therein). This means that there is only one P_b center on an area of about $100 \times 100\,\text{nm}^2$ if assuming a density of surface bonds of $5 \cdot 10^{14}\,\text{cm}^{-2}$ and a distribution of P_b centers over about $0.5\,\text{eV}$ within the band gap of c-Si.

Atomically flat hydrogenated facets at c-Si(111) crystals were roughened during subsequent etching in concentrated hydric fluoride (HF) solution. $D_{it,\text{min}}$ increased to about $3 \cdot 10^{11}$, $1 \cdot 10^{12}$ and $2 \cdot 10^{12}\,\text{eV}^{-1}\text{cm}^{-2}$ for an increase of the effective surface micro-roughness to about one, two and three monolayers, respectively (see, Angermann, 2014 and references therein).

Silicon surfaces can be oxidized by anodic oxidation in an oscillating regime in diluted NH$_4$F solution (see, for example, Rappich and Dittrich, 2002 and references therein). In this regime, the c-Si surface is permanently oxidized and the oxide is permanently etched. After interrupting anodic oxidation, the disappearance of a current transient gives evidence for the formation of a hydrogenated c-Si surface (Bitzer *et al.*, 1993). Hydrogenated c-Si surfaces obtained after anodic oxidation in the oscillating regime in diluted NH$_4$F solution (pH 4.0) are relatively rough with very smooth variations of the morphology and without atomically flat hydrogenated facets (Dittrich *et al.*, 1999). The value of $D_{it,\text{min}}$ can be as low as $1 \cdot 10^{10}\,\text{eV}^{-1}\text{cm}^{-2}$ for corresponding electrochemically hydrogenated c-Si surfaces (Rauscher *et al.*, 1995). Therefore, the local roughness, i.e. steep changes in the local morphology, is essential for the formation of defect states.

Figure 5.5 shows examples of SPV$_{\text{max}}$–U_{bias} cycles for an electrochemically hydrogenated c-Si surface and for a HF treated c-Si surface. The maximum slope of the SPV$_{\text{max}}$–U_{bias} cycle is much steeper and

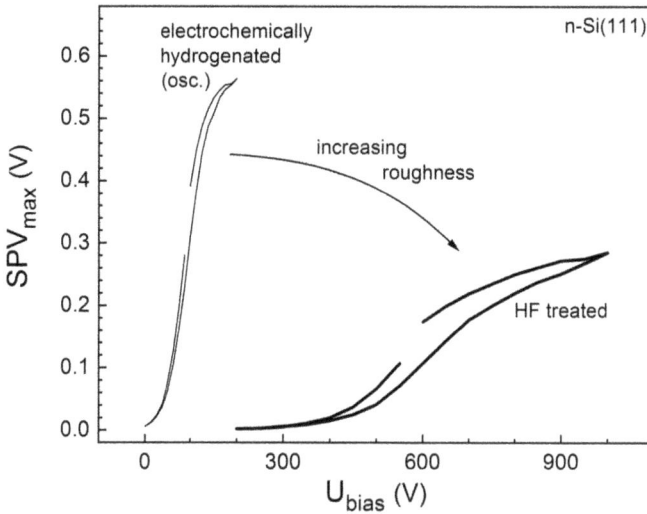

Figure 5.5. SPV_{max}–U_{bias} cycles for an electrochemically hydrogenated n-Si(111):H surface (thin line) and for a surface hydrogenated by HF treatment (thick line). The line with the arrows marks the direction of increasing (local) surface roughness. Data from Dittrich *et al.* (1999).

the hysteresis and the fixed positive charge are much lower for the electrochemically hydrogenated surface than for the HF treated c-Si surface. Therefore, P_b centers and adsorbed acceptor molecules are related to each other and are located at or near reactive surface sites.

5.2.3. *Distribution of the density of surface states*

The distribution of the density of surface states ($D_{it}(E_{F,s})$) can be deduced from SPV_{max}–U_{bias} cycles, whereas one has to keep in mind that the distributions are influenced by all charge transfer processes within the time of switching on U_{bias}. As an example, Figure 5.6 shows $D_{it}(E_{F,s})$ for a hydrogenated c-Si surface and for c-Si surfaces covered with one, two and three monolayers of anodic oxide. Incidentally, an advantage of anodic oxidation is that the integrated charge flown during anodization is precisely related to the amount of oxidized Si atoms.

For the given hydrogenated c-Si surface, $D_{it,min}$ was about $2 \cdot 10^{11}\,eV^{-1}cm^{-2}$. $D_{it,min}$ increased toward $8 \cdot 10^{11}$ and $5 \cdot 10^{11}\,eV^{-1}cm^{-2}$ close to $E_{F,s} - E_V = E_{F,bulk}$ (flat band) and close to $E_{F,s} - E_V = E_C - E_{F,bulk}$ (strong inversion), respectively. After anodic oxidation of one

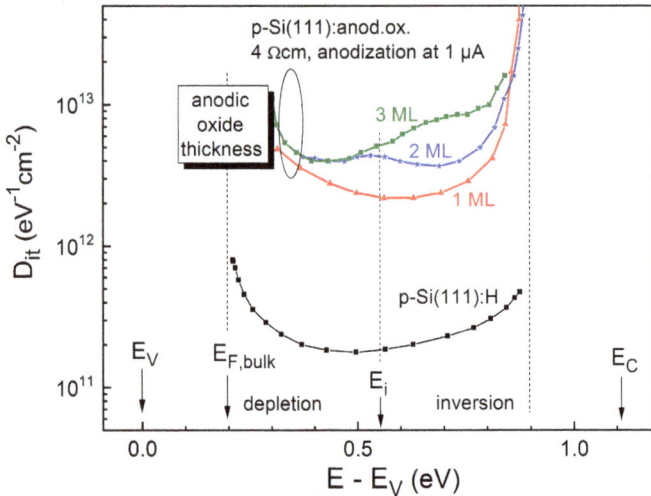

Figure 5.6. Distributions of the density of surface states for a hydrogenated p-Si(111):H surface (black squares) and at p-Si(111) surfaces covered with anodic oxide layers with thicknesses of one, two and three monolayers (red triangles, blue stars and green squares, respectively). Data from Rappich and Dittrich (2002).

monolayer, $D_{it,\text{min}}$ increased to about $2 \cdot 10^{12}\,\text{eV}^{-1}\text{cm}^{-2}$. Anodic oxidation of two and three monolayers resulted in a very similar $D_{it,\text{min}}$ equal to about $3 \cdot 10^{12}\,\text{eV}^{-1}\text{cm}^{-2}$, i.e. the density of P_b centers increased during anodic oxidation of the first two monolayers. D_{it} around $E_{F,s} - E_V = 0.7\,\text{eV}$ (inversion) increased from about $3 \cdot 10^{12}$ to about $7 \cdot 10^{12}\,\text{eV}^{-1}\text{cm}^{-2}$ for anodic oxidation of two and three monolayers, respectively, i.e. the density of acceptor states close to the c-Si surface increased with ongoing anodic oxidation.

SPV$_{\text{max}}$–U_{bias} cycling allows for the investigation of defect states under depletion and inversion conditions. Furthermore, if the density of states of the fixed charge are very high, the range of $E_{F,s} - E_V$ is strongly reduced due to Fermi-level pinning. However, often, information about localized states closer to the edges of the conduction and/or valence bands is of interest. In such a case, transient SPV spectroscopy can be helpful.

Figure 5.7 shows plots of transient SPV spectroscopy for a highly n-type doped c-Si (a) and for a p-type doped c-Si surface (b) covered with a native oxide. These measurements were performed at a constant photon flux. The SPV signals were positive and of the order of 20–30 mV at higher photon energies for the highly n-type doped sample. At the given light intensity,

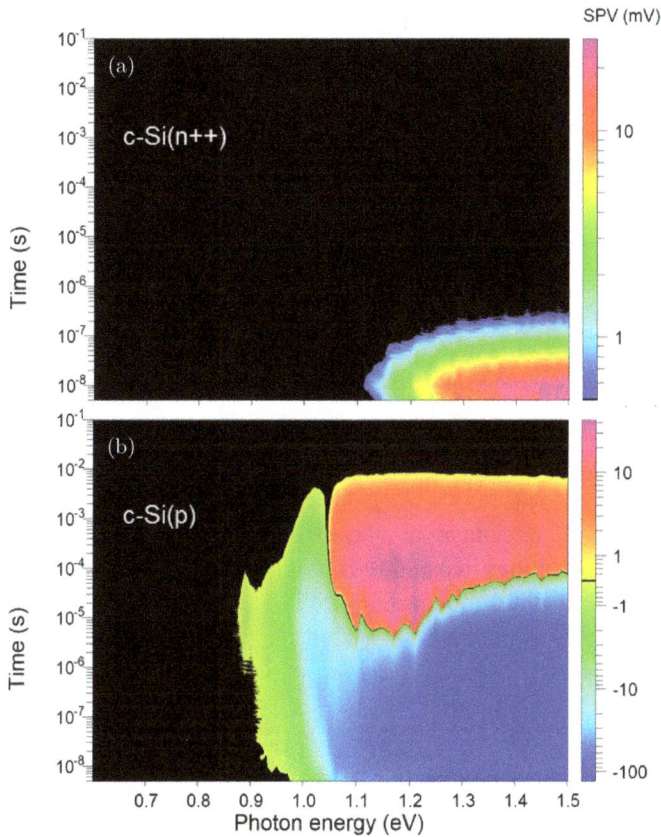

Figure 5.7. Transient SPV spectroscopy at constant photon flux for highly n-type doped c-Si (a) and p-type doped c-Si with surface defects (b).

the sensitivity of the SPV measurements was not high enough to detect surface states on the highly n-type doped sample.

The SPV signals were negative and of the order of $-120\,\mathrm{mV}$ at higher photon energies for the p-type doped sample. The plot of transient SPV spectroscopy for the p-type doped c-Si surface showed an onset of small negative SPV signals between 0.8 and 0.9 eV. These signals can be assigned to the excitation of electrons from the valence band into unoccupied acceptor states at the c-Si surface (Figure 5.8(a)). The electrons were excited from the valence band having a high density of states into unoccupied acceptor states with a lowest energy of about 0.25 eV below the conduction band edge.

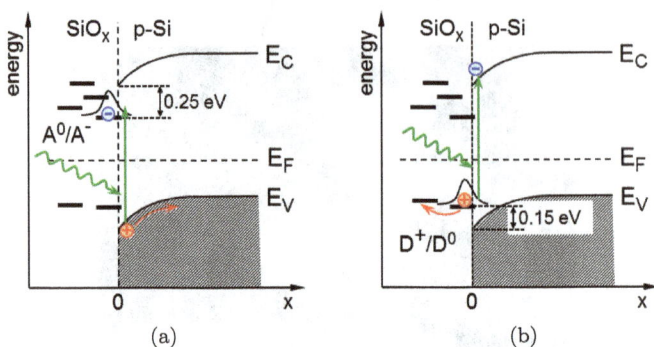

Figure 5.8. Models for separation of a hole photogenerated by excitation of an electron into an unoccupied acceptor surface state toward the bulk of p-type doped c-Si (a) and of a hole photogenerated by excitation of an electron from an occupied donor surface state toward the external surface of a sub-oxide surface layer (b).

Furthermore, a change of the sign from negative to positive was observed between about five and tens of microseconds depending on photon energy for the p-type doped sample. The positive SPV signals reached a maximum of the order of 30 mV at photon energies between 1.1 and 1.2 eV. The change of the sign set on at photon energies around 1.05 eV, i.e. this photon energy was required for exciting a hole moving from the c-Si surface toward the surface of the sub-oxide layer. The change of the sign of the SPV signal can be explained by a relatively fast relaxation of holes separated toward the bulk and a relatively slow release of photogenerated holes trapped at surface states and separated toward the surface of the sub-oxide layer (Figure 5.8(b)). The electrons were excited from occupied donor states with lowest energy of about 0.15 eV above the valence band edge into the conduction band having a high density of states. Incidentally, charge separation after excitation from an occupied donor into an unoccupied acceptor state was not observed. Incidentally, doping of mesoporous silicon is possible by slow predominant donor (HF treatment) or acceptor (moderate oxidation) surface states due to compensation of trapped charge by free charge carriers (Dittrich and Duzhko, 2003).

The analysis of surface states by SPV_{max}–U_{bias} cycling and by transient SPV spectroscopy is complementary in the sense of sensitivity. SPV_{max}–U_{bias} cycling is highly sensitive to defects with low and very low densities near the middle of the band gap, whereas transient SPV spectroscopy is sensitive to defects with higher density closer to the conduction and valence band edges. However, one shall keep in mind that transient SPV spectroscopy does not provide directly the density of states. Incidentally, transient SPV spectroscopy is also sensitive to bulk defects which have to be distinguished from surface defects by additional experiments in which the properties of the bulk or of the surface region are varied separately.

5.2.4.   *SPV at electrolyte–silicon interfaces*

Transient SPV measurements can be applied for *in situ* investigations of photoactive materials during chemical and electrochemical processing in electrolytes if those are accompanied by changes of the amount and/or distribution of charge near a semiconductor surface.

Figure 5.9 shows an idealized structure and band diagram of a semiconductor–electrolyte interface in thermal equilibrium. The semiconductor–electrolyte interface is separated into two regions which are the Helmholtz layer, also called Helmholtz double layer or dipole layer, and a reaction layer. The thickness of the Helmholtz double layer is given by the dimensions of polar molecules in an electrolyte. Surface chemical

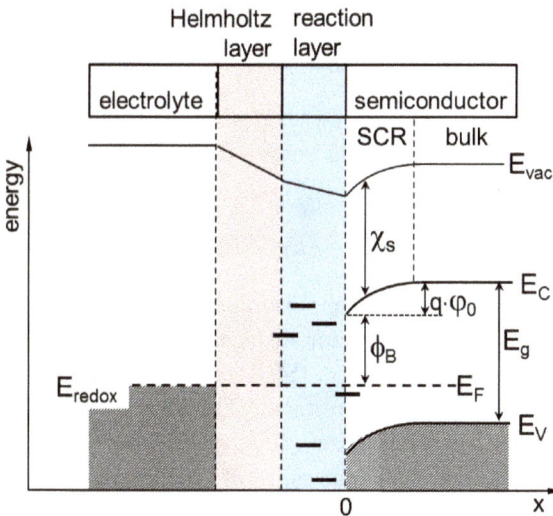

Figure 5.9. Idealized band diagram in thermal equilibrium of a semiconductor–electrolyte interface including the Helmholtz dipole layer and a reaction layer. $E_{vac}, E_{redox}, E_g, E_F, E_V, E_C$, SCR, ϕ_0, φ_B and χ_s denote and the vacuum energy, the redox potential of the electrolyte, the band gap, Fermi energy, valence and conduction band edges, space charge region, surface band bending, barrier height at the semiconductor surface and electron affinity of the semiconductor, respectively.

and electrochemical reactions cause changes in the bond configuration and electronic states near the semiconductor surface. The effective thickness of the reaction layer can vary from less than a monolayer to several monolayers depending on the reactions. A reaction layer can be, for example, a thin anodic oxide or a sub-oxide layer at the surface of a silicon crystal.

The main characteristic of the electrolyte is its redox potential (E_{redox}) which describes the equilibrium between reduction and oxidation reactions in the electrolyte. Due to the high concentration of reduced and oxidized species, the redox potential is fixed in electrolytes. In this sense, the redox potential of an electrolyte is equivalent to the Fermi energy of a metal. However, one shall keep in mind that a redox potential may change near a semiconductor–electrolyte interface due to transport limitation in the electrolyte under vigorous electrochemical reactions.

The vacuum energy (E_{vac}) serves as the reference energy in the band diagram. The difference between E_{vac} and E_{redox} is equivalent to metal work function. At a semiconductor–electrolyte interface, E_{vac}–E_{redox} is equal to the sum of the barrier height at the semiconductor surface (ϕ_B), the electron affinity of the semiconductor (χ_s), the potential drop across the Helmholtz layer ($\Delta\varphi_{Hl}$) and the potential drop across the reaction layer ($\Delta\varphi_{rl}$):

$$E_{vac} - E_{redox} = \phi_B + \chi_s + q \cdot \Delta\varphi_{Hl} + q \cdot \Delta\varphi_{rl} \qquad (5.1)$$

The surface band bending (φ_0) is described by the difference between the band gap (E_g) and the Fermi energy in the bulk of the semiconductor (($E_F - E_V)_{bulk}$) and on ϕ_B:

$$q \cdot \varphi_0 = E_g - (E_F - E_V)_{bulk} - \phi_B \qquad (5.2)$$

An SPV transient at short times is mainly sensitive to the change of the band bending at the semiconductor surface, i.e. to the change of the charge in the surface space charge region (SCR). Charge neutrality is considered at semiconductor–electrolyte interfaces. The change of the charge in the surface SCR of the semiconductor (ΔQ_{SC}) is balanced by the changes of the charge influenced at the capacitance of the Helmholtz layer (ΔQ_i), the charge injected into deep localized states in the reaction layer (ΔQ_{ds}) and the charge caused by trapping at surface states (ΔQ_{ss}):

$$\Delta Q_{SC} = -(\Delta Q_{SS} + \Delta Q_{ds} + \Delta Q_i) \qquad (5.3)$$

The changes of the surface band bending of $\Delta\varphi_{Hl}$ and of $\Delta\varphi_{rl}$ depend on the changes of the different components of charge. Therefore,

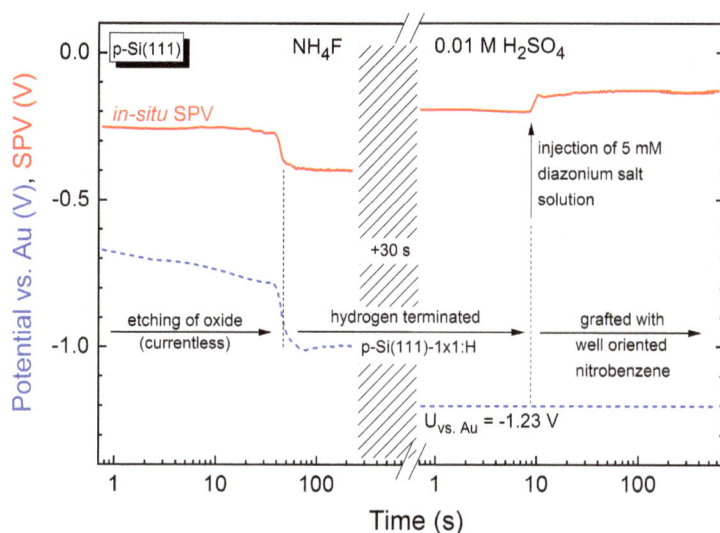

Figure 5.10. Applied potential (blue) and *in situ* monitoring of SPV signals of p-type doped c-Si(111) obtained from SPV transients at short times (red) for a sequence of currentless etching of a thin anodic oxide and hydrogen termination in NH_4F followed by an exchange of the electrolyte to 0.01 M H_2SO_4 under cathodic potential preserving hydrogen termination and subsequent grafting of a monolayer of oriented nitrobenzene molecules by injecting a diluted solution of 4-nitrobenzene-diazonium salt (see Hartig *et al.*, 2002a, 2002b for more details).

the application of Equations (5.1)–(5.3) marks the direction toward the analysis of SPV signals during chemical and electrochemical reactions at semiconductor surfaces. Furthermore, the reaction layer can be treated within models of small systems with localized states.

As an example, Figure 5.10 shows the SPV signal and the applied potential (U_{bias}) during a sequence of chemical and electrochemical treatments at the surface of a p-type doped c-Si(111) crystal. In the beginning of the sequence, a thin anodic oxide layer was removed by currentless etching in ammonium fluoride solution (NH_4F). The SPV signal stayed constant at about $-0.25\,V$ until about $40\,s$ and increased to $-0.4\,V$ within the following $10–20\,s$. This change of the SPV signal was a signature for completion of etching of the oxide and reaching of hydrogen termination.

The SPV signals at -0.25 and $-0.4\,V$ corresponded to Fermi energies at the c-Si(111) surface near the middle of the band gap and closer to the conduction band edge, respectively. During the removal of the oxide, U_{bias} was changed from about -0.67 to $-0.8\,V$ (U_{bias} was measured against a gold reference electrode) in order to keep the current at 0. After finishing the hydrogen termination, U_{bias} stabilized at about $-1.0\,V$. The change of U_{bias} during etching of the anodic oxide gave evidence for a reduction of charge in the anodic oxide. At the same time, the Fermi energy was fixed near midgap, i.e. the amount of charge removed with the oxide was less than the amount of charge staying in surface states. The change of the SPV and of U_{bias} during the formation of a hydrogenated c-Si(111)-1 × 1:H surface by about -0.15 and $-0.2\,V$, respectively, gave evidence for an accumulation of positive charge at the c-Si surface.

Hydrogen termination of c-Si surfaces is preserved in diluted hydrosulfuric acid (0.01 M H_2SO_4) under slight cathodic potential keeping the density of oxidizing holes at the c-Si surface low. The redox potential changed during the exchange of NH_4F solution by 0.01 M H_2SO_4 and the SPV signal change to $-0.2\,V$, i.e. the Fermi energy was not pinned at the c-Si surface.

Between 9 and 10 s, a diluted solution of 4-nitrobenzene-diazonium tetrafluoroborate was injected into the electrolyte. Electrons were injected from c-Si into the electrolyte and captured by diazonium ions. As a result, N_2 and nitrobenzene radicals were formed. The nitrobenzene radicals reacted with ≡S–H bonds at the c-Si(111) surface and formed a monolayer of chemically bonded nitrobenzene molecules. Incidentally, such a process is also called grafting. Due to the orientation of the nitrobenzene molecules, a permanent surface dipole appeared at the c-Si(111) surface which caused a stabilization of the SPV signals at about $-0.13\,V$ within a few seconds.

The change of the SPV signals after grafting of different benzene derivatives on c-Si(111)-1 × 1:H surfaces correlated very well with the dipole moments of the molecules (Hartig *et al.*, 2002a). Incidentally, a change of SPV signals during grafting of nitrobenzene molecules was not observed on c-Si surfaces covered with a very thin oxide layer or on rough hydrogenated c-Si surfaces (see, also Hartig *et al.*, 2002b).

5.3.   Charge Separation in a Layer System with Gallium Arsenide

5.3.1.   *GaO_x/n-GaAs/SI GaAs layer system*

Gallium arsenide (GaAs) is one of the best known semiconductors which is widely applied in photovoltaics (the highest solar energy conversion efficiency of a single-gap solar cell was achieved with GaAs; Alta Devices (2018)), electronics (metal semiconductor field effect transistor, MES-FET,

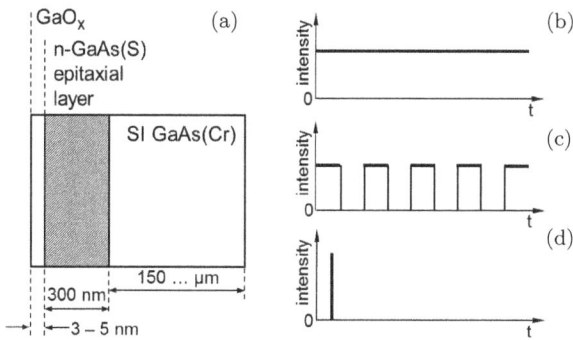

Figure 5.11. Structure of a n-type GaAs epitaxial layer grown on a SI GaAs(Cr) substrate (a) and schematic time-dependent light intensities of applied Kelvin probe (b), modulated (c) and transient (d) SPV measurements.

see, for example, Rocchi, 1989) and optoelectronics such as laser diodes (see, for example, Horiuchi, 2010 and references therein). Surface states of different GaAs surfaces were investigated by spectroscopy of ΔCPD in the 1970s (see, for example, Lagowski *et al.*, 1973).

As an example, the detection of electronic states with SPV techniques is shown on a sample of an n-GaAs epitaxial layer grown on a semi-insulating substrate compensated with chromium (SI GaAs(Cr)). Figure 5.11 shows a schematic diagram of the structure. The substrate was about $200\,\mu$m thick. The epitaxial layer was doped with sulfur (density of free electrons $10^{17}\,\text{cm}^{-3}$). The thicknesses of the epitaxial layer and of the substrate wafer were $300\,$nm and $200\,\mu$m, respectively. Related systems are applied, for example, for the production of MES-FETs. The sample was stored in air over a longer time so that a native oxide layer (GaO$_x$) with a thickness of $3.5\,$nm (Ruske, 2019) was formed on top (Lukeš, 1972).

Depending on the excitation conditions, SPV techniques can show preferential sensitivity to different processes. This will be demonstrated for SPV spectroscopy with continuous, modulated and pulsed illumination (schematic in Figure 5.11). A quartz prism monochromator with a halogen lamp and a tunable pulsed laser were used as light sources (see also Chapter 3).

Charge separation in built-in electric fields with different directions and charge transfer across internal interfaces shall be considered for the given sample (see Figure 5.12). Charge carriers can be photogenerated in the n-GaAs layer, the SI GaAs substrate and the GaO$_x$ layer. The densities and distributions of free and trapped charge carriers in space

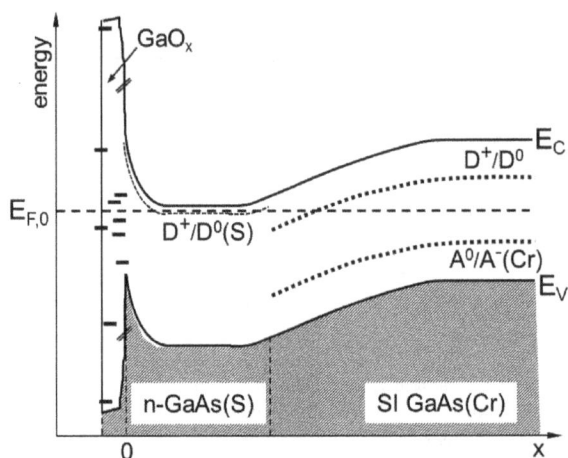

Figure 5.12. Schematic band diagram of n-GaAs/SI GaAs.

depend on the distributions of electronic states in space and energy and on the photogeneration rates. The band gap of undisturbed GaAs is 1.42 eV. The native oxide of GaAs consists mainly of Ga_2O_3 and suboxides in the interface region. The band gaps of Ga_2O_3 and amorphous a-GaO_x:H are 4.9 eV (Orita *et al.*, 2000) and 3.4–4.0 eV (Kobayashi *et al.*, 2018). Defect states are present in GaO_x and in SI GaAs(Cr), for example, the deep Cr acceptor with an activation energy of 0.57 eV (Kato *et al.*, 1979).

The bands are bended upward at the GaO_x and n-GaAs interface and upward with opposite direction of the built-in electric field at the n-GaAs/SI GaAs interface. The maximum bending is similar for both built-in electric fields if we assume that E_F is pinned near midgap at the external n-GaAs surface and that E_F is near midgap in SI GaAs. Sulfur forms shallow donors in GaAs. SI GaAs is characterized by deep donor and acceptor states in the band gap.

5.3.2. *SPV spectroscopy of n-GaAs/SI GaAs with a Kelvin probe*

SPV measurements with a Kelvin probe are limited by the slowest processes. Figure 5.13 shows the time dependence of the ΔCPD of the n-GaAs/SI GaAs(Cr) sample after switching on and off illumination (wavelength 700 nm). The sign of the light-induced ΔCPD was negative, which is typical for an n-type doped semiconductor with a depletion region

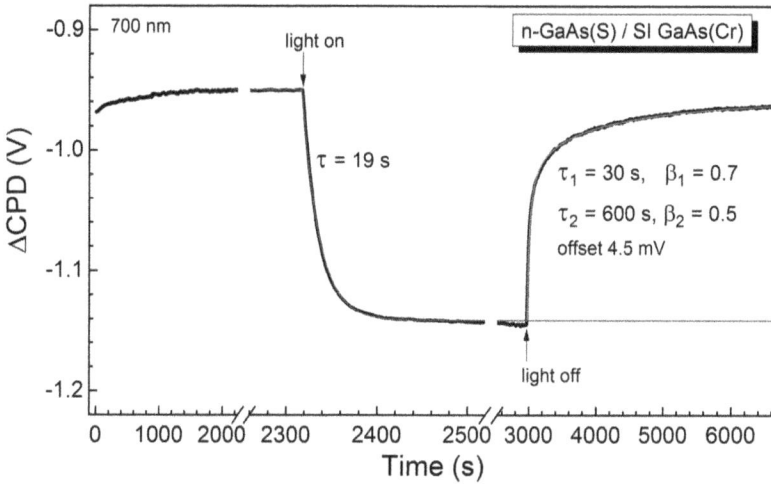

Figure 5.13. Time dependence of the ΔCPD under illumination (700 nm, 4.5 μW/cm^2) and in the dark of n-type GaAs/SI GaAs. Charging was well described by an exponential increase time constant 19 s). Discharging was fitted with two stretched exponentials (τ_1, τ_2, β_1 and β_2 equal to 30 s, 600 s, 0.7 and 0.5, respectively).

at the surface. Therefore, ΔCPD was dominated by charge separation at the GaO$_x$/n-GaAs interface.

For the given conditions, the increase of ΔCPD in time could be well fitted with a logistic growth function (τ_{on} less than 1 s) and an exponential growth function with a time constant of 19 s corresponding to the time needed for establishing a stationary state at the low light intensity. After switching off illumination, the decrease of ΔCPD could be fitted with two stretched exponentials (τ_1, τ_2, β_1 and β_2 equal to 30 s, 600 s, 0.7 and 0.5, respectively) and an offset of 4.5 mV. The need for an offset in the fit gave evidence for processes with much longer time constants than the time of measurement. Therefore, a significant part of positive charge separated toward the surface was trapped at deep states in the interface region between the n-GaAs and GaO$_x$ layers and in the GaO$_x$ layer.

A ΔCPD spectrum of n-GaAs/SI GaAs(Cr) is depicted in Figure 5.14. ΔCPD started to increase in the range between 0.6 and 0.7 eV and continued to increase continuously up to the range near the band gap. The increase became steeper between about 1.32 and 1.38 eV (see also the spectrum of the first derivative) and reached a minimum at 1.41 eV and a second maximum around 1.44 eV. ΔCPD decreased with increasing photon energies above 2 eV. The decrease of ΔCPD was a bit reduced in the range

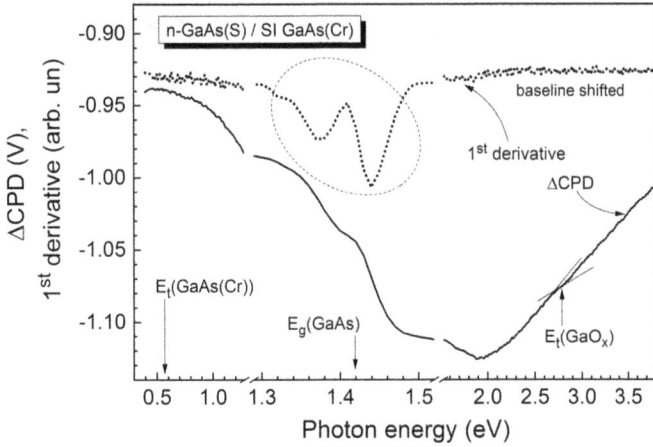

Figure 5.14. Spectra of ΔCPD and first derivative of n-GaAs/SI GaAs. The arrows mark the acceptor state of Cr, the band gap of GaAs and a defect state in the native oxide. The baseline of the first derivative was shifted for clarity. The peculiarities in the range of the band gap cannot be assigned to certain transitions.

between 2.56 and 2.58 eV. The very slight change in the decrease of ΔCPD around 2.56 and 2.58 eV may be a hint to a transition related to excitation from defect states in GaO_x.

Despite the relatively large number of possible optical transitions, only very few transitions could be detected in the ΔCPD spectrum. Furthermore, several processes contributed differently to the ΔCPD around E_g. There is practically no chance to define a procedure for the precise measurement of E_g from the ΔCPD spectrum. The number and the origin of processes contributing to charge separation in the n-GaAs/SI GaAs sample cannot be separated from the ΔCPD spectrum.

5.3.3. *Modulated SPV spectroscopy of n-GaAs/SI GaAs*

Figure 5.15 shows the spectra of the in-phase and phase-shifted by 90° signals of the n-GaAs/SI GaAs(Cr) sample in the spectra regions below the band gap of GaAs (Figure 5.15(a)), around the band gap (Figure 5.15(b)) and in the region with high absorption coefficients (Figure 5.15(c)). In contrast to the ΔCPD spectrum, the modulated SPV spectra were rich in structures and peculiarities Furthermore, the X-signals were negative and became positive at about 1.8 eV. Therefore, modulated SPV signals at photon energies below E_g were strongly influenced by charge separation in

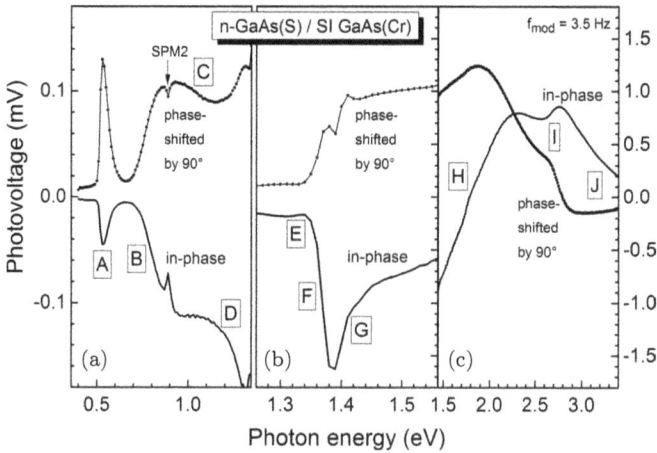

Figure 5.15. Spectra of modulated SPV in-phase and phase-shifted by 90° signals (thick and thin lines, respectively) of n-GaAs/SI GaAs in the ranges between 0.4 and 1.3 eV (a), 1.26 and 1.56 eV (b) and 1.57 and 3.8 eV (c). The letters denote characteristic features (see text). The feature around 0.9 eV was caused by the monochromator.

the region of the built-in electric field at the interface between n-GaAs and SI GaAs. Modulated charge separation in the region of the built-in electric field at the interface between GaO_x and n-GaAs became dominating only at higher photon energies when most of the light was absorbed in the n-GaAs layer.

At first glance, 10 features and one peculiarity at about 0.9 eV caused by the light source could be well distinguished in the modulated SPV spectra. The features are denoted by A–J and belong to the approximate spectral ranges 0.5–0.54, 0.68–0.83, 0.93–1.14, 1.1–1.3, 1.3–1.34, 1.34–1.38, 1.38–1.41, 2.5–3.0 and >3.0 eV, respectively.

Feature A was characterized by an increase of the X- and Y-signals to the maxima of -45 and $75\,\mu V$, respectively, and a decrease of both at higher photon energies. Feature A seems to be caused by the excitation of electrons form occupied donor states into the conduction band of SI GaAs near the n-GaAs/SI GaAs interface. The precise determination of the transition energy is not possible due to the increase of the light intensity in the same range. The decrease of the X- and Y-signals was dominated by the decrease of the light intensity in this range (see also Figure 5.16).

Feature B was characterized by an increase of the X- and Y-signals to -75 and $100\,\mu V$, respectively. Future B can be assigned to excitation of electrons from occupied acceptor states in SI GaAs.

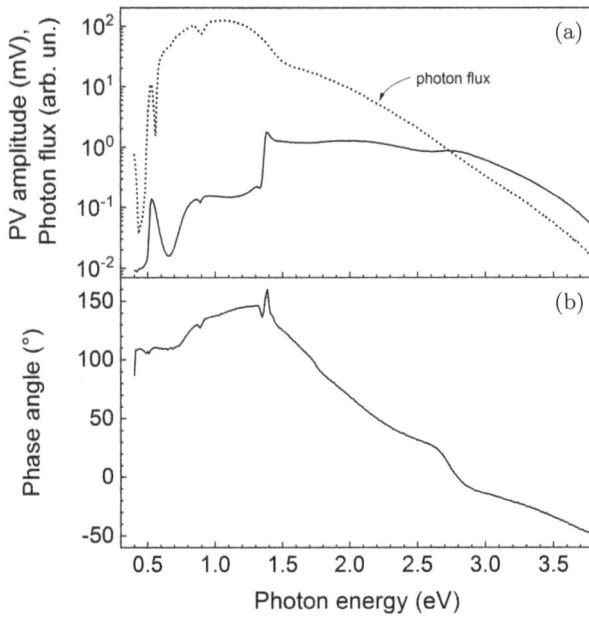

Figure 5.16. Spectra of the (a) PV amplitude and the photon flux (solid and dashed lines, respectively) and (b) the phase angle of n-GaAs/SI GaAs.

Feature C was characterized by a constant X-signal, whereas the Y-signals decreased to $90\,\mu V$. Therefore, feature C is caused by a slow process, probably at the GaO_x/n-GaAs interface. Feature D was characterized by a practically exponential increase of the X- and Y-signals to -185 and $120\,\mu V$, respectively. Feature D can be assigned to photogeneration of charge carriers from defect states near the conduction and valence bands of the SI GaAs. Feature E was characterized by a decrease of the X-signal to $-160\,\mu V$ and a constant Y-signal. The origin of feature E is not clear. Feature F was characterized by an increase of the X and Y-signals to -1.6 and $0.62\,mV$, respectively. Feature F was caused by photogeneration of charge carriers from shallow defect states in SI GaAs.

Feature G was characterized by a constant X-signal up to $1.39\,eV$ and a decrease of the X-signal to $-1.1\,mV$ at $1.41\,eV$. In the same range, the Y-signal first decreased to $0.6\,mV$ and then increased to $0.96\,mV$. Feature G is an overlap of signals caused by photogeneration of charge carries from shallow defect states in the space charge regions near the GaO_x/n-GaAs and n-GaAs/SI GaAs interfaces.

Feature H was characterized by a continuous change of the X-signal from negative to positive signals reaching 0.8 mV at 2.34 eV and a slight decrease at slightly higher photon energies. The Y-signal increased to 1.2 mV at 1.87 eV and decreased to 0.47 mV at 2.5 eV. The absorption coefficient of GaAs increases strongly to up to $>10^5$ cm^{-1} in the range of feature H. Therefore, feature H was caused by the change of the dominating spatial region for charge separation from the range of the n-GaAs/SI GaAs interface to the range of the GaO$_x$/n-GaAs interface.

Feature I was mainly characterized by an increase of the X-signal from 0.74 mV at 2.57 eV to 0.84 mV at 2.76 eV and a continuous change of the Y-signal from positive to negative sign with values of 0.35 mV at 2.65 eV and -0.14 mV at 2.94 eV. Feature I was caused by photogeneration of electrons from occupied states in GaO$_x$ near the GaO$_x$/n-GaAs interface followed by an electron transfer into the space charge region of the n-GaAs layer. Feature J was characterized by a continuous decrease of the X- and Y-signals due to the decrease of the light intensity (see also Figure 5.16(a)).

Most of the features A–J can be also well distinguished in the amplitude spectrum (Figure 5.16(a)). Especially the increase of the amplitude near E_g and near features A and B are very well pronounced.

Some features which were not so well pronounced in the amplitude spectrum dominate the spectrum of the phase angle (Figure 5.16(b)), for example, feature I. Vice versa, some features which were not well pronounced in the spectrum of the phase angle became very well pronounced in the amplitude spectrum, for example, feature A.

A phase angle–PV amplitude plot gives a profound overview of the features in modulated SPV spectra (Figure 5.17). Each feature has its specific fingerprint.

The phase angle of feature A remained practically constant, whereas the amplitude changed over one order of magnitude. Therefore, feature A is influenced by one process independent of Fermi-level splitting. This is the case for modulated charge separation in the region with occupied and unoccupied donor states in the space charge region at the side of SI GaAs.

The phase angle of feature H changed over a huge range, whereas the amplitude remained quasi-unchanged. This is only possible for a variation of the dominating charge-selective region, whereas both regions should have similar maximum band bending.

For feature J, the phase angle changed continuously from about $-15°$ at about 0.8 mV to $-55°$ at about 80 μV, i.e. the processes became slow with decreasing light intensity. This behavior is typical for modulated charge

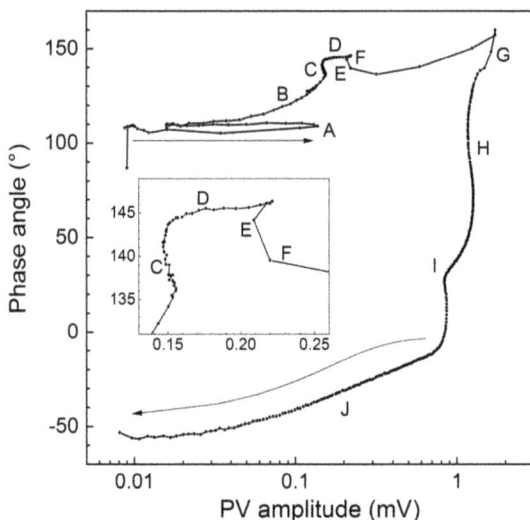

Figure 5.17. Phase angle–PV amplitude plot of n-GaAs/SI GaAs. The letters mark characteristic features (see text).

separation in a system with localized states (see Figure 4.29) and gives additional evidence for the nature of localized trap states in the GaO_x layer.

The behavior of the other features can be discussed in a similar way as for features A, H and J.

Phase angle–amplitude plots are very useful since they give immediately a clear idea about the number of processes involved in modulated charge separation and their nature. Incidentally, the modulated SPV spectra depend sensitively on the modulation frequency. Depending on the modulation frequency, features can appear or disappear. The analysis of the influence of f_{mod} on the modulated SPV spectra is beyond the scope of this book.

Modulated SPV spectroscopy on n-GaAs/SI GaAs gave some evidence for charge separation including also surface states at the GaO_x/n-GaAs interface. However, it was not possible to separate the energy range of this process from the modulated SPV spectra due to the dominating modulated charge separation in the region of the n-GaAs/SI GaAs interface. Higher resolution times toward milliseconds, microseconds and nanoseconds are required in order to separate the influence of surface states at the GaO_x/n-GaAs interface. However, this is not possible with modulated SPV spectroscopy.

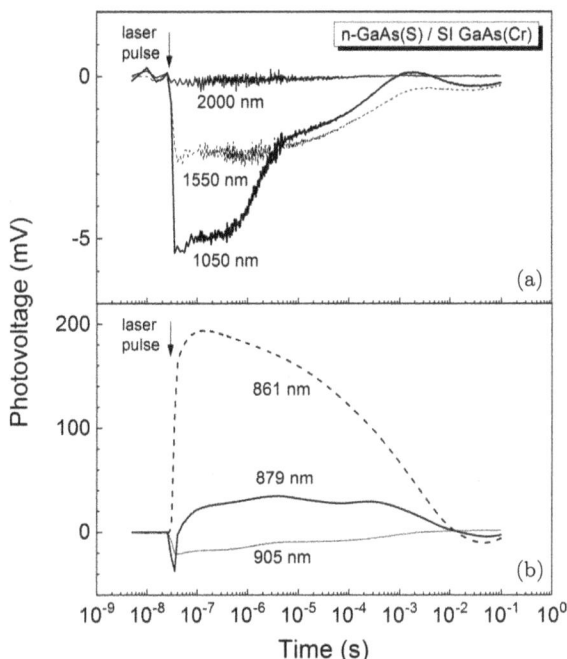

Figure 5.18. SPV transients of n-GaAs/SI GaAs excited at 2,000, 1,550 and 1,050 nm (thin, dashed and thick lines, respectively, (a)) and at 905, 879 and 861 nm (thin, thick and dashed lines, respectively, (b)). The transients were measured at a constant photon flux. The repetition rate of the laser pulses was 1 Hz. Ten transients were averaged in one. The arrows mark the laser pulses.

5.3.4. *Transient SPV spectroscopy of n-GaAs/SI GaAs*

Characteristic SPV transients excited at different wavelengths are presented for n-GaAs/SI GaAs in Figure 5.18. Extremely weak negative SPV signals with a maximum of the order of -0.2 mV were obtained for the excitation wavelength of 2,000 nm. This gives some evidence for very weak photogeneration from occupied deep donor states in the space charge region at the n-GaAs/SI GaAs interface and will not considered in the following.

Excitation at 1550 nm, i.e. at 0.8 eV, leads to a well-pronounced negative SPV transient setting on within the resolution time of the system and having a maximum of about -2.4 mV. The SPV transient decayed to -0.4 mV within several milliseconds and increased slightly to -0.5 mV in the following milliseconds and started to decrease at times longer than 20–30 ms. This behavior gives evidence that at least two processes were involved in the relaxation of the charge distribution. It is reasonable to suppose that the relaxation of charge separated in space is different for photogeneration from the valence band to unoccupied donor states and from occupied acceptor states to the conduction band.

Excitation at 1,050 nm, i.e. at 1.18 eV, resulted in an SPV transient with a maximum of about −5.3 mV. First, the SPV signal decreased to −5.0 mV within 30–40 ns. Second, the SPV signal stayed nearly constant until 300–400 ns and decreased to −2 mV within the next 4–5 μs. For comparison, the SPV signal excited at 1,550 nm was higher at this time, i.e. an additional process with the opposite direction of charge separation started. Third, the sign of the SPV signal became positive at about 1 ms and became again negative at about 4 ms. This gives additional evidence for a process with opposite direction of charge separation, i.e. photogeneration from surface states at the GaO_x/n-GaAs interface. Fourth, the SPV transient excited at 1,050 nm started to become closer to that excited at 1,550 nm at longer times.

Excitation at 905 nm, i.e. at 1.37 eV, resulted in an SPV transient with a maximum of about −20 mV. The shape of the SPV transient excited at 905 nm was rather similar to that excited at 1,050 nm at times up to the millisecond range. At longer times, the sign of the SPV transient excited at 905 nm did not change to negative again.

Excitation at 879 nm, i.e. at 1.41 eV, resulted in an SPV transient with a signal of about −36 mV at 10 ns. The sign changed to positive within the following 10–15 ns and increased to 35 mV at 3–4 μs. The SPV signal reached 28 mV after 60–70 μs, increased to about 32 mV at 200 μs decreased in the following time, changed the sign to negative at 15 ms, reached −5 mV at 50 ms and decreased at longer times. The very rapid change of the sign gave evidence for the onset of photogeneration from the valence band into unoccupied shallow donor states in the surface space charge region of n-GaAs. The change to negative SPV signals at longer times gave evidence that the relaxation of charge separated in space was slower in SI GaAs. Incidentally, one has to keep in mind that the repetition rate of the laser pulses was 1 Hz, i.e. the transients were not sensitive to processes slower than 0.1–1 s.

Excitation at 861 nm, i.e. at 1.44 eV, resulted in a positive SPV transient with a signal of about 170 mV at about 100 ns. Therefore, fundamental absorption in n-GaAs dominated the charge separation. The fact that the maximum was not reached within the resolution time of the system gives evidence for a superposition of positive and negative SPV signals, i.e. 170 mV was only the difference. The SPV transient decreased

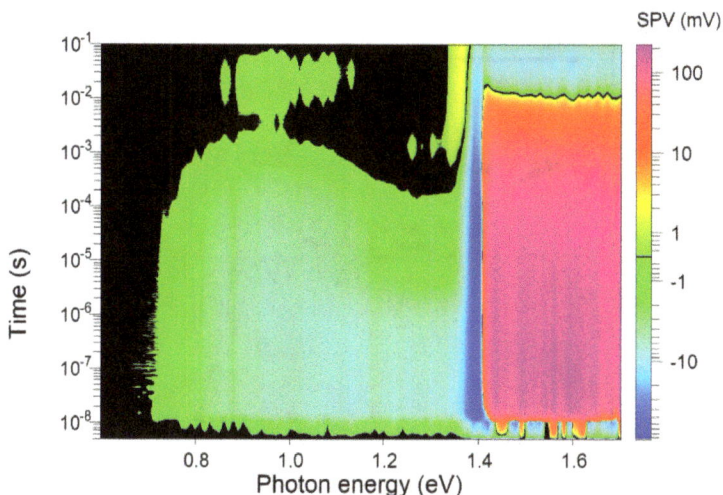

Figure 5.19. Transient SPV spectroscopy of n-GaAs/SI GaAs(Cr).

monotonously and changed the sign at about 15 ms and reached a negative maximum of −10 mV at 50 ms. Therefore, the amount of charge separated into electronic states in SI GaAs with long relaxation times increased with increasing photogeneration, probably due to trapping.

The SPV transients excited at numerous wavelengths were transformed to SPV spectra measured as a function of time after switching off the laser pulses. The plot given in Figure 5.19 summarizes the measurements of transient SPV spectroscopy. Regions with changing sign and dynamics can be clearly distinguished.

It is useful to focus on some details in separate spectra measured at certain times after switching on the laser pulses (Figure 5.20). After 10 ns, the maximum negative and positive SPV signals of −83 and 162 mV were reached at 1.39–1.40 eV and 1.44 eV, respectively. The SPV signals set on at about 0.7 eV and reached about −6 mV at 0.9 eV. The SPV signal was −7.5 mV at 1.2 eV and remained nearly constant until 1.33 eV. The SPV signals increased steeply between 1.33 and 1.39 eV. The steep increase between 1.33 and 1.39 eV of the negative SPV signal is equivalent to the behavior of feature F of the modulated SPV spectrum.

With increasing time up to about 10 μs, the SPV signals between 0.7–0.75 and 0.95 eV remained unchanged and vanished practically after several milliseconds. After 1 μs, the SPV signals decreased between 1.05 and 1.2 eV from about −6.5 mV to −4 mV and remained constant between

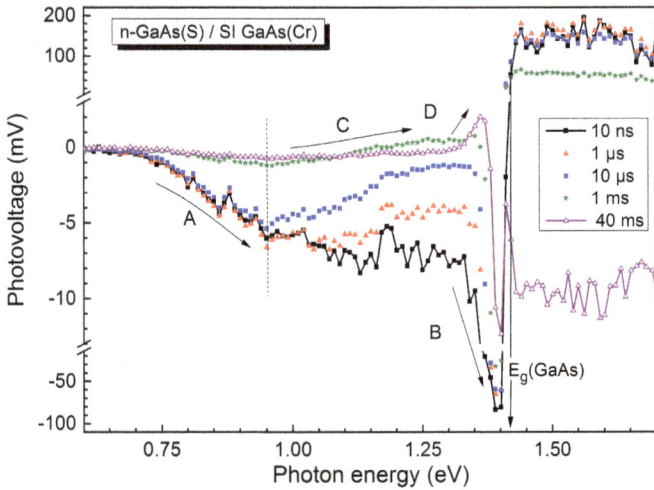

Figure 5.20. Transient SPV spectra of the n-GaAs/SI GaAs(Cr) sample measured after 10 ns, 1 μs, 10 μsm, 1 ms and 40 ms (black square, triangles, blue square, stars and empty triangles, respectively). The letters and the arrows mark characteristic transitions and the band gap of GaAs.

about 1.2 and 1.33 eV. After 10 μs, the SPV signals decreased between 0.95 and 1.25 eV from about -5 to -1.4 mV and remained practically constant between about 1.25 and 1.33 eV.

After 1 ms, the SPV signals changed between 0.95 and 1.25 eV from about -1.2 to $+0.5$ mV and remained constant between about 1.25 and 1.33 eV. The behavior of the SPV signals between 0.95 and 1.25 eV gave evidence for the appearance of charge separation within the space charge region at the surface of n-GaAs. These SPV signals correspond to surface states at n-GaAs between 0.9 and 1.0 eV below the conduction band edge (measured by ΔCPD, Lagowski *et al.*, 1973). After 40 ms, these SPV signals between 0.95 and 1.25 eV practically disappeared.

A pronounced increase of the SPV signals from 0 to 2 mV appeared between 1.3 and 1.37 eV after 40 ms. These signals were caused by photogeneration from surface states close to the valence band edge followed by an electron transfer into the surface space charge region of n-GaAs and a transfer of the positive charge deeper into the GaO$_x$ layer.

After 40 ms, the SPV signals between 1.37 and 1.40 eV, between 1.40 and 1.41 eV and between 1.41 and 1.43 eV changed from 1.8 to -12.3 mV, decreased from -12.3 to -3.8 mV and increased from -3.8 to -9.5 mV, respectively. The change of the SPV signals between 1.40 and 1.41 eV gave evidence for strong photogeneration due to absorption into the large number of unoccupied shallow sulfur donors in n-GaAs in contrast to photogeneration in SI GaAs which does not have that high density of shallow donors or acceptors. Holes photogenerated in n-GaAs were partially transferred into surface states with a lifetime longer than 40 ms.

The SPV signals above 1.43 eV were practically unchanged up to about 10 μs, decreased to 60 mV after 1 ms and changed to about -10 mV after 40 ms, i.e. charge separation within the n-GaAs layer relaxed faster than within SI GaAs for the given time window.

5.4. Charge Separation in Systems with Quantum Dot Layers

5.4.1. *Colloidal quantum dot layers*

QDs are semiconductor nanoparticles in which the extension of the electron and hole wave functions is limited by the dimension of the nanoparticle or crystallite (quantum confinement). Electrons and holes form excitons in QDs. Due to quantum confinement, the transition energy of excitons (E_{eh}) is larger than the band gap of the semiconductor. Furthermore, the value of E_{eh} increases with decreasing size of QDs. Light absorption is not possible at photon energies lower than E_{eh}. Therefore, the absorption edge of QDs is shifted to higher energies with regard to E_{eh} (quantum size effect). The diameters of QDs are of the order of several nanometers.

Colloidal QDs (called QDs in the following) can be prepared by precipitation of nanoparticles, for example, of CdSe or CdTe, in solutions. At a certain point, the surfaces of QDs shall be passivated chemically in order to stop further growth and in order to avoid agglomeration. Organic molecules are used for the stabilization of QDs. Therefore, QDs consist of a semiconductor core and a passivating shell of organic molecules (Figure 5.21). The passivating molecules are also called surfactants or terminating molecules.

The electronic properties of QDs depend also on the surfactant molecules and on chemical surface bonds. Surfactant molecules can be, for example, TGA (thioglycolic acid), dithiol, pyridine, OA (oleic acid) or TOP (trioctylphosphine).

The energies of bonding and antibonding states of surfactant molecules are large so that the organic shell acts as a barrier layer. For charge transfer, the surfactant molecules shall be very short in order to enable tunneling between neighboring QDs and between QDs and substrates.

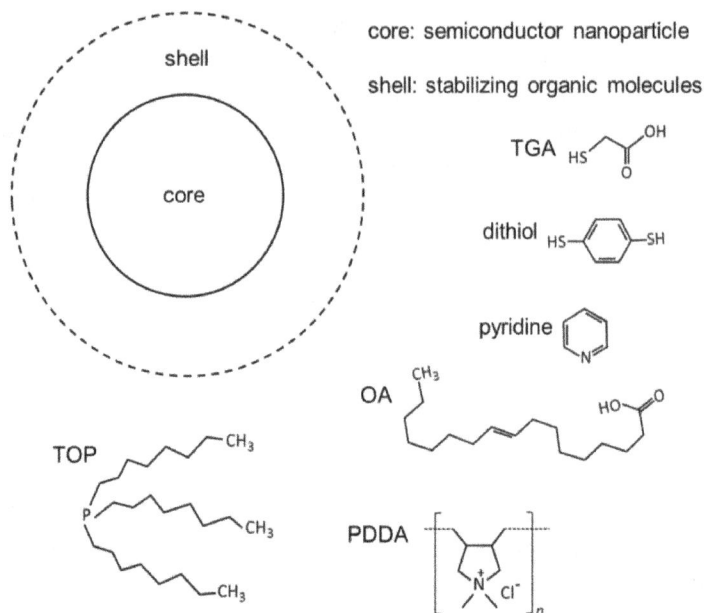

Figure 5.21. Scheme of a colloidal QD consisting of a semiconductor nanoparticle and a shell of stabilizing organic molecules and structures of TGA (thioglycolic acid), dithiol (1,4-benzenedithiol), pyridine, OA (oleic acid), TOP (trioctylphosphine) and PDDA (poly(diallyldimethylammonium chloride)).

There are two strategies of realizing very thin shells in colloidal QDs. First, QDs can be directly stabilized in the solution with very short molecules such as TGA. An advantage of TGA molecules is that they bind directly to the QD via the thiol groups (SH) and that the carboxyl groups (COOH) point toward the solution. CdSe QDs (Rogach *et al.*, 1999) and CdTe QDs (Rogach *et al.*, 1996, 2007) can be prepared with surfaces passivated by TGA. Second, stabilizing molecules with longer chains such as OA and TOP can be exchanged by very short molecules such as pyridine during additional steps (Celik *et al.*, 2012). This is demonstrated in Figure 5.22 for CdSe QDs with diameters of 4.5 nm.

QD layers can be deposited on different kinds of substrates. Carboxyl groups can bind directly to the surface of metal oxides. Molecules such as TGA are also called linker molecules since they bind QDs onto a substrate surface. An advantage of the use of TGA as a linker molecule is that the thickness of a QD layer is limited to one monolayer since agglomeration via carboxyl groups is hindered.

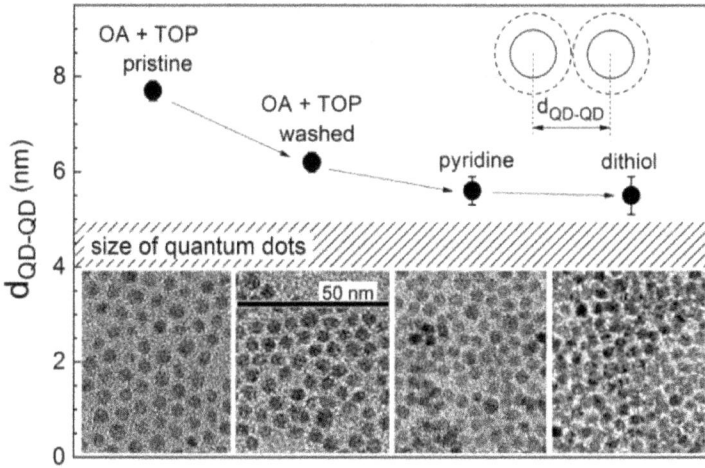

Figure 5.22. Dependence of the average distance between the centers of CdSe QDs (inset) on the treatments of washing residual molecules of OA and TOP, subsequent exchange of OA and TOP by pyridine and subsequent exchange of pyridine by dithiol molecules. The photographs show the corresponding micrographs (transmission electron microscopy). See, for details, also Zillner *et al.* (2012) and Zillner (2013).

For the deposition of multilayers of QDs passivated with TGA, an intermediate treatment in a polyelectrolyte such as PDDA (poly(diallylmethylammonium), see also Figure 5.21) is required. Due to electrostatic attraction, negatively charged QDs can adsorb on long chains of positively charged PDDA molecules. This effect is used in the so-called layer-by-layer deposition of QDs in which concentrations and treatment times have to be carefully adjusted (Rogach *et al.*, 2002).

QD layers can be deposited rather fast by the very robust dip-coating technique, which is suitable, for example, for the deposition of films with a well-defined thickness of QDs passivated with OA and TOP or pyridine (see, for example, Zillner and Dittrich, 2011; Zillner *et al.*, 2012). Furthermore, QDs can be linked by chemical bonding to each other if for example, pyridine surfactants are exchanged by bidentate linker molecules such as dithiol (Figure 5.22).

5.4.2. The $TiO_2/CdSe$ quantum dot layer system

In the following, CdSe QDs are linked to the surface of a compact TiO_2 layer with TGA molecules (see, also Mora-Seró *et al.*, 2007). The electron affinities of CdSe and TiO_2 are about 4.3 and 4.4 eV (at pH 1, Hagfeldt and

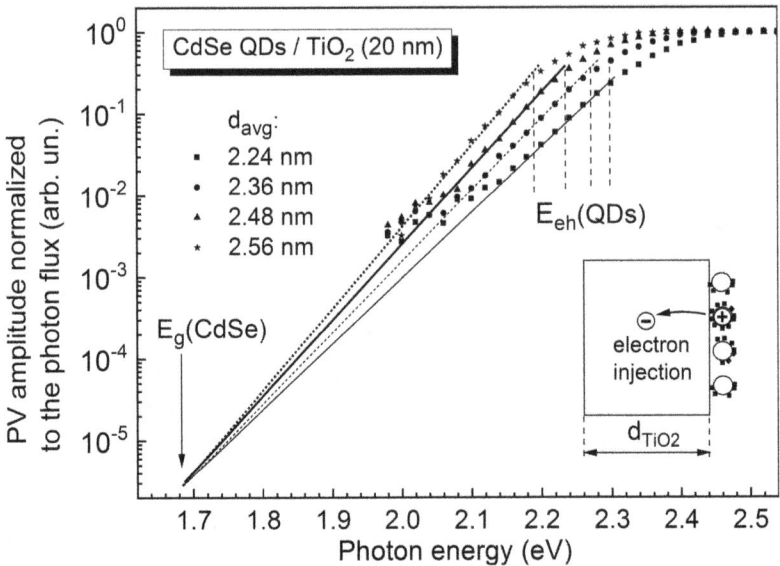

Figure 5.23. Spectra of the modulated SPV amplitudes normalized to the photon flux on a semi-logarithmic scale for CdSe QDs with average diameters of 2.24, 2.36, 2.48 and 2.56 nm (squares, circles, triangles and stars, respectively). The QDs were bonded to the surface of a TiO_2 layer with TGA linker molecules. The dashed lines mark the first excitonic transition energies. The exponential tails have a common intersection point at the band gap of CdSe (reevaluation of data given by Mora-Seró *et al.* (2007)).

Grätzel, 1995), respectively. Therefore, electrons photogenerated in CdSe QDs are injected into the conduction band of a TiO_2 layer if the linker molecules are very short so that the electron transfer is not limited by the barrier of the shell, whereas the photogenerated holes stay in the CdSe QDs (inset of Figure 5.23).

If the TiO_2 layer is much thicker than the radius of the CdSe QDs, the SPV signals are dominated by the TiO_2 layer and the SPV signals are proportional to the thickness of the TiO_2 layer (Mora-Seró *et al.*, 2007). Furthermore, in the small signal case, the SPV signals are proportional to the density of photogenerated charge carriers, i.e. to the light intensity. Therefore, a TiO_2 layer can serve as an excellent sensor for the analysis of isolated CdSe QDs linked to the surface of the TiO_2 layer with very short molecules.

Figure 5.23 shows modulated SPV spectra normalized to the photon flux for CdSe QDs linked to a 20 nm thin TiO_2 layer. At higher photon energies, the SPV signals were very similarly independent of the average diameter of the CdSe QDs (d_{avg}). At lower photon energies, the spectra follow an exponential dependence. The values of an E_{eh} can be obtained from the energy between the exponential dependence and the ongoing non-exponential increase. The values of E_{eh} decreased with increasing diameter of the CdSe QDs. The values of E_{eh} were 2.30, 2.27, 2.23 and 2.19 eV for d_{avg} equal to 2.24, 2.36, 2.48 and 2.56 nm, respectively.

Furthermore, the SPV spectra correlated with the absorption spectra via a power law (not shown in Figure 5.23, see Mora-Seró *et al.*, 2007). The power law is a signature for the role of dispersive transport of electrons injected into TiO_2.

The values of the exponential tails of the SPV spectra (E_t) were 54, 50, 47 and 43 meV for d_{avg} equal to 2.24, 2.36, 2.48 and 2.56 nm, respectively, i.e. E_t increased with increasing E_{eh}. The data shown by Mora-Seró *et al.* (2007) were re-examined in Figure 5.13 in the sense that the exponential tails resulted in a common intersection point at the band gap of CdSe. Therefore, the exponential tails in the SPV spectra of the CdSe QD layers are defined by exponential distributions of their size and the influence of surface and/or other defect states on E_t can be neglected.

5.4.3.   *Charge separation at quantum dot heterojunctions*

CdSe and CdTe form a charge-selective heterojunction which may be used for charge separation in nanocrystal solar cells (Gur *et al.*, 2005). The similar concept will be proven for CdSe QD/CdTe QD heterojunctions in the following. For this purpose, different layer systems of CdTe QDs and CdSe QDs passivated with TGA were deposited onto a SnO_2:F electrode by the layer-by-layer technique with PDDA molecules as polyelectrolyte, whereas the numbers and the order of the layers were changed (Gross *et al.*, 2010). Investigations were performed by modulated SPV spectroscopy and by transient SPV measurements (Gross *et al.*, 2010). As an example, the ΔCPD spectra are compared for a system with six monolayers of CdTe QDs and one monolayer of CdSe QDs on top and a system with six monolayers of CdSe QDs and one monolayer of CdTe QDs on top in Figure 5.24.

The photoinduced responses set on at photon energies of about 1.84 and 2.03 eV for the samples with 6 ML of CdTe QDs and 6 ML of CdSe QDs, respectively. Therefore, the absorption was dominated by the thicker QD layers.

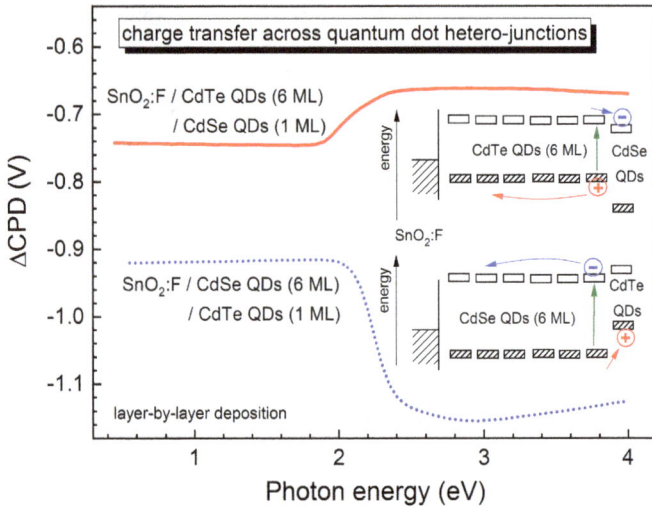

Figure 5.24. Spectra of ΔCPD for six monolayers of CdTe QDs and one monolayer of CdSe QDs on top (red line) and for six monolayers of CdSe QDs and one monolayer of CdTe QDs on bottom (blue line). The baselines were shifted by 5 V. The inset shows the configuration of the layer system. The colloidal QDs were stabilized with TGA molecules. The layer systems were successively deposited layer-by-layer with polyelectrolyte PDDA molecules. More details and modulated SPV spectra and SPV transients are given for more systems by Gross *et al.* (2010).

The ΔCPD signals became more positive (by about +80 mV) for the system with six monolayers of CdTe QDs and one monolayer of CdSe QDs on top. Therefore, electrons photogenerated in the CdTe QD layers were separated into the CdSe QD layer at the surface. On the other side, the ΔCPD signals became more negative (by about −230 mV) for the system with six monolayers of CdSe QDs and one monolayer of CdTe QDs. Therefore, holes photogenerated in the CdSe QD layers were separated into the CdTe QD layer at the surface. Furthermore, the ΔCPD and modulated and transient SPV signals increased with increasing numbers of QD layers (not shown, see, also Gross *et al.*, 2010). This was a direct proof that QD heterojunctions can be used as charge-selective contacts.

5.4.4. *Initial charge separation across a CdSe quantum dot monolayer*

Layers of CdSe QDs with different surfactants and different thickness were deposited onto ITO (SnO$_2$:In) by dip coating (Zillner *et al.*, 2012). SPV

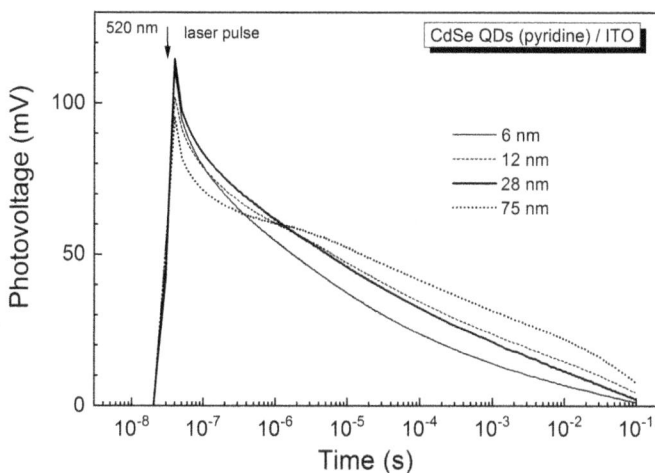

Figure 5.25. SPV transients of CdSe QD (pyridine)/ITO layers with thicknesses of 6, 12, 28 and 75 nm (thin solid, thin dashed, thick solid and thick dotted lines, respectively). The average diameter of the CdSe QDs was about 4.5 nm. The repetition rate of the laser pulses was 1 Hz. The OA and TOP surfactant molecules were exchanged by pyridine molecules in the solution. The layers were deposited with a dip robot (see Zillner and Dittrich, 2011; Zillner, 2013 for more details).

signals were not observed for CdSe QDs with untreated OA and TOP surfactant molecules (Zillner *et al.*, 2012), i.e. there was no charge transfer across such QD layers. SPV signals were observed for layers based on CdSe QDs with OA and TOP surfactant molecules after washing and with successively exchanged pyridine and dithiol surfactant molecules. As an example, Figure 5.25 shows SPV transients of CdSe QD (pyridine) layers with thicknesses of 6, 12, 28 and 75 nm. For comparison, 6 nm correspond to about one monolayer.

The SPV signals appeared within the laser pulse and reached the maximum at 10–20 ns which corresponded to the resolution time of the given measurements (100 MS/s). The maximum SPV signals were positive and amounted to about 120, 105, 125 and 90 mV for the CdSe QD (pyridine) layers with thicknesses of 6, 12, 28 and 75 nm, respectively. Therefore, the initial charge separation was practically independent of the layer thickness and photogenerated electrons were separated toward the ITO surface. This means that the interface between the first monolayer of CdSe QDs and the ITO was decisive for the charge transfer and that a strong built-in electric field resulting in additional charge separation was absent in the thicker layers.

The SPV transients decayed within 0.1–1 s, i.e. the decays were limited by the repetition rate at the longest times. The SPV transients tended to decay faster for thinner layers of CdSe QDs, i.e. charge transport from the near-surface region into deeper regions did play a certain role for the decays.

ITO is highly doped. Therefore, the influence of charge transfer across the space charge region of ITO can be neglected in comparison to the thickness of a monolayer of CdSe QDs. Charge transfer across the monolayer of CdSe QDs at the ITO surface dominated the SPV transients.

With regard to Equation (1.45) and taking into account that d and ε_r are about 3 nm and 5, respectively, an SPV signal of 100 mV corresponds to a density of separated charge carriers of the order of $10^{12}\,\mathrm{cm}^{-2}$. This means that electrons were transferred toward the ITO surface from about 1/4–1/3 of the CdSe QDs in the monolayer at the ITO surface.

The SPV signals at 20 ns were proportional to the intensity of the laser pulses for SPV signals up to about 60–70 mV (Zillner and Dittrich, 2011). At higher intensities, the SPV signals at 20 ns tended to saturate at values of the order of 110–130 mV.

Different processes influence initial charge separation across a QD layer. In the simplest model, a QD layer consists of QDs with surface states (energy E_{st}). The Fermi energy (E_{F}) is above E_{st} so that the surface states act as traps for holes (Figure 5.26(a)). The thickness of the surfactant layer is denoted by δ.

One photon can be absorbed in one QD resulting in the formation of an exciton (Figure 5.26(b)). SPV signals at 20 ns were sensitive to those excitons which dissociated by trapping of holes at surface states and transfer of electrons toward the ITO surface (Figure 5.26(c)). Directed charge separation was limited by radiative recombination (Figure 5.26(d)) and surface recombination, for example, at the ITO surface after desorption of surfactant molecules (δ reduced to δ', Figure 5.26(e)). At low light intensity, the number of QD in which one photon was absorbed, and therefore also the SPV signals at 20 ns, increased linearly with the light intensity. At high intensities, the number of QDs in which two photons were absorbed increased with increasing light intensity (Figure 5.26(f)). Due to the very high density of charge carriers in one QD, Auger recombination (Figure 5.26(g)) was much faster than trapping and charge transfer. Therefore, the second photon absorbed in CdSe QDs did not contribute to the SPV signals and the SPV signals at 20 ns tended to saturate at high light intensities.

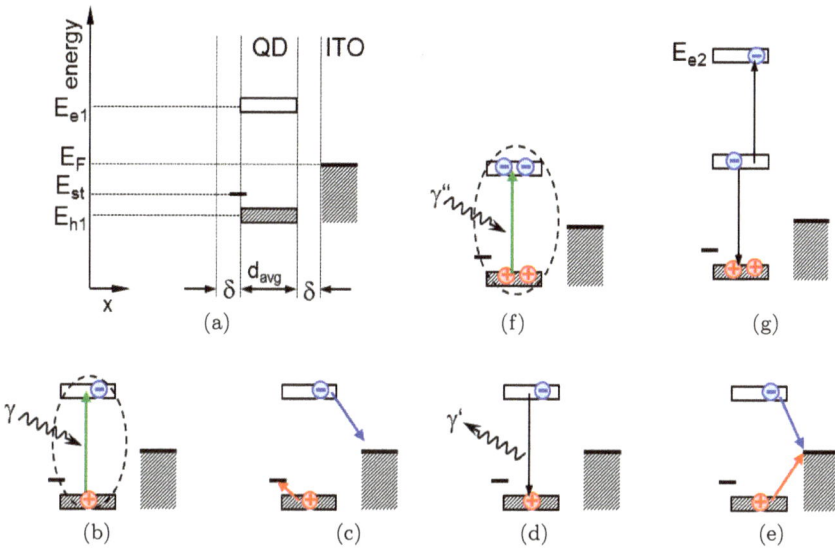

Figure 5.26. Schematic diagram of the energy levels of a QD with a surface state and an ITO surface (a) and elementary processes playing a role for initial charge separation (absorption of one photon in one QD and formation of an exciton (b), dissociation of an exciton and charge separation by trapping of a hole at a surface state and transfer of an electron toward the ITO surface (c), radiative recombination (d), surface recombination at the ITO surface (e), absorption of a second photon in one QD (f) and Auger recombination (g). E_{h1}, E_{e1}, E_{e2}, E_{st}, E_F, d_{avg}, δ, γ, γ', and γ'' denote the energy of the hole and electron in the exciton, the energy of an excited electron, the energy of the surface state, the Fermi energy, the average diameter of the QDs, the thickness of the surfactant layer, the first photon absorbed by one QD, a photon emitted due to radiative recombination and a second photon absorbed by one QD, respectively.

Incidentally, typical photoluminescence lifetimes of QDs are of the order of nanoseconds (see, for example, Jones *et al.*, 2003). Auger recombination rates in CdSe QDs are of the order of $3\,\mathrm{ns}^{-1}$ (Pelton *et al.*, 2017), i.e. Auger recombination lifetimes are about 300 ps. Therefore, with respect to the sign and the slow decay of SPV transients, it is reasonable to suppose that the dissociation of an exciton was caused by trapping of the hole at a surface state in the CdSe QD within a time of 0.1–1 ns.

Figure 5.27 shows (*in situ*) SPV transients of a CdSe QD (pyridine) monolayer measured at different temperatures during and after heating at increasing maximum temperatures. The initial value of the SPV signal at 20 ns (45°C) was 65 mV and the transient decreased within 0.1 s.

The SPV signal at 20 ns decreased to 48 mV at 120°C and increased subsequently to 56 mV at 60°C, whereas the SPV transients decayed

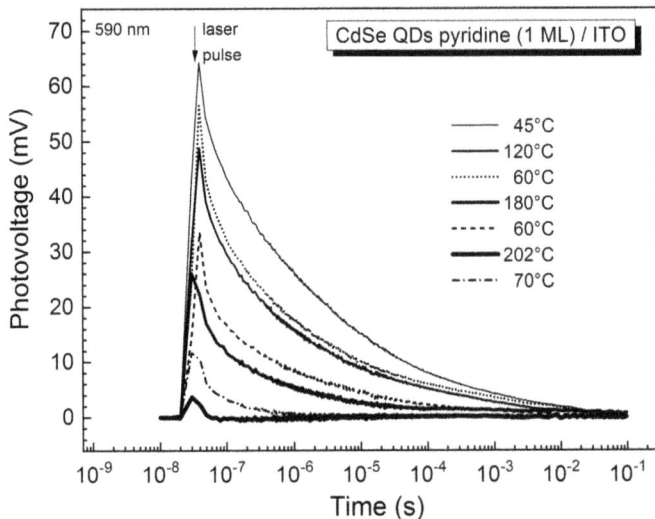

Figure 5.27. SPV transients of a CdSe QD (pyridine) monolayer measured at 45°C, 120°C, 60°C after heating in high vacuum to 120°C, 180°C, 60°C after heating to 180°C, 202°C and 70°C after heating to 202°C (thin solid, thicker solid and dotted, medium solid and dashed and thick solid and dash-dot lines, respectively).

to about half of the initial value within a similar time as the initial SPV transient. During the second heating cycle, the SPV signal at 20 ns decreased to 25 mV at 180°C and increased subsequently to 33 mV at 60°C, whereas the SPV transients decayed very similarly as the transients measured before. During the third heating cycle, the SPV signal at 20 ns decreased to 4 mV at 202°C and increased subsequently to 11 mV at 70°C, whereas the SPV transients decayed within a much shorter time than the transients measured before.

The processes limiting the relaxation of the SPV transients of the CdSe QD (pyridine) monolayer remained practically unchanged up to temperatures of about 180°C. This means that heating up to about 180°C did not change significantly the nature of the hole traps at the surface of CdSe QDs.

The behavior of the SPV signals at 20 ns is summarized in Figure 5.28 for the heating and cooling cycles of the CdSe QD (pyridine) monolayer, whereas the SPV signals were correlated with time and temperature.

The correlation with time showed that the SPV signals at 20 ns increased by up to about 5–8%, 40% and 300% for maximum temperatures of 120°C, 180°C and 202°C, respectively, during cooling. Therefore, the

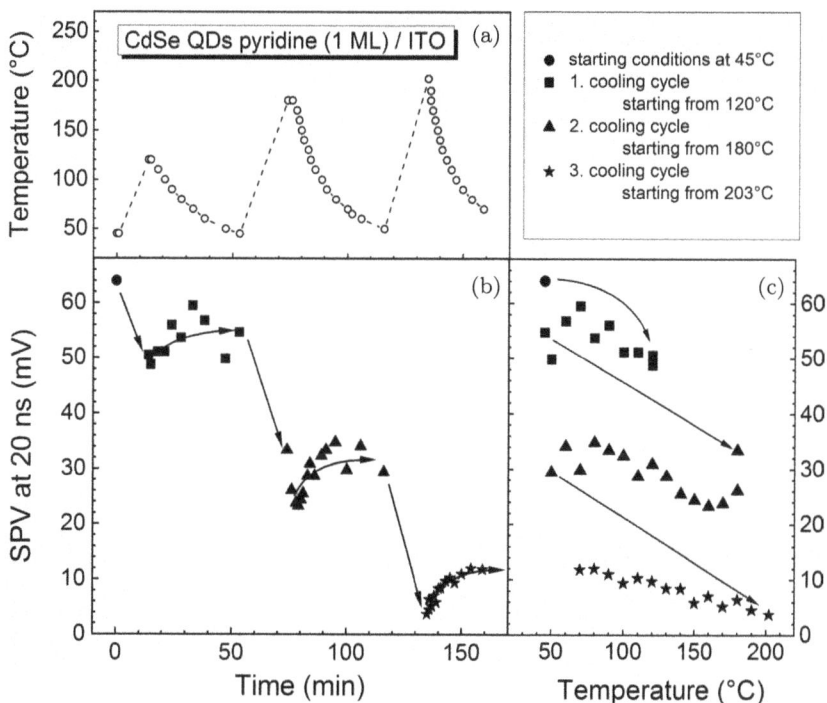

Figure 5.28. Sequence of the temperature variation in time (a), sequence of the corresponding SPV signals at 20 ns after switching on the laser pulses (b) and correlation between the SPV signals at 20 ns and the temperature (c). The sequence consisted of cooling cycles after heating to 120°C, 180°C and 202°C (squares, triangles and stars, respectively).

temperature dependence of initial charge separation of the CdSe QD (pyridine) monolayer increased strongly with increasing maximum temperature. This can be explained by a decrease of the relative influence of tunneling for the limitation of initial charge separation.

The initial charge separation across a CdSe QD (pyridine) monolayer depended strongly on preheating cycling. This can be explained within the model shown in Figure 5.26(e) if taking into account that pyridine molecules did not form chemical bonds at the surface of CdSe QDs. Pyridine molecules evaporated and the value of δ decreased during heating, and therefore surface recombination at the ITO electrode increased.

Figure 5.29 shows (*in situ*) SPV transients of a CdSe QD (dithiol) monolayer measured at 50°C, 100°C, 180°C and 230°C during heating and at 180°C, 100°C and 50°C during cooling within one cycle.

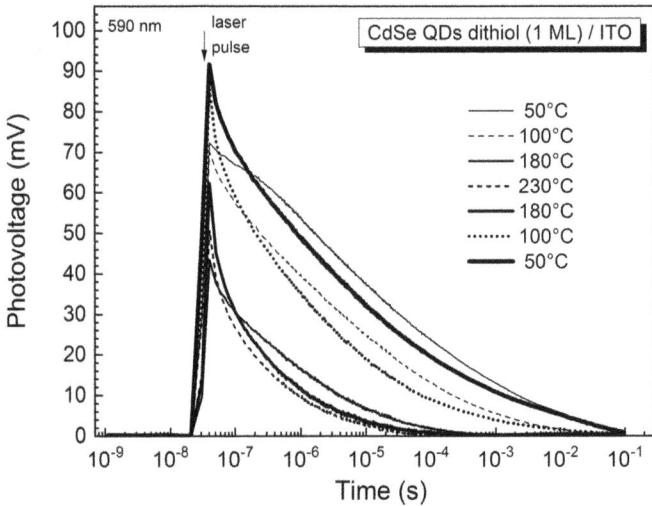

Figure 5.29. SPV transients of a CdSe QD (dithiol) monolayer measured at 50°C, 100°C, 180°C and 230°C during heating and at 180°C, 100°C and 50°C during cooling (thin solid and dashed, thicker solid and dashed, medium solid and dotted and thick solid lines, respectively).

The initial value of the SPV signal at 20 ns (50°C) was 72 mV and the transient decreased within 0.1–1 s. The SPV signal at 20 ns remained at 72 mV at 100°C and decreased to 43 mV at 180°C. In contrast to the CdSe QD (pyridine) monolayer, the SPV signal at 20 ns of a CdSe QD (dithiol) monolayer increased with increasing temperature for temperatures above 180°C and reached a value of 50 mV at the maximum temperature of 230°C. Furthermore, the SPV signals at 20 ns increased to 62, 88 and 92 mV for cooling to 180°C, 100°C and 50°C, respectively, i.e. the initial charge separation was stronger after heating to 230°C.

For the CdSe QD (dithiol) monolayer, the SPV transients decayed also within 0.1–1 s after the heating and cooling cycle. But the SPV transient decayed much more homogeneously after heating and cooling, i.e. without the shoulder around a microseconds which was observed before the heating and cooling cycle.

Dithiol forms stable bonds at the surface of CdSe QDs and links neighboring CdSe QDs with each other. Heating of a CdSe QD (dithiol) monolayer led to a remarkable reduction of the influence of processes limiting initial charge separation across the CdSe QD (dithiol) layer and to a pronounced homogenization of the hole traps at the surface of CdSe QDs (dithiol) which limit the relaxation of the charge separated in space. The reason is probably some reordering of bonds and evaporation of residual solvent molecules during heating.

5.4.5. *Single quantum dot approximation for random walk analysis*

Monolayers of QDs are practically ideal systems for random walk simulations due to the small number of available electronic states (see Chapter 4) and limited charge transfer into neighboring QDs. One isolated QD is considered in the single QD approximation (Fengler *et al.*, 2013). In the following it will be shown that fits by random walk simulations of temperature-dependent SPV transients of one monolayer (ML) CdSe QDs deposited on ITO (1 ML CdSe QDs) allow for the extraction of energy parameters of electronic states in systems with monolayers of QDs.

The SPV transients of a monolayer of CdSe QDs deposited onto ITO showed that photogenerated electrons were separated toward the ITO surface within the resolution time of the measurements. Furthermore, only QDs containing one hole contributed to SPV signals. Therefore, initial charge separation of a delocalized electron at the ITO surface and a localized hole at the QD surface is considered as the starting condition for random walk simulations (Figure 5.30(a)).

The electron and the hole separated in space can both undergo charge transfer steps and therefore random walks in a system with a QD at a surface. However, it is not possible to distinguish the motion of electrons from that of holes by the transient SPV measurements on CdSe QDs shown in this paragraph. Therefore, it is reasonable to assume only the motion of one of the charge carriers in the random walk simulations. In the given model, the random walk of holes in the QD was considered (Figure 5.30(b)).

A hole at a localized state can be transferred to other available localized states by tunneling. In addition, a hole at a localized state can be emitted into a delocalized state and randomly trapped at any available localized state. The random walk of the hole is finished by a charge transfer step of the hole to the ITO surface where the hole and the electron recombine (Figure 5.30(c)).

For the random walk simulations, the energy of the delocalized state was set as a sharp reference energy ($E_{dl} = 0\,\text{eV}$). The random walk of a hole in the QD was described by the number of localized states at the surface

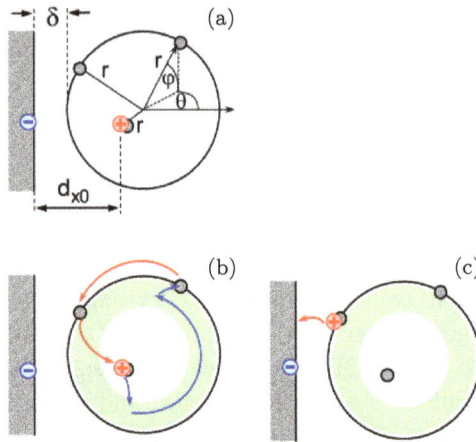

Figure 5.30. Schematic diagram of the single QD approximation consisting of the starting condition with an electron separated onto the metal surface and a hole trapped at a surface state of the QD (a), a random walk of the hole under consideration of localized and a delocalized states (b) and a final recombination step (c). See, for details, also Zillner *et al.* (2012) and Fengler *et al.* (2013).

of a QD (N_{ts}), the distribution of energies of localized states (standard deviation: σ_d), the difference between E_{dl} and the energy of the maximum of the energy distribution (E_d) and by the Fermi energy or cut-off energy (E_F) limiting the number of available states. E_F was introduced in order to consider charged deep trap states with detrapping times longer than the reciprocal repetition rate of the laser pulses (1 s). The value of N_{ts} defined the distribution of distances between localized states by assigning randomly azimuthal (φ) and horizontal (θ) angles to each localized state at a constant distance to the center of the sphere (radius r) and calculating the distances between localized states.

The probability of the hole transfer to the ITO surface was limited by the difference between E_{dl} and the energy of the electron at the ITO surface (E_s) and by the distance between the hole and the ITO surface (d_{x0}). The value of d_{x0} depended on the position of the hole at the QD and the distance between the ITO surface and the sphere (δ) was the main limiting parameter.

The tunneling probabilities of the charge transfer steps were calculated for Miller–Abrahams tunneling (Equation (4.2)) with a constant reciprocal tunneling length (α_{tunn}). The bouncing frequency ($\nu_{0,tunn}$) was set to 10^{13} 1/s.

The electric field between the positive and negative charges was considered in the model (Equation (4.3)) by an approximation for the dielectric constant. For this purpose, an effective dielectric constant ($\varepsilon_{r,\mathrm{eff}}$) was introduced as an additional free parameter in the system. The value of $\varepsilon_{r,\mathrm{eff}}$ was assumed to be independent of the position of the hole, i.e. the variable volume of CdSe remaining between the hole and the ITO surface was not taken into account.

The minimum number of parameters for random walk simulations in the single QD approximation was given by r, δ, N_{ts}, α_{tunn}, $\nu_{0,\mathrm{tunn}}$, $\varepsilon_{r,\mathrm{eff}}$, E_{dl}, E_{d}, σ_{d}, E_{s} and E_{F}. For multi-parameter fitting of SPV transients with random walk simulations, the values of δ, N_{ts}, $\varepsilon_{r,\mathrm{eff}}$, E_{d}, σ_{d}, E_{s} and E_{F} were variable and the parameters r (0.225 nm), α_{tunn} (1/0.7 nm), $\nu_{0,\mathrm{tunn}}$ (10^{13} 1/s) and E_{dl} (0 eV) were kept constant.

As an example, Figure 5.31 shows measured and simulated SPV transients of 1 ML CdSe QDs (dithiol) at different temperatures during cooling for one common set of fitting parameters. At time longer than about 100 ns, the measured and simulated SPV transients were the same within the errors of measurement and simulation. Deviations between measured and simulated experiments at short times depended on temperature and could be reduced or eliminated by adding an exponential distribution to the states at the ITO surface (not shown).

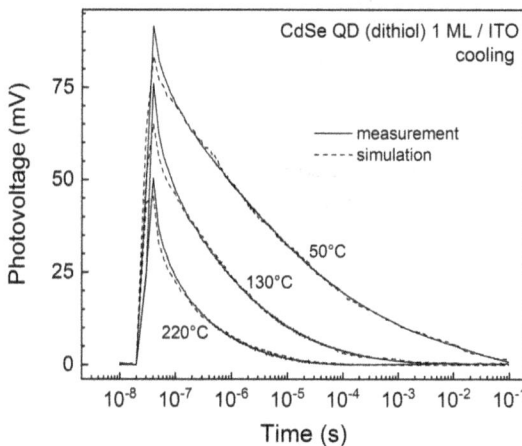

Figure 5.31. Measured (solid lines) and fitted by random walk simulations (dashed lines) SPV transients for 1 ML CdSe QDs (dithiol)/ITO measured at different temperatures during cooling.

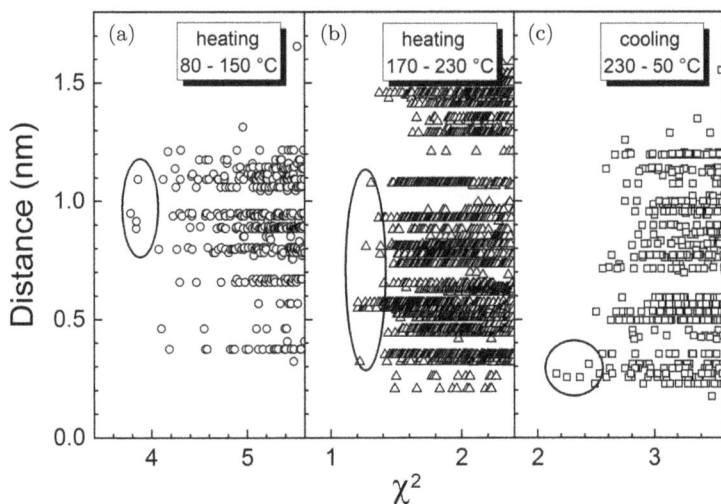

Figure 5.32. Plots of the fitted distance between the ITO surface and 1 ML CdSe QD (dithiol)/ITO as a function of χ^2 for heating from 80°C to 150°C (a) and from 170°C to 230°C (b) and for cooling from 230°C to 50°C (c).

The fact that measured SPV transients of 1 ML CdSe QDs (dithiol) could be well fitted with one common set of parameters for cooling starting at high temperatures gave evidence for a very stable configuration of distances and electronic states in the 1 ML CdSe QD (dithiol)/ITO system. This was not the case during heating. It followed from the fits of SPV transients during heating by random walk simulations that the distance between the QDs and the ITO surface was the most critical parameter in the system (Figure 5.32). The multi-parameter fits of SPV transients of 1ML CdSe QDs (dithiol) measured during heating did not result in a sharp distribution for the values of δ at low values of χ^2.

In order to get an impression about the evolution of δ during heating, SPV transients measured during heating between 80°C and 150°C and between 170°C and 230°C were fitted in two separate groups. During heating between 80°C and 150°C, the values of δ resulting in low values of χ^2 ranged between 0.85 and 1.1 nm (Figure 5.32(a)). During heating between 150°C and 230°C, the values of δ resulting in low values of χ^2 ranged between 0.3 and 1.1 nm with a tendency more to 0.3–0.55 nm (Figure 5.32(b)). In contrast, during cooling between 230°C and 50°C, the values of δ resulting in low values of χ^2 were distributed much sharper and amounted to about 0.28 nm (Figure 5.32(c)). Therefore, the average

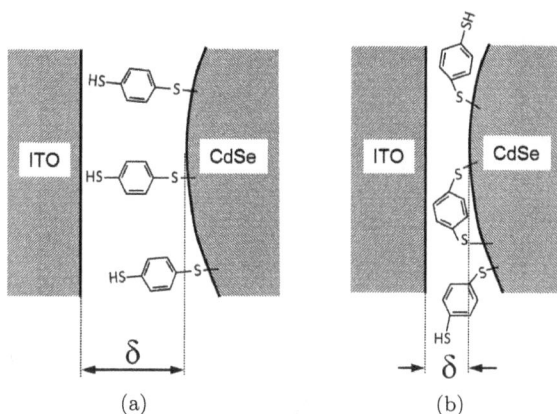

Figure 5.33. Schematic diagram of the bond configuration at an ITO/CdSe QD (dithiol) before (a) and after (b) heating to higher temperatures.

distance between CdSe QDs (dithiol) and ITO decreased strongly during heating by bending dithiol molecules out from the contact area and/or by forming two Cd-S bonds between dithiol molecules and the surface of CdSe QDs. The reduction of δ during heating to higher temperatures of 1 ML CdSe QDs (dithiol)/ITO is schematically illustrated in Figure 5.33.

Figure 5.34 shows the plots of E_d, σ_d, E_s and E_F as a function of χ^2 for cooling of 1 ML CdSe QDs (pyridine)/ITO after heating to 120°C (a) and for cooling of 1 ML CdSe QDs (dithiol)/ITO after heating to 230°C (b). The fitted values of E_d, σ_d, E_s and E_F resulted in sharp distributions at low values of χ^2. Incidentally, a value of δ of 0.4 nm resulted from the simulations of 1 ML CdSe QDs (pyridine) which was of the order of the dimension of a pyridine molecule. The values of $\varepsilon_{r,eff}$ were about 3.3 for 1 ML CdSe QDs (pyridine) and 3.0 for 1 ML CdSe QDs (dithiol).

The values of E_d, σ_d, E_s and E_F at the lowest χ^2 were equal to -0.21, 0.26, -0.02 and -0.91 eV for 1 ML CdSe QDs (pyridine)/ITO and -0.37, 0.18, 0.15 and -0.73 eV for 1 ML CdSe QDs (dithiol)/ITO, respectively.

The difference between E_s and E_F was practically the same for both samples (0.89 and 0.88 eV for 1 ML CdSe QDs (pyridine)/ITO and 1 ML CdSe QDs (dithiol)/ITO, respectively). It can be assumed that the Fermi energy of the highly doped ITO was in quasi-equilibrium with the cut-off energy in the CdSe QDs before photogeneration of one electron transferred toward the ITO and one hole remaining in the CdSe QD. Furthermore, it

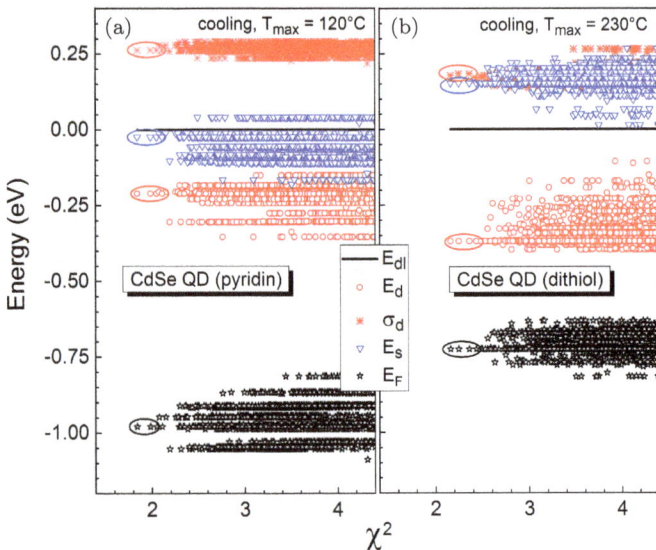

Figure 5.34. Plots of the fitted energy of the defect states (circles) and their standard deviation (double crosses), the energy states at the ITO surface (triangles) and Fermi energy (stars) as functions of χ^2 for 1 ML CdSe QDs (pyridine) during cooling after heating to 120°C (a) and for 1 ML CdSe QDs (dithiol)/ITO during cooling after heating to 230°C (b).

is reasonable to assume that E_F was pinned at the ITO surface. Therefore, it is not surprising that $E_s - E_F$ was unchanged.

Due to the pinning of E_F at the ITO surface, the bands were bended upward in energy and a depletion region with a width of 1–2 nm was formed which was thicker than the distance between the recombination states and the CdSe QDs. Therefore, the delocalized recombination state at the ITO surface was formed by the electrons in the valence band at the ITO surface. Incidentally, the consideration of the width of the depletion region for the charge separation length would result in a surface density of separated charge carriers reduced by 30–40%.

The difference between E_{dl} and E_s was −0.02 and 0.15 eV for 1 ML CdSe QDs (pyridine)/ITO and 1 ML CdSe QDs (dithiol)/ITO, respectively. This means that E_{dl} was 0.02 eV below and 0.15 eV above the valence band edge of ITO in the corresponding band diagram (Figure 5.35). The difference between E_s of the 1 ML CdSe QDs (pyridine)/ITO and 1 ML CdSe QDs (dithiol)/ITO systems was caused by the different surface work functions of both systems due to the different dipole moments of pyridine and dithiol.

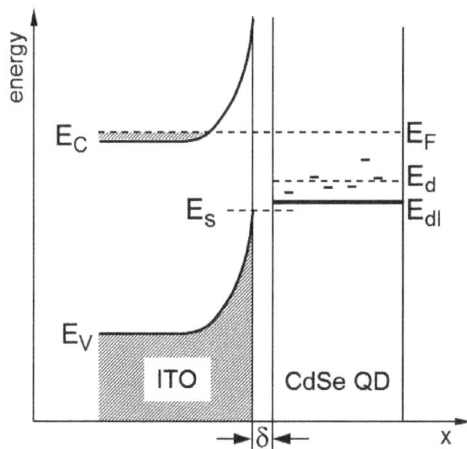

Figure 5.35. Schematic band diagram of the CdSe QD/ITO system with a delocalized recombination state at the ITO surface and localized and a delocalized states for hole transfer in the QD (Fengler and Dittrich, 2019).

The energy and the distributions of the localized states were different for the 1 ML CdSe QDs (pyridine)/ITO and 1 ML CdSe QDs (dithiol)/ITO systems. This was caused by different chemical interactions of different molecules at the CdSe surface.

5.5. Charge Separation in Doped Tin Oxide and Titanium Oxide

5.5.1. *Electron transfer over a barrier at highly doped tin oxide*

Highly doped metal oxide layers play an important role as transparent conductive oxides (TCOs), for example, in thin-film solar cells. Typical TCOs are indium-doped tin oxide (ITO), fluorine-doped tin oxide (SnO_2:F) or aluminum-doped zinc oxide. Furthermore, reference electrodes with SnO_2:F are often used in SPV measurement with a fixed capacitor (see Chapter 2). The density of ionized donors (N_D^+) in TCOs is of the order of 10^{21} cm^{-3}. For such a highly doped semiconductor, the width of the space charge region (SCR) at the surface of a TCO is expected to be of the order of 1–2 nm (see, for example, Sze, 1981).

The back transfer of electrons injected, for example, from dye molecules adsorbed at the surface of a TCO into the conduction band of a TCO is limited by tunneling through the SCR and/or by thermionic emission over

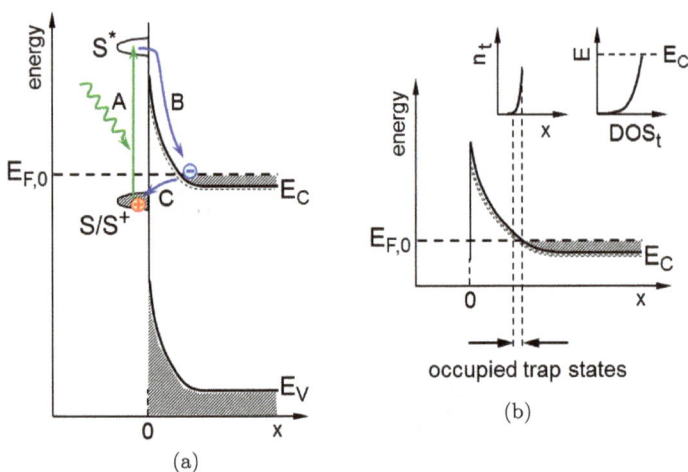

Figure 5.36. Schematic band diagram of the surface of a TCO with adsorbed dye molecules (a) and detail of the region of the conduction band of the band diagram with the distributions of occupied traps in space and of trap states in energy close to the conduction band edge (b). A, B and C denote photogeneration of electrons by injection from dye molecules, separation of electrons in the SCR and electron back transfer, respectively.

the barrier (Figure 5.36(a)). The tunneling length is of the order of the width of the space charge region.

Trap states in the SCR close to E_C and close to the bulk of a TCO are occupied. The (local) tunneling or charge separation length is reduced by the distance between an occupied trap and the edge of the SCR (Figure 5.36(b)). Due to the cut of the Fermi energy through defect states in space and due to the very low density of photogenerated charge carriers in relation to the density of ionized donors in the TCO, the distribution of trap states in energy is transformed to a distribution of occupied trap states in space.

Figure 5.37 shows SPV transients of a SnO_2:F sample sensitized with N3 dye molecules (Nazeeruddin *et al.*, 1993) measured at 28°C and 320°C. In contrast to the highly n-type doped c-Si, the SPV transients decayed much more slowly for the TCO. This can be caused by a higher barrier and/or by a reduced density of ionized donors near the surface due to chemical reactions with ambient gas molecules. The SPV transients depended strongly on temperature and could be well described by a power law (Equation (4.22)). The values of α_p amounted to 0.26 and 0.47 for 28°C and 320°C, respectively.

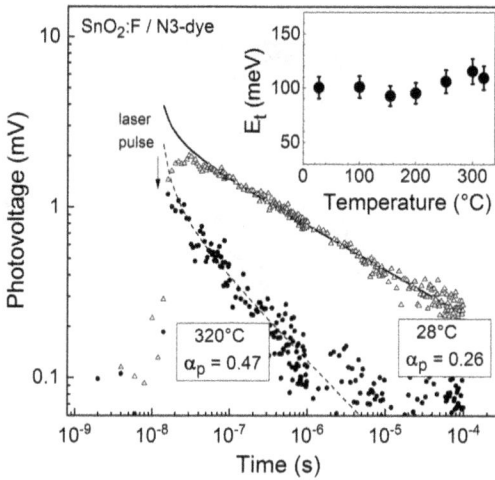

Figure 5.37. SPV transients on a double-logarithmic scale of a SnO_2:F /N3-dye sample measured at 28°C (triangles) and 320°C (circles) and temperature dependence of E_t (inset). Reevaluation of data from partially unpublished data and partially from Mora-Seró *et al.* (2006a).

A decay of SPV transients by a power law is characteristic for charge transfer limited by an exponential distribution. The corresponding values of E_t were obtained by using Equation (4.16) and are given as a function of temperature in the inset of Figure 5.37. The values of E_t were of the order of 100 meV.

As an example for photogeneration and charge separation in a SnO_2:F reference electrode, a plot of transient SPV spectroscopy is given in Figure 5.38 for a highly n-type doped c-Si sample in an extended range of photon energies of up to 4.4 eV.

Features caused by photogeneration and charge separation in the SnO_2:F reference electrode occurred in the plot of transient SPV spectroscopy at photon energies around 2.5–2.8 eV and at 3.5 eV. For comparison, the band gap of highly doped SnO_2:F is about 3.6 eV. The signs of the SPV signals related to SnO_2:F were negative and of the order of -1 mV (2.8 eV, 10^{-6}–10^{-4} s) and -3 to -4 mV (3.7–4.4 eV, 10^{-7}–10^{-1} s). The SPV signals related to fundamental absorption in the highly n-type doped c-Si sample decreased with increasing SPV signals, causing fundamental absorption in the SnO_2:F reference electrode.

The fact that SPV signals related to fundamental absorption in SnO_2:F did not decay within 0.1 s gives evidence for additional charge

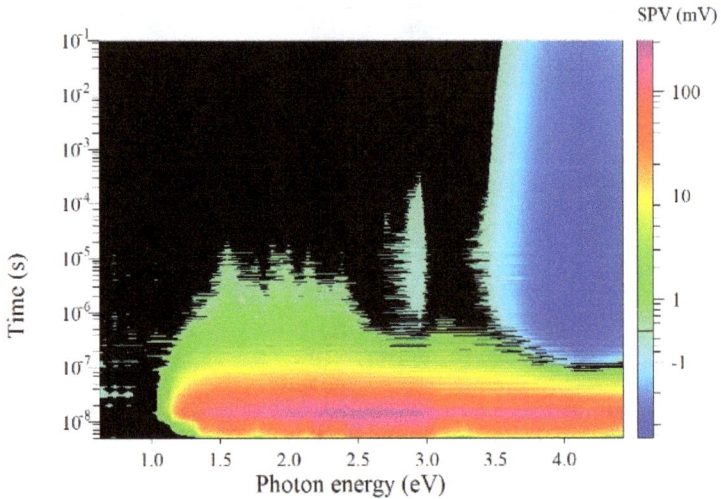

Figure 5.38. Plot of transient SPV spectroscopy of a highly n-type doped c-Si sample in a wide range of photon energies. Features setting on above 2.4 eV were caused by photogeneration and charge separation in the SnO_2:F reference electrode.

transfer into deep defect states at the SnO_2:F/SiO_2 interface. Therefore, possible charging of TCO reference electrodes under UV illumination has to be considered in SPV measurements. In principle, one can get rid of the SPV signals related to TCO electrodes since these signals are well reproducible. However, one shall be careful with such corrections since signals of investigated samples cannot be corrected to the light intensity reduced by absorption in the TCO layer.

5.5.2. *Charge transport in titanium oxide layers (anatase)*

Injection of electrons from dye molecules into TiO_2 was applied in order to study the temperature-dependent transport of electrons. Figure 5.39 shows SPV transients measured in high vacuum on a 100 nm thick compact TiO_2 (anatase) layer. The measurement cycle started with the virgin TiO_2 layer and the temperature was increased up to 300°C. The SPV signals increased within 10–20 ns, which corresponded to the resolution time of the given measurements, continued to increase for longer times and decayed after reaching the maximum.

The time at which the maximum SPV signals was reached (t_{peak}) is related to the dielectric relaxation time (Dittrich *et al.*, 2006). Incidentally, a direct correlation between t_{peak} and the dielectric relaxation time

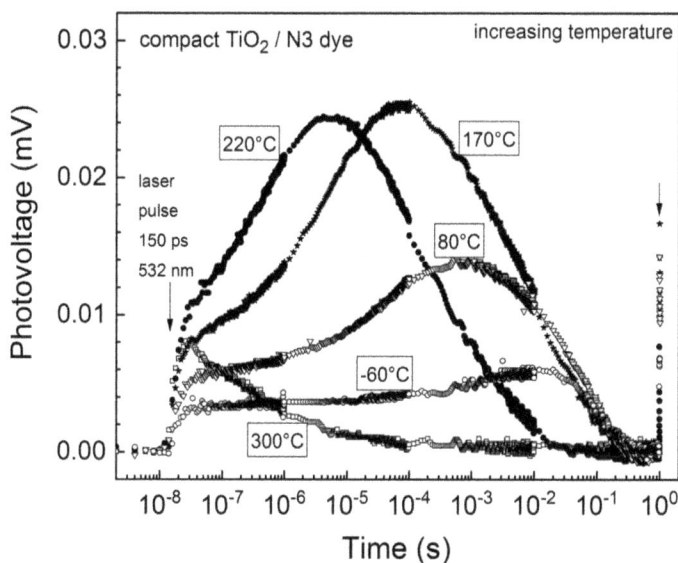

Figure 5.39. SPV transients for electrons injected from N3 dye molecules into a compact layer of TiO_2 (anatase, 100 nm thick) measured at different temperatures in high vacuum during a heating cycle starting from the virgin sample.

(see also Chapter 1) was shown for temperature-dependent SPV and current voltage measurements on mesoporous silicon (Duzhko *et al.*, 2002). The time at which the transients decayed the half of the maximum SPV signal ($t_{0.5}$) and t_{peak} decreased with increasing temperature corresponded to an increase of the conductivity.

Figure 5.40 shows the Arrhenius plot of $1/\tau_{0.5}$ for the compact TiO_2/N3 layer during the heating cycle starting from the virgin sample. For values of $1/\tau_{0.5}$ larger than about $10^2\,s^{-1}$, the activation energy (E_A) was of the order of 1.3 eV and tended to increase toward 1.5 eV at higher temperatures. For a Gaussian defect distribution, the activation energy of $1/\tau_{0.5}$ corresponds to the energy at the maximum (E_d, see Chapter 4). Therefore, the conductivity in the compact TiO_2 layer was limited by thermal emission from deep defect states around the middle of the band gap of anatase. These defects can be located, for example, at grain boundaries between anatase crystallites in the layer.

For $1/\tau_{0.5}$ below $10^2\,s^{-1}$, the activation energy was of the order of 0.1 eV. This low activation energy was limited by the repetition rate of the laser pulses which did not allow measurements of $\tau_{0.5}$ at times

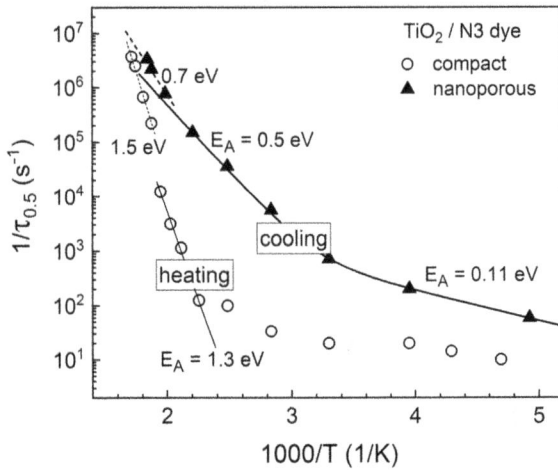

Figure 5.40. Arrhenius plot of $1/\tau_{0.5}$ for a compact TiO_2 (anatase)/N3 dye layer during a heating cycle starting from the virgin sample (circles) and for a nanoporous TiO_2 (anatase)/N3 dye layer during a cooling cycle starting from the maximum temperature of 270°C (triangles). The measurements were performed in high vacuum. Some values of activation energies are given. Data partially from Dittrich *et al.* (2006).

longer than 0.5–1 s and which caused a stationary occupation of deep trap states. Therefore, a low activation energy gives evidence for limitation of electron transport by emission from trap states closer to the conduction band edge under the condition that deep trap states are permanently occupied.

At higher temperatures, the SPV signals at short times were practically independent of temperature (8–10 mV). This means that the conditions for a fast initial charge separation, for example, across a grain boundary, did not change. However, the signals at short times decreased with decreasing temperature at the lower temperatures (about 6 and 3 mV at 80°C and −60°C, respectively). This additionally points to an increasing stationary reduction of charge in deep trap states with decreasing temperature.

Heating of metal oxides in vacuum caused their partial reduction. The positive charge in oxygen vacancies is compensated by free electrons. Therefore, heating of TiO_2 in vacuum causes non-reversible changes of the conductivity and therefore of SPV transients. This is demonstrated in Figure 5.41 for SPV transients measured at 200°C during the heating and during the cooling (maximum temperature was 300°C) cycles. The values of $\tau_{0.5}$ were $9 \cdot 10^{-4}$ and $1.2 \cdot 10^{-6}$ s for the measurements during heating

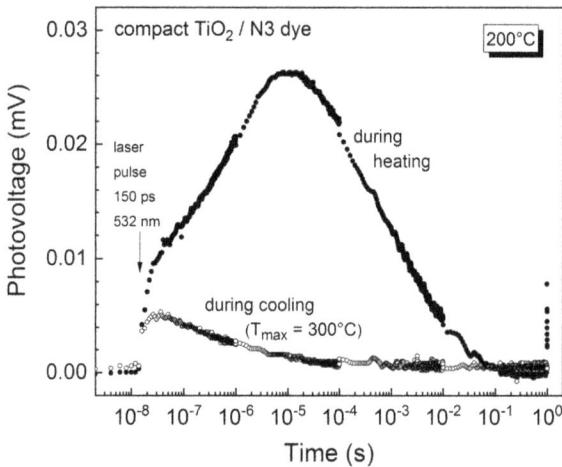

Figure 5.41. SPV transients of a compact TiO_2/N3 dye layer measured at $200°C$ during heating (filled circles) and cooling from $300°C$ (open circles) as an example for the non-reversible change of TiO_2 after heating in vacuum.

and cooling, respectively. Therefore, metal oxides shall be preconditioned at a temperature higher than the highest temperature in a temperature-dependent measurement.

Figure 5.40 also shows the Arrhenius plot of $1/\tau_{0.5}$, a nanoporous TiO_2 (anatase)/N3 dye layer for cooling after heating up to $270°C$ in high vacuum. During cooling, the values of E_A were about $0.8\,eV$ at higher temperatures, $0.5\,eV$ at intermediate temperatures and $0.11\,eV$ at lower temperatures. For comparison, preconditioning at $450°C$ in high vacuum resulted in an activation energy of $0.8\,eV$ for electronic transport in porous TiO_2 over the whole temperature range independent of the measurement technique, particle size and phase (Dittrich *et al.*, 2000). Therefore, a surface region with a high density of defects ($E_d = E_C - 0.8\,eV$) was formed during prolonged heating of porous TiO_2 at $450°C$ in high vacuum. Similar defects were also formed during heating at $270°C$ but at lower density.

5.5.3. *Role of oxidation and reduction in titanium oxide (rutile)*

The densities of titanium and oxygen vacancies in TiO_2 increase with increasing and decreasing $p(O_2)$, respectively. Vacancies of titanium and oxygen form acceptor and donor states in TiO_2, respectively (see

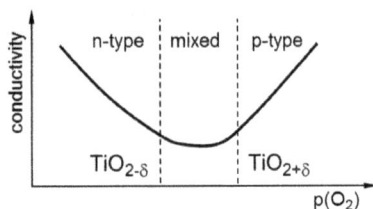

Figure 5.42. Schematic diagram of the dependence of the conductivity on $p(O_2)$ of a metal oxide with a n–p transition.

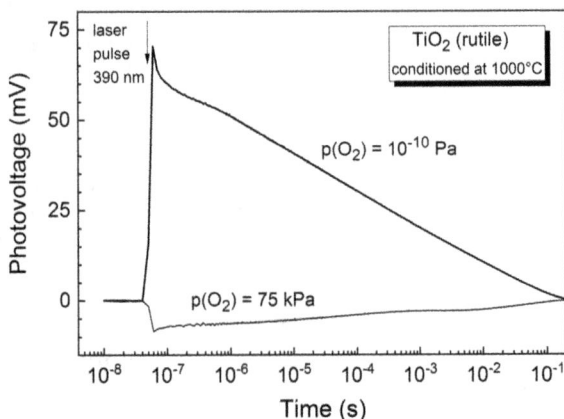

Figure 5.43. SPV transients of TiO_2 (rutile) samples conditioned at $1000°C$ and partial pressures of oxygen of 10^{-10} Pa (thick line) and $75\,kPa$ (thin line). Data from Nowotny and Dittrich (2009).

Nowotny *et al.*, (2006a and 2006b) and refs. therein). Therefore, TiO_2 can undergo an n-p transition depending on the $p(O_2)$ as shown schematically in Figure 5.42.

The rutile phase of TiO_2 is stable at high temperatures. Dense pellets of polycrystalline TiO_2 (rutile) can be therefore well preconditioned at high temperatures in a wide range of partial pressures of oxygen ($p(O_2)$). Figure 5.43 shows SPV transients of polycrystalline TiO_2 (rutile) samples preconditioned at $1000°C$ for $24\,h$ in high vacuum at $10^{-10}\,Pa$ (heavily reduced) and in oxygen atmosphere at $75\,kPa$ (oxidized). The transients were excited with a photon energy above the band gap of rutile (E_g(rutile) $= 3.06\,eV$). The SPV signals increased within the laser pulse and reached the maximum within the resolution time of the system. The maximum SPV signals were about $70\,mV$ and $-8\,mV$ for the sample

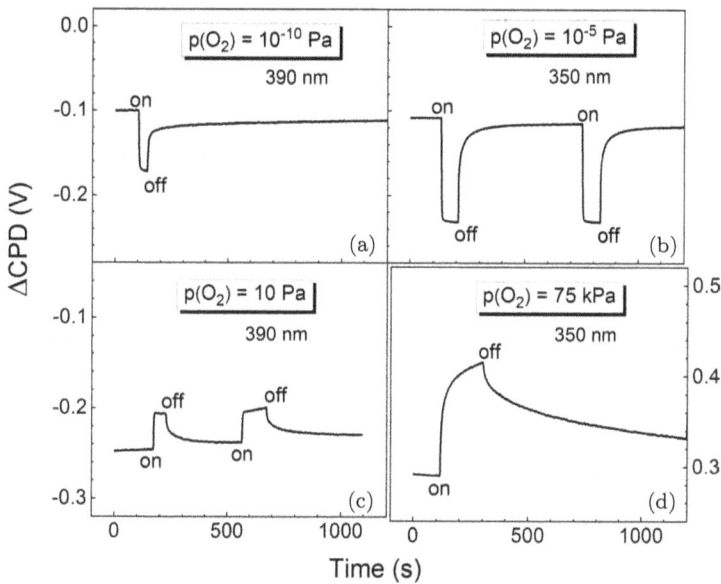

Figure 5.44. Illumination on–off cycles of ΔCPD of TiO$_2$ (rutile) samples conditioned at 1000°C and partial pressures of oxygen of 10^{-10}, 10^{-5} and 10 Pa and 75 kPa ((a)–(d), respectively). Data from Nowotny and Dittrich (2009).

preconditioned at 10^{-10} Pa and 75 kPa, respectively. Positive and negative SPV signals are typical for n- and p-type semiconductors with a depletion region near the surface due to surface band bending.

The decays of the SPV transients were limited by the repetition rate of the laser pulses, i.e. charging significantly influenced the values of the SPV signals. The influence of charging on maximum SPV signals can reduced by measuring ΔCPD during on–off cycling of illumination. This is demonstrated in Figure 5.44 for p(O$_2$) of 10^{-10} Pa, 10^{-5} Pa (moderately reduced), 10 Pa (slightly oxidized) and 75 kPa. The values of the change of ΔCPD were -75 mV (10^{-10} Pa, illumination at 390 nm), -120 mV (10^{-5} Pa, illumination at 350 nm), 42 mV (10 Pa, illumination at 390 mV) and 130 mV (75 kPa, illumination at 350 nm). Therefore, the SPV signals were positive for 10^{-10} and 10^{-5} Pa and negative for 10 Pa and 75 kPa which can be assigned to positive and negative band bending at the surface of n-type and p-type doped TiO$_2$, respectively. This is in agreement with the observations of p-type conductivity in rutile at room temperature (Nowotny *et al.*, 2010). Incidentally, a change of the sign of SPV signals was not observed for polycrystalline rutile films preconditioned in high vacuum or dry atmospheric air at 450°C and 400°C, respectively (Rothschild *et al.*, 2006). This is not surprising since high temperatures and long equilibration times are required for the extremely slow diffusion of titanium vacancies in TiO_2 (Nowotny *et al.*, 2006c).

The ΔCPD signals of the samples preconditioned at 10^{-10}, 10^{-5} and 10 Pa increased rapidly and reached the nearly maximum change within seconds after switching on illumination. After switching off illumination, the ΔCPD signals decreased rapidly up to a certain value but decreased very slowly within tens of minutes and hours in the following. The increase and decrease of ΔCPD after switching on and off illumination, respectively, were much slower for the sample preconditioned at 75 kPa than for the other samples. The relaxation of the ΔCPD signals after switching off illumination can be approximated by a superposition of a power law and stretched exponentials. Therefore, detrapping of electrons from a broad distribution of electronic states limited the relaxation of the disturbance of the space charge at long times.

Figure 5.45(a) shows the ΔCPD spectrum of the sample preconditioned at 10^{-10} Pa. For this sample, an onset of deep defect states around 2.2 eV and exponential tails with E_t equal to about 90 meV were found. Despite the sample being dark brownish only little signatures appeared in ΔCPD below E_g, i.e. charge separation from absorption via deep states was nearly absent in highly reduced rutile.

In contrast to the heavily reduced sample, numerous transitions were observed below E_g in the ΔCPD spectrum of the moderately reduced sample (Figure 5.45(b)). ΔCPD changed to more positive values (holes toward the surface) around 0.7, 1.2, 1.7 and 2.7 eV and to more negative values (electrons toward the surface) at 1.1, 1.5, 2.1 (strong) and 2.8 (pronounced) eV. The occupation of defect states can change within a surface space charge region. Therefore, it is reasonable to suppose that the transitions around 1.2 and 1.7 eV belong to the same deep defect state with energy of about 1.2 eV below E_C which is close to the ionization energy of the doubly ionized oxygen vacancy (Nowotny *et al.*, 2006d and references therein). The transition around 1.5 eV can be ascribed to the ionization energy of the titanium interstitials (Nowotny *et al.*, 2006d and references therein) the concentration of which depends only slightly on $p(O_2)$ (Nowotny *et al.*, 2006a). The change of ΔCPD setting on around 2.8 eV was ascribed to the pronounced onset of tail states. The fact for opposite directions of changes in ΔCPD gave evidence for

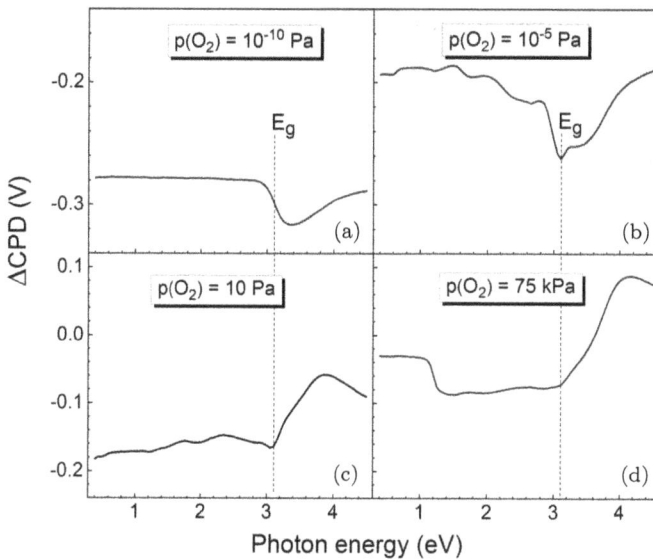

Figure 5.45. ΔCPD spectra of TiO$_2$ (rutile) samples conditioned at 1000°C and partial pressures of oxygen of 10^{-10}, 10^{-5} and 10 Pa and 75 kPa ((a)–(d), respectively). Illumination was performed with a Xe lamp. Measurements were performed in oxygen-free wet atmosphere. The dashed lines mark E_g of rutile.

charge separation due to photogeneration from defect states in the region of the built-in electric field and from defect states in a compensated surface layer.

At photon energies below E_g, the ΔCPD spectrum of the slightly oxidized sample (Figure 5.45(c)) changed to more negative values around 1.1 eV (slight change) and around 1.7–1.75, 2.3–2.4 and 2.95 eV and to more positive values around 1.2 and 1.9–2.0 eV. It is reasonable to suppose that the transitions around 1.2 and 1.9–2.0 eV belong to the same deep defect state with energy of about 1.2 eV above E_V which is very close to the energy of the titanium vacancy (Nowotny *et al.*, 2006d and references therein).

After preliminary illumination of the oxidized sample, the change of ΔCPD to more negative values was very pronounced at 1.1 eV and the change to more positive values around 2 eV appeared as well. Slight changes of ΔCPD to more positive and negative values were observed around 1.5 and 2.9 eV and around 1.7 and 2.7 eV, respectively. Therefore, charging of defect states caused a quite strong change of the distribution of the electric field in the compensated surface region.

5.6.　Local Charge Separation at Bismuth Vanadate Crystallites

5.6.1.　*Bismuth vanadate crystallites as photocatalysts*

The oxidizing and reducing chemical actions of photogenerated holes and electrons, respectively, are explored in photocatalytic systems for photocatalytic water splitting (see, for example, the review of Chu *et al.*, 2017). A photocatalyst is a photoactive material in which excess electrons and holes can be photogenerated and at the surface of which photocatalytic reactions can take place. Photocatalytic materials are, for example, TiO_2 (Fujishima and Honda, 1972) and bismuth vanadate ($BiVO_4$, Abdi *et al.*, 2013a; Ren *et al.*, 2019).

The surface area of photocatalysts is usually increased in order to increase the number of reactive surface sites. Furthermore, the access of photogenerated charge carriers to reactive surface sites is limited by the diffusion and/or drift lengths of photogenerated electrons and holes. The diffusion length in $BiVO_4$ layers is of the order of 70 nm (Abdi *et al.*, 2013b). The reduction of the size of photocatalytic crystallites such as in inorganic nanostructures (see, for example, Osterloh, 2013) and the implementation of cocatalytic nanoparticles (see, for example, Chen *et al.*, 2018) are common measures for increasing the photocatalytic activity and for reducing recombination losses of photogenerated electrons and holes.

Noble metal nanoparticles (gold (Au), platinum (Pt) and silver (Ag)) and metal oxide nanoparticles (manganese oxide (MnO_x), lead oxide (PbO_2) and cobalt oxide (CoO_x)) can be selectively deposited onto $\{010\}$ and $\{011\}$ facets of $BiVO_4$ crystallites, respectively (Li *et al.*, 2014). Figure 5.46 shows a schematic cross section of a $BiVO_4$ crystallite (a) and a micrograph of the top view of a $BiVO_4$ crystallite with Pt and MnO_x nanoparticles deposited selectively onto $\{010\}$ and $\{011\}$ facets. It was found that the amount of photocatalytic oxygen evolution increased by more than one order of magnitude after facet-selective deposition of Pt and CoO_x nanoparticles at the $\{010\}$ and $\{011\}$ facets of $BiVO_4$ crystallites, respectively (Li *et al.*, 2014).

The direct imaging of charge carriers separated in space at surfaces and interfaces of photocatalysts with SPV microscopy implemented into KPFM (see Chapter 2) is a powerful tool for studying the role of cocatalysts in photocatalysis (Chen *et al.*, 2018). This will be illustrated for $BiVO_4$ crystallites, the $\{010\}$ and $\{011\}$ facets of which were selectively covered

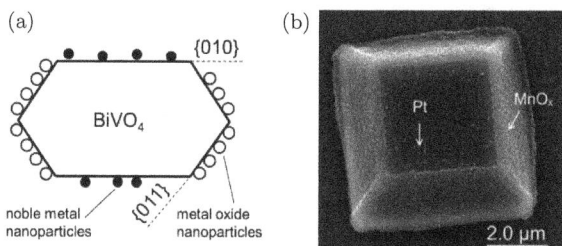

Figure 5.46. Schematic diagram of the cross-section of a $BiVO_4$ crystallite with noble metal and metal oxide nanoparticles deposited selectively onto {010} and {011} facets, respectively (a) and micrograph of the top view of a $BiVO_4$ crystallite with Pt and MnO_x nanoparticles deposited selectively onto {010} and {011} facets, respectively (b). (b) reprinted with permission from Zhu *et al.* (2017), Copyright (2017) American Chemical Society.

with Pt and MnO_x cocatalyst nanoparticles, respectively (Zhu *et al.*, 2015, 2017).

5.6.2. *Spatially resolved SPV on BiVO$_4$ crystallites with cocatalysts*

Figures 5.47(a) and 5.47(b) show KPFM images of a $BiVO_4$ crystallite with Pt and MnO_x nanoparticles deposited selectively onto {010} and {011} facets, respectively, in the dark (a) and under illumination (b) (Zhu *et al.*, 2017). Under illumination, the gray tones became darker at the {010} facet and brighter at the {011} facets, i.e. the CPD changed to more negative or positive values at the {010} and {011} facets, respectively, due to separation of photogenerated electrons (toward the {010} facet) and holes (toward the {011} facets).

A line scan of ΔCPD along the $BiVO_4$ crystallite is presented in Figure 5.47(c). ΔCPD was positive (up to about $100\,\text{mV}$) at the {011} facets and negative (up to about $-60\,\text{mV}$) at the {010} facet. The SPV spectra measured at the {010} and {011} facets are given in Figure 5.47(d). The SPV signals were positive ($100\,\text{mV}$ in the maximum) at the {011} facet and negative ($-75\,\text{mV}$ in the maximum) at the {010} facet, i.e. the total difference of the SPV signals across the $BiVO_4$ crystallite was about $175\,\text{mV}$. The onset energies (E_{on}, see also Chapter 2) of the SPV signals were roughly $2.5\,\text{eV}$ which was equal to the band gap of $BiVO_4$ (Stoughton *et al.*, 2013). The energy parameter characterizing the exponential tails

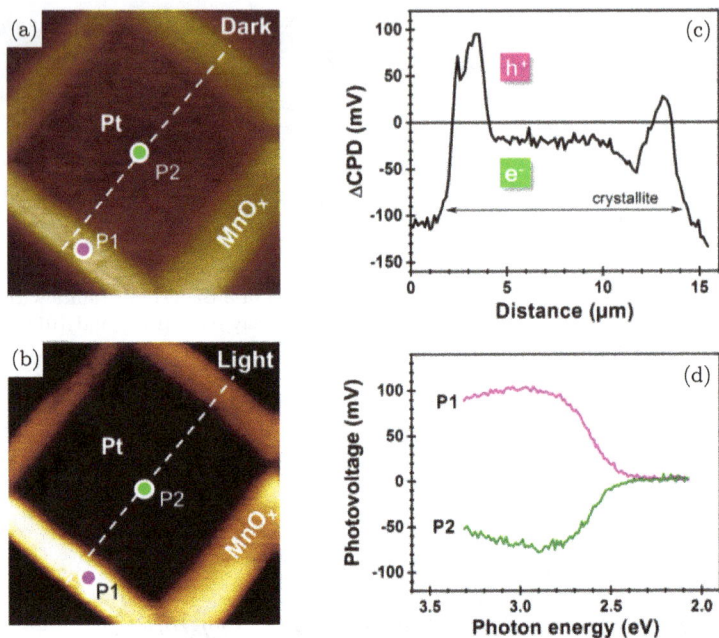

Figure 5.47. KPFM images of a $BiVO_4$ crystallite with Pt and MnO_x nanoparticles selectively deposited onto {010} and {011} facets, respectively, in the dark (a) and under illumination (b) and line scan of ΔCPD (c) and SPV spectra (d) measured along the dashed line and in the points P2 and P1 at the {010} and {011} facets, respectively, (green and pink lines, respectively) marked in (a) and (b). The scales of the gray tones in (a) and (b) ranged from 120 to 260 and from 100 to 300 mV, respectively. Adapted with permission from (Zhu *et al.*, 2017), Copyright (2017) American Chemical Society.

was about 60 meV which was relatively large in comparison to crystalline semiconductors and of the same order as for amorphous silicon (Cody *et al.*, 1981). The relatively large value of E_t gave evidence for the formation of defects close to the conduction and valence bands, for example, due to the evolution of stress near the surface during facet-selective deposition of Pt and/or MnO_x nanoparticles.

The SPV signals measured on the {011} and {010} facets of $BiVO_4$ crystallites depended strongly on the surface conditioning with MnO_x and Pt nanoparticles (Figure 5.48). At the bare and homogeneously covered with MnO_x nanoparticles {010} and {011} facets, the SPV signals were positive and low (about 3 and 6 mV for the {010} and 7 and 8 mV for the {011} facets, respectively). After deposition of MnO_x nanoparticles at {011} facets and after additional deposition of Pt nanoparticles at

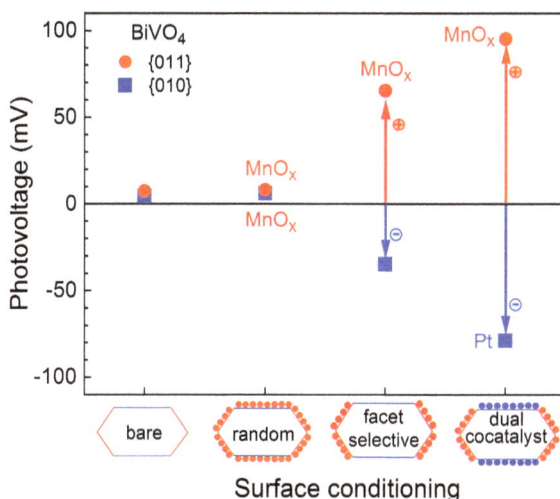

Figure 5.48. Dependence of the SPV signals at the {011} (circles) and {010} (squares) facets of $BiVO_4$ crystallites on the surface conditioning with MnO_x and Pt nanoparticles. Data values were taken from Zhu *et al.* (2017).

{010} facets, the SPV signals increased to about -35 and -80 mV at the {010} facets and to about 65 and 95 mV at the {011} facets, respectively. Therefore, pronounced local charge separation across $BiVO_4$ crystallites was only possible after local deposition of MnO_x nanoparticles at {011} facets and strongly enhanced after additional local deposition of Pt nanoparticles at {010} facets.

The SPV signals measured on the {011} facets of $BiVO_4$ crystallites depended strongly also on the diameter of the selectively deposited MnO_x nanoparticles (Figure 5.49). For MnO_x nanoparticles smaller than 30 nm, the SPV signals increased strongly with increasing size. The SPV signals measured on the {011} facets saturated around 85–90 mV for sizes of MnO_x nanoparticles equal or larger than about 50 nm.

The dependence of the O_2 evolution rate on the size of MnO_x nanoparticles selectively deposited onto {011} facets is also shown in Figure 5.49. The SPV signals and the O_2 evolution rates correlate very well for sizes of MnO_x nanoparticles smaller than 30–50 nm. The O_2 evolution rates decreased for larger sizes of MnO_x nanoparticles. A possible reason for the decrease of the O_2 evolution rate can be, in contrast to the SPV measurements, an increasing limitation by transport in the $BiVO_4/MnO_x/$electrolyte system.

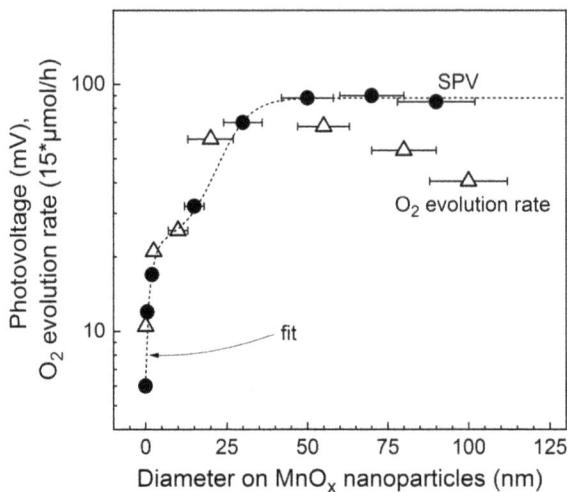

Figure 5.49. Dependencies of the SPV signals at {011} facets of BiVO$_4$ crystallites (circles) and of the O$_2$ evolution rate (triangles) on the diameter of MnO$_x$ cocatalyst nanoparticles selectively deposited onto {011} facets. The fit function is given by Equation (5.4). Re-evaluation of experimental data from Zhu *et al.* (2017).

The general behavior of the dependencies of the SPV and of the O$_2$ evolution rate (without the decrease) on the diameter of the MnO$_x$ nanoparticles (d_{MnO_x}) can be described by the following function:

$$
\text{SPV}(d_{MnOx}) = \text{SPV}(0) + \text{SPV}(1) \cdot \left[1 - \exp\left(-\frac{d_{MnOx}}{d_{MnOx,0}} \right) \right]
$$

$$
+ \frac{\text{SPV}(2)}{1 + \exp\left(-\frac{d_{MnOx} - d_{MnOx,1}}{d_{MnOx,2}} \right)} \tag{5.4}
$$

The three terms in SPV(d_{MnOx}) give evidence for the influence of three different phenomena on ΔCPD. The parameters SPV(0), SPV(1), SPV(2), $d_{MnOx,0}$, $d_{MnOx,1}$ and $d_{MnOx,2}$ were obtained by fitting and amounted to about 6 mV, 17 mV, 65 mV, 2 nm, 25 nm and 5 nm, respectively.

5.6.3. *Mechanism of charge separation on BiVO$_4$ with cocatalysts*

The spatial distribution of the SPV signals on the BiVO$_4$ crystallite with selectively deposited Pt and MnO$_x$ nanoparticles at the {010} and {011} facets, respectively, is displayed in Figure 5.50(a). Photogenerated electrons

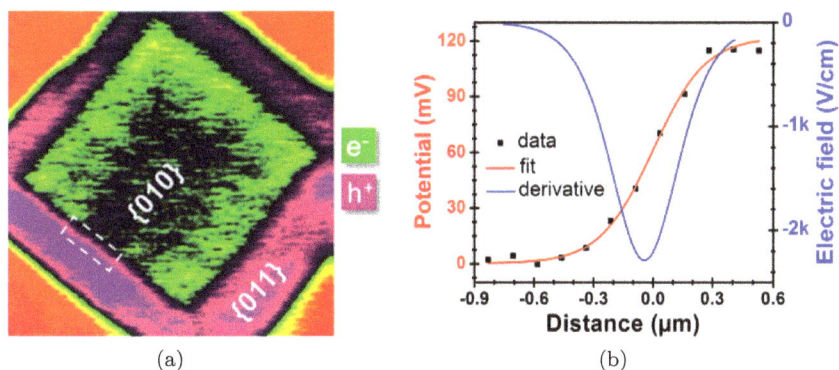

(a)   (b)

Figure 5.50. (a) Spatial distribution of the SPV signals at a $BiVO_4$ crystallite (a) shown in Figure 5.47 and (b) plots of the CPD (squares and red line) and the electric field (blue line) across the boundary between the {011} and {010} facets marked by the dashed square in (a). The color scale in (a) ranged from $-50\,\text{mV}$ (green) to $100\,\text{mV}$ (pink). Adapted with permission from Zhu *et al.* (2017), Copyright (2017) American Chemical Society.

and holes were separated toward the {010} and {011} facets, respectively, in a relatively sharp cut at the boundary between the {010} and {011} facets. Therefore, the boundary between the areas coated with Pt and MnO_x nanoparticles marks a charge-selective contact comparable to a pn-junction in a solar cell.

The potential across the boundary between {010} and {011} facets dropped within a range of 0.5–1 μm resulting in a maximum electric field of more than $2\,\text{kV/cm}$ (Figure 5.50(b)). Built-in electric fields of about half of this value can be assumed at the $BiVO_4$ surfaces on the areas of the {010} and {011} facets (see also Figure 5.48) due to opposite surface band bending.

Figure 5.51 shows schematic band diagrams of a $BiVO_4$ crystallite before (a) and after (b) facet-selective deposition of MnO_x and Pt nanoparticles. The increase of the SPV signals with increasing size of MnO_x nanoparticles deposited on {011} facets of $BiVO_4$ crystallites (see Figure 5.49) can be explained by increasing the surface band bending, i.e. MnO_x nanoparticles act as acceptor states the negative charge of which increases with increasing volume of MnO_x (see also Zhu *et al.*, 2017). Incidentally, Equation (5.4) gave evidence for two different kinds of acceptor states. The low value of $d_{\text{MnOx},0}$ of 2 nm showed that the initial growth of MnO_x leads to the formation of acceptor states at the $BiVO_4$ surface, whereas the values of $d_{\text{MnOx},1}$ and $d_{\text{MnOx},2}$ of 25 and 5 nm gave evidence

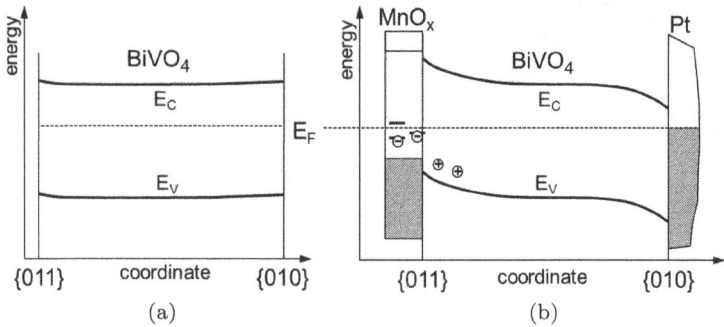

Figure 5.51. Schematic band diagrams of a $BiVO_4$ crystallite before (a) and after (b) facet-selective deposition of MnO_x and Pt nanoparticles.

for a logistic growth of negative charge trapped in acceptor states inside the MnO_x nanoparticles.

The work function of Pt is relatively high (5.7 eV, see, for example, Kiskinova *et al.*, 1983). The additional increase of the negative SPV signals after selective deposition of Pt nanoparticles on {010} facets of $BiVO_4$ crystallites (see Figure 5.48) gave evidence for the formation of local Schottky barriers with downward surface band bending between the Pt nanoparticles and the $BiVO_4$ surface (see also Zhu *et al.*, 2017).

Photogenerated electrons and holes drift through electric fields without recombining if the drift time is shorter than the lifetime. The drift length (d_{dr}) across the boundary between {010} and {011} facets can be estimated if taking into account a lifetime (τ) and a mobility (μ) of charge carriers in $BiVO_4$ thin films of 40 ns and 0.044 cm²/(Vs) (Abdi *et al.*, 2013b):

$$d_{dr} = \sqrt{\mu \cdot U \cdot \tau} \tag{5.5}$$

Depending on the potential drop (U), d_{dr} is of the order of 0.1–0.2 µm which is much larger than the diffusion length (0.07 µm; Abdi *et al.*, 2013). Furthermore, surface band bending acts in a similar way as back surface fields reducing recombination losses at contacts for the collection of majority charge carriers in solar cells. Therefore, the alignment of built-in electric fields at {010} and {011} facets of $BiVO_4$ crystallites by selectively deposited Pt and MnO_x nanoparticles drastically enhances separation and collection of photogenerated charge carriers which is crucial for the efficiency of photocatalytic reactions.

5.7. Summary

SPV techniques based on Kelvin probes and fixed capacitors have been applied for the characterization of numerous photoactive materials and for the investigation of different phenomena in those materials. Usually, SPV experiments demand an individual approach depending on the kind of samples and the method applied. Therefore, the application of SPV techniques for the characterization and analysis of photoactive materials could be demonstrated only for several selected examples. The focus of the selected examples was set to the characterization of surface and bulk states in semiconductors and in nanoparticles, to locally resolved SPV and to *in situ* SPV measurements under electrochemical surface engineering.

The occupation of surface states on hydrogenated and oxidized Si surfaces was varied with charge influenced by different bias potentials (U_{bias}). The resulting band bending was obtained from SPV transients in the large signal case. The density of surface states (D_{it}) was extracted as a function of the surface Fermi energy from the dependence of surface band bending on U_{bias}. This approach works well for relatively low values of D_{it} and in the range of depletion and weak inversion at a silicon surface. In case of E_F pinning at the silicon surface, a distribution of D_{it} cannot be obtained from SPV measurements under variation of U_{bias}. Surface states close to the conduction and valence band edges cannot be characterized by applying U_{bias}. Information about surface states closer to the conduction and valence band edges and in case of E_F pinning at semiconductor surface in general can be obtained by transient SPV spectroscopy.

"Fast" surface states were related to defect states directly at the silicon surface, for example, $-Si\equiv Si_3$ dangling bonds, so that photogenerated charge carriers could be directly captured without limitation by tunneling. Etching of silicon surfaces in hydrofluoric acid or mild oxidation led to preferential formation of "slow" donor (related to protonation of water molecules) or acceptor (related to hydroxide radicals) states, respectively, at a certain distance to the surface. Changes of surface band bending during chemical and electrochemical reactions were detected during electrochemical grafting of nitrobenzene molecules from diazonium salt solutions on silicon.

The specific influence of the SPV technique on the characterization of electronic states was shown for a $GaO_x/n\text{-}GaAs(S)/SI\ GaAs(Cr)$ sample for continuous, modulated and transient SPV spectroscopy. Numerous peculiarities, for example, were observed in the phase angle–PV amplitude plot of modulated SPV. In such plots, each peculiarity is caused by one specific process with its individual frequency and intensity dependence. Due to the limited frequency range in modulated SPV spectroscopy, continuous and transient SPV spectroscopy give complementary information about processes at time scales much longer or much shorter than the modulation period, respectively.

Electronic transitions can be changed over wide ranges in colloidal QDs. Electron injection from CdSe QDs into a very thin TiO_2 layer was used as a probe for the investigation of the absorption edge of QDs deposited from solutions with different dispersions in size by low-signal modulated SPV spectroscopy. Charge transfer across charge-selective CdTe QD/CdSe QD heterojunctions was demonstrated. Also, SPV experiments were also performed on charge separation in QD/dye molecules systems (Mora Seró *et al.*, 2008; 2010) and PbSe QDs/CdSe QD/TiO_2 systems (Gonzáles-Pedro *et al.*, 2014).

Due to their limited dimension and number of electronic states, colloidal QDs are nearly ideal systems for analysis with random walk simulations. This was shown for the extraction of the energy of states by multi-parameter fits of temperature-dependent SPV transients on monolayers of CdSe QD (pyridine) and CdSe QD (dithiol) deposited onto ITO.

Incidentally, random walk simulations were also applied for the analysis of modulated and transient SPV measurements on $BiVO_4$ powders and cold gas-sprayed layers (Fengler *et al.*, 2019). The role and the change of localized states at and near the surface were shown qualitatively by random walk simulations. However, a fitting of those data by random walk simulations was not possible since delocalized states in large crystalline regions could not be adequately considered in a model of a small system. The adequate combination of the worlds of localized and delocalized states is still challenging and reliable models are not available yet.

SPV signals can be caused by charge transfer across space charge regions in highly doped metal oxide layers such as SnO_2:F. Therefore, the influence of reference electrodes shall be considered in SPV measurements in the fixed capacitor arrangement. In metal oxides such as TiO_2 (anatase), reactions of oxidation and reduction depend strongly on the partial pressures of oxidizing and/or reducing molecules and temperature. This can cause, for example, non-reversible changes in SPV measurements and shall be taken into account for planning a series of measurements. Reversible conditioning of oxidized and reduced TiO_2 (rutile) was performed at fixed partial pressures of oxygen during prolonged heating at high temperature. Electronic states were distinguished in oxidized and reduced TiO_2 by measurements with continuous SPV spectroscopy.

Spatially resolved SPV measurements with a Kelvin probe force microscope were performed in order to analyze the specific role of cocatalysts for local charge separation across $BiVO_4$ crystallites. Photogenerated electrons and holes were separated toward {010} and {011} facets, respectively, onto which Pt and MnO_x nanoparticles were selectively deposited. The charge selectivity was caused by the high work function of Pt and the formation of acceptor states due to deposition of MnO_x nanoparticles. A correlation between lateral charge separation and the O_2 evolution rate was shown and the distribution of the electric field in the sample was estimated. Incidentally, lateral separation of photogenerated electrons and holes was also achieved across Cu_2O crystallites by removing locally a part of a surface region which had electrochemically modified defect states (Chen *et al.*, 2019).

The selected examples covered only a relatively small part of experiments in which photoactive materials were characterized by SPV measurements. Figure 5.52 gives an overview of some phenomena and materials, in part combined with each other, investigated by SPV techniques.

Hybrid organic–inorganic metal halide perovskites (MHP) can be applied in solar cells with very high efficiency and attracted a lot of attention for this reason. MHP of high electronic quality can be well deposited from precursor salt solutions and SPV techniques have been applied, for example, for the investigation of passivating (Supasai *et al.*, 2013) and charge-selective interfaces (Prajongtat and Dittrich, 2015; Kegelmann *et al.*, 2017; Prajongtat *et al.*, 2018; Brus *et al.*, 2018), of the temperature dependence of the bandgap of $CH_3NH_3PbI_3$ (Dittrich *et al.*, 2015), of the limitation of the diffusion lengths by grain boundaries (Shargaieva *et al.*, 2017), of degradation under illumination (Dittrich *et al.*, 2016b) and correlation of phase transitions with doping under annealing (Naikaew *et al.*, 2015).

Complex investigations of photocatalysts in combination with continuous SPV spectroscopy, for example, succinic acid-stabilized IrO_2 nanocrystals (Frame *et al.*, 2011), modulated SPV spectroscopy, for example, hematite layers prepared by a sol–gel process (Hermann-Geppert *et al.*, 2013), or transient SPV spectroscopy, for example, mesoporous $BiVO_4$ decorated with $V_{13}O_{16}$ (Ren *et al.*, 2019), contribute to a better understanding of processes in photocatalysis.

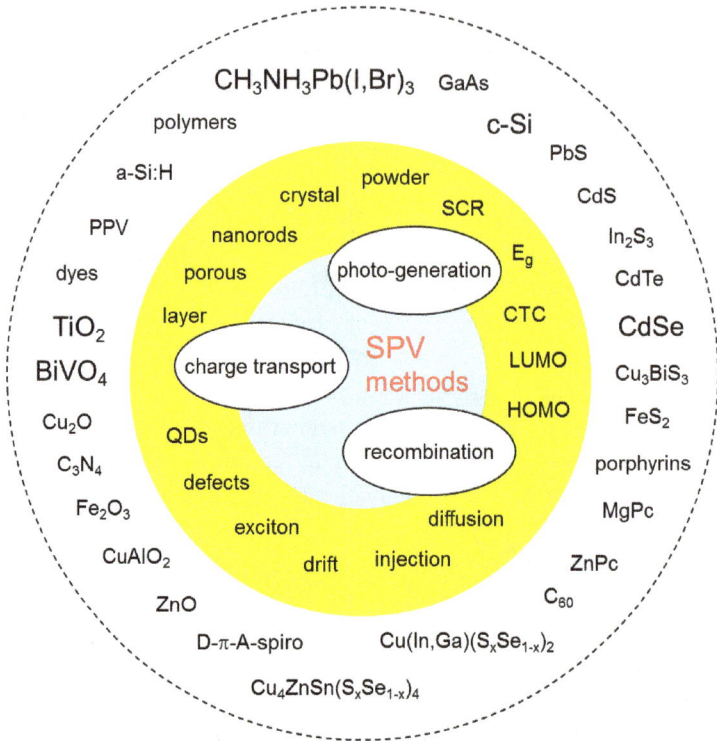

Figure 5.52. Overview of some phenomena and materials studied by SPV methods.

Furthermore, information about the dissociation of excitons, electronic transitions, charge transfer states and shifts of transitions due to interactions with substrates was obtained for different organic layers and molecules by SPV measurements. For example, continuous and modulated SPV spectroscopy measurements were performed on layers of conjugated organic molecules such as tetraphenyl-porphyrin evaporated onto metal oxide, metal and semiconductor surfaces (Zidon *et al.*, 2007a, 2007b, 2007c, 2008), layer systems with evaporated conjugated acceptor and donor molecules such as C_{60} and SubPc, respectively (Fengler *et al.*, 2015), layer systems of conjugated donor and acceptor electropolymers (for example, Durantini *et al.*, 2012; Belén Suarez *et al.*, 2014; Durantini *et al.*, 2015), on monolayers of directed donor-acceptor dyads implemented into spiro compounds (for example, Macor *et al.*, 2012; Dittrich *et al.*, 2013) and on electropolymer–QD layer systems (Otero *et al.*, 2015).

Bibliography

Abdi, F. F., Han, L., Smets, A. H. M., Zeman, M., Dam, B., van de Krol, R. (2013a), Efficient solar water splitting by enhanced charge separation in a bismuth vanadate-silicon tandem photoelectrode, *Nat. Commun.* 4, pp. 2195-1–2195-7.

Abdi, F. F., Savenije, T. J., May, M. M., Dam, B., van de Krol, R. (2013b), The origin of slow carrier transport in $BiVO_4$ thin film photoanodes: A time-resolved microwave conductivity study, *J. Phys. Chem. Lett.* 79, pp. 2752–2757.

Agmon, N. (1995), The Grotthuss mechanism, *Phys. Chem. Lett.* 244, pp. 456–462.

Ahuja, R., Arwin, H., Ferreira da Silva, A., Persson, C., Osorio-Guillién, J. M., Souza de Almeda, J., Moyses Araujo, C., Veje, E., Veissid, N., An, C. Y., Pepe, I., Johansson, B. (2002), Electronic and optical properties of lead iodide, *J. Appl. Phys.* 92, pp. 7219–7224.

Alex, V., Finkbeiner, S., Weber, J. (1996), Temperature dependence of the indirect energy gap in crystalline silicon, *J. Appl. Phys.* 79, pp. 6943–6946.

Alta Devices (2018), Alta Devices sets 29.1% solar efficiency record; NASA selects Alta Devices for International Space Station Test, *Press Release*, December 12, 2018, pp. 1–2.

Angermann, H., (2014), Conditioning of Si-interfaces by wet-chemical oxidation: Electronic interface properties study by surface photovoltage measurements, *Appl. Surf. Sci.* 312, pp. 3–16.

Anta, J. A., Mora-Seró, I., Dittrich, Th., Bisquert, J. (2007), Dynamics of charge separation and trap-limited electron transport in TiO_2 nanostructures, *J. Phys. Chem. C* 111, pp. 13997–1400.

Anta, J. A., Mora-Seró, I., Dittrich, Th., Bisquert, J. (2008), Interpretation of diffusion coefficients in nanostructured materials from random walk numerical simulation, *Phys. Chem. Chem. Phys.* 10, pp. 4478–4485.

261

Anta, J. A., Nelson, J., Quirke, N. (2002), Charge transport model for disordered materials: Application to sensitized TiO_2, *Phys. Rev. B* 65, pp. 125324-1–125324-10.

Barford, W. (2013), Excitons in conjugated polymers: A tale of two particles, *J. Phys. Chem. A* 117, pp. 2665–2671.

Bastide, S., Gal, D., Cahen, D., Kronik, L. (1999), Surface photovoltage measurements in liquids, *Rev. Sci. Instrum.* 70, pp. 4032–4036.

Belén Suarez, M., Durantini, J., Otero, L., Dittrich, Th., Santo, M., Milanesio, M. E., Durantini, E., Gervaldo, M. (2014), Electrochemical generation of porphyrin-porphyrin and porphyrin-C_{60} polymeric photoactive organic heterojunctions, *Electrochim. Acta* 133, pp. 399–406.

Beranek, R., Neumann, B., Sakthivel, S., Janczarek, M., Dittrich, Th., Tributsch, H., Kisch, H. (2007), Exploring the electronic structure of nitrogen-modified TiO_2 photocatalysts through photocurrent and surface photovoltage studies, *Chem. Phys.* 339, pp. 11–19.

Besocke, K., Berger, S. (1976), Piezoelectric driven Kelvin probe for contact potential difference studies, *Rev. Sci. Instrum.* 47, pp. 840–842.

Besocke delta phi, data sheet, http://www.besocke-delta-phi.de/Kelvin_probe_s_htm.

Binnig, G., Rohrer, H., Gerber, Ch., Weibel, E. (1982), Surface studies by tunneling microscopy, *Phys. Rev. Lett.* 49, pp. 57–61.

Bitzer, T., Gruyters, M., Lewerenz, H. J., Jacobi, K. (1993), Electrochemically prepared Si(111) 1x1-H surface, *Appl. Phys. Lett.* 63, pp. 397–399.

Blakemore, J. S., (1982), Semiconducting and other major properties of gallium arsenide, *J. Appl. Phys.* 53, pp. R123–R181.

Bönisch, S., Namaschk, B., Wittmaack, H.-J. (2004), The high impedance buffer V79 based on the OPA656 was developed, optimized and manufactured at the former Hahn-Meitner-Institute within the years between 2004 and 2008, unpublished.

Bolotin, K. I., Sikes, K. J., Stormer, H. L., Kim, P. (2008), Temperature-dependent transport in suspended graphene, *Phys. Rev. Lett.* 101, pp. 096802-1-096802-4.

Bröcker, D., Gießel, T., Widdra, W. (2004), Charge carrier dynamics at the $SiO_2/Si(100)$ surface: a time-resolved photoemission study with combined laser and synchrotron radiation, *Chem. Phys.* 299, pp. 247–251.

Brus, L. (1984), Electron-electron and electron-hole interactions in small semiconductor crystallites: The size dependence of the lowest excited state, *J. Chem. Phys.* 80, pp. 4403–4409.

Brus, V. V., Lang, F., Fengler, S., Dittrich, Th., Rappich, J., Nickel, N. H. (2018), Doping effects and charge transfer dynamics at hybrid perovskite/graphene interfaces, *Adv. Mats. Interfaces* 5, pp. 1800826-1–1800826-7.

Celik, D., Krueger, M., Veit, C., Schleiermacher, H. F., Zimmermann, B., Allard, S., Dumsch, I., Scherf, U., Rauscher, F., Niyamakom, P. (2012), Performance enhancement of CdSe nanorods-polymer based hybrid solar cells utilizing a novel combination of post-synthetic nanoparticle surface treatments, *Sol. En. Mat. Sol. Cells* 98, pp. 433–440.

Chen, R., Fan, F., Dittrich, Th., Li, C. (2018), Imaging photogenerated charge carriers on surfaces and interfaces pf photocatalysts with surface photovoltage microscopy, *Chem. Soc. Rev.* 47, pp. 8238–8262.

Chen, R., Pang, S., An, H., Dittrich, Th., Fan, F., Li, C. (2019), Giant defect-induced effects in nanoscale charge separation in semiconductor photocatalysts, *Nano Lett.* 19, pp. 426–432.

Chu, S., Yan, Y., Hamann, T., Shih, I., Wang, D., Mi, Z. (2017), Roadmap on solar water splitting: current status and future prospects, *Nano Futures* 1, pp. 022001-1–022001-29.

Clabes, J., Henzler, M. (1980), Determination of surface states on Si(111) by surface photovoltage spectroscopy, *Phys. Rev. B* 21, pp. 625–631.

Cody, G. D., Tiedje, T., Abeles, B., Brooks, B., Goldstein, Y. (1981), Disorder and the optical-absorption edge of hydrogenated amorphous silicon, *Phys. Rev. Lett*, 47, pp. 1480–1483.

Crowell, C. R. (1965), The Richardson constant for thermionic emission in Schottky barrier diodes, *Solid-State Electron.* 8, pp. 395–399.

Dember, H. (1932), Forward motion of electrons induced by light, *Physikalische Zeitschrift* 33, pp. 207–208.

De Wolf, S., Holovsky, J., Moon, S.-J., Löper, P., Niesen, B., Ledinsky, M., Haug, F.-J., Yum, J.-H., Ballif, C. (2014), Organometallic halide perovskites: sharp optical absorption edge and its relation to photovoltaic performance, *J. Phys. Chem. Lett.* 5, pp. 1035–1039.

Dittrich, Th. (2018), *Materials Concepts for Solar Cells*, 2nd ed., World Scientific, London, Singapore.

Dittrich, A., Angermann, H., Füssel, W., Flietner, H. (1993), Electronic properties of the HF-passivated Si(111) surface during the initial oxidation in air, *Phys. Stat. Sol. (a)* 149, pp. 463–470.

Dittrich, Th., Awino, C., Prajongtat, P., Rech, B., Lux-Steiner, M. Ch. (2015), Temperature dependence of the band gap of $CH_3NH_3PbI_3$ stabilized with PMMA: a modulated surface photovoltage study, *J. Phys. Chem. C* 119, pp. 23668–23972.

Dittrich, Th., Bitzer, T., Rada, T., Timoshenko, V. Yu., Rappich, J. (2002), Non-radiative recombination at reconstructed Si surfaces, *Solid-State Electronics* 46, pp. 1863–1872.

Dittrich, Th., Bönisch, S., Zabel, P., Dube, S. (2008), High precision differential measurement of surface photovoltage transients on ultrathin CdS layers, *Rev. Sci. Intrum.* 79, pp. 113903-1–113903-6.

Dittrich, Th., Burke, Th., Koch, F., Rappich, J. (2001), Passivation of an anodic oxide/p-Si interface stimulated by electron injection, *J. Appl. Phys.* 89, pp. 4636–4642.

Dittrich, Th., Dloczik, L., Guminskaya, T., Lux-Steiner, M. Ch. (2004), Photovoltage characterization of $CuAlO_2$ crystallites, *Appl. Phys. Lett.* 85, pp. 742–744.

Dittrich, Th., Duzhko, V. (2003), Photovoltage in free-standing mesoporous silicon layers, *Phys. Stat. Sol. (a)* 197, pp. 107–112.

Dittrich, Th., Duzhko, V., Koch, F., Kytin, V., Rappich, J. (2002), Trap-limited photovoltage in ultrathin metal oxide layers, *Phys. Rev. B* 65, pp. 155319-1–155319-5.

Dittrich, Th., Fengler, S., Franke, M. (2017), Transient surface photovoltage measurement over 12 orders of magnitude in time, *Rev. Sci. Intrum.* 88, pp. 053904-1–053904-7.

Dittrich, Th., Fiechter, S., Thomas, A. (2011), Surface photovoltage study of carbon nitride powder, *Appl. Phys. Lett.* 99, pp. 084105-1–084105-3.

Dittrich, Th., Garcia Vera, O., Pineda, S., Fengler, S., Bönisch, S. (2019), Extraction of source functions of surface photovoltage transients at very short times, *Rev. Sci. Instrum.* 90, pp. 026102-1–026102-3.

Dittrich, Th., Gonzáles, A., Rada, T., Rissom, T., Zillner, E., Sadewasser, S., Lux-Steiner, M. (2013), Comparative study of $Cu(In,Ga)Se_2/CdS$ and $Cu(In,Ga)Se_2/In_2S_3$ systems by surface photovoltage techniques, *Thin Solid Films* 535, pp. 357–361.

Dittrich, Th., Lang, F., Shargaieva, O., Rappich, J., Nickel, N. H., Unger, E., Rech, B. (2016a), Diffusion length of photo-generated charge carriers in layers and powders of $CH_3NH_3PbI_3$ perovskite, *Appl. Phys. Lett.* 109, pp. 073901-1–073901-3.

Dittrich, Th., Macor, L., Gervaldo, M., Fungo, F., Otero, L., Lin, C.-Y., Chi, L.-C., Fang, F.-C., Lii, S.-W., Wong, K.-T., Tsai, T.-H., Wu, C.-C. (2013), Charge separation in donor-acceptor spiro compounds at metal and metal oxide surfaces investigated by surface photovoltage, *J. Nanosci. Nanotechnol.* 13, pp. 5158–5163.

Dittrich, Th., Mora-Seró, I., García-Belmonte, G., Biquert, J. (2006), Temperature dependent normal and anomalous electron diffusion in porous TiO_2 studied by transient surface photovoltage, *Phys. Rev. B* 73, pp. 045407-1–045407-8.

Dittrich, Th., Muffler, H.-J., Vogel, M., Guminskaya, T., Ogacho, A., Belaidi, A., Strub, E., Bohne, W., Röhrich, J., Hilt, O., Lux-Steiner, M. Ch. (2005), Passivation of TiO_2 by ultra-thin Al-oxide, *Appl. Surf. Sci.* 240, pp. 236–243.

Dittrich, Th., Neumann, B. (2005), unpublished.

Dittrich, Th., Neumann, B., Tributsch, H. (2007), Sensitization via reversible $Ru(dcbpyH_2)_2(NCS)_2$-TiO_2 charge transfer complex, *J. Phys. Chem. C* 111, pp. 2265–2269.

Dittrich, Th., Schwartzkopff, M., Hartmann, E., Rappich, J. (1999), On the origin of the positive charge on hydrogenated Si surfaces and its dependence on the surface morphology, *Surf. Sci.* 437, pp. 154–162.

Dittrich, Th., Shargaieva, Lang, F., Nickel, N. H., Rech, B., Rappich, J. (2016b), Dependence of the transport length in $CH_3NH_3PbI_3$ powders on light soaking: a surface photovoltage study, *Proc. 32nd Europ. Photovolt. Sol. En. Conf. and Exhib. (EU-PVSEC)*, pp. 1104–1107.

Dittrich, Th., Valle Rios, L. E., Kapil, S., Gurieva, G., Rujisamphan, N., Schorr, S. (2017), Temperature dependent transient surface photovoltage spectroscopy of a $Cu_{1.95}Zn_{1.1}Sn_{0.96}Se_4$ kesterite single phase powder, *Appl. Phys. Lett.* 110, pp. 023901-1–023901-3.

Dittrich, Th., Weidmann, J., Timoshenko, V. Yu., Petrov, A. A., Koch, F., Lisachenko, M. G., Lebedev, E. (2000), Thermal activation of the electronic transport in porous titanium dioxides, *Mat. Sci. Engin. B* 69–70, pp. 489–493.

Dloczik, L., Tomm, Y., Könenkamp, R., Lux-Steiner, M. C., Dittrich, Th. (2004), CuAlO$_2$ prepared by ion exchange from LiAlO$_2$, *Thin Solid Films* 451–452, pp. 116–119.

Dumas, P., Chabal, Y. J., Jakob, P. (1992), Morphology of hydrogen-terminated Si(111) and Si(100) surfaces upon etching in HF and buffered-HF solutions, *Surf. Sci.* 269–270, pp. 867–878.

Durantini, J., Morales, G. M., Santo, M., Funes, M., Durantini, E. N., Fungo, F., Dittrich, Th., Otero, L., Gervaldo, M. (2012), Synthesis and characterization of porphyrin electrochromic and photovoltaic electropolymers, *Org. Electron.* 13, pp. 604–614.

Durantini, J., Suarez, M. B., Santo, M., Durantini, E., Dittrich, Th., Otero, L., Gervaldo, M. (2015), Photoinduced charge separation in organic-organic heterojunctions based on porphyrin electropolymers. Spectral and time dependent surface photovoltage study, *J. Phys. Chem. C* 119, pp. 4044–4051.

Duzhko, V., Dittrich, Th. (2003), Photovoltage in free-standing mesoporous silicon layers, *Phys. Stat. Sol. (a)* 197, pp. 107–112.

Duzhko, V., Dittrich, Th., Kamenev, B., Timoshenko, V. Yu., Brütting, W. (2001), Diffusion photovoltage in poly(p-phenylenevinylene), *J. Appl. Phys.* 89, pp. 4410–4412.

Duzhko, V., Koch, F., Dittrich, Th. (2002), Transient photovoltage and dielectric relaxation time in porous silicon, *J. Appl. Phys.* 91, pp. 9432–9434.

Duzhko, V., Timoshenko, V. Yu., Koch, F., Dittrich, Th. (2001b), Photovoltage in nanocrystalline TiO$_2$, *Phys. Rev. B* 64, pp. 075204-1–075204-7.

Eckhardt, H., Shacklette, L. W., Jen, K. Y., Eisenbaumer, R. L. (1989), The electronic and electrochemical properties of poly(phenylene vinylenes) and poly(thienylene vinylenes): An experimental and theoretical study, *J. Phys. Chem.* 91, pp. 1303–1315.

Eichberger, R., Willig, F. (1990), Ultra-fast electron injection from excited dye molecules into semiconductor electrodes, *Chem. Phys.* 141, pp. 159–173.

Fengler, S., Dittrich, Th. (2016), Algorithm for random walk simulation of modulated surface photovoltage signals in nanostructured systems with localized states, *J. Phys. Chem. C* 120, pp. 17777–17783.

Fengler, S., Dittrich, Th., Rusu, M. (2015), Electronic transitions and band offsets in C$_{60}$:SubPc and C$_{60}$:MgPc on MoO$_3$ studied by modulated surface photovoltage spectroscopy, *J. Appl. Phys.* 118, pp. 035501-1–035501-7.

Fengler, S., Dittrich, Th., Schieda, M., Gutzmann, H., Emmler, T., Villa-Vidaller, M., Klassen, T. (2019), Characterization of BiVO$_4$ powders and cold gas sprayed layers by surface photovoltage techniques, *Catalysis Today* 321–322, pp. 34–40.

Fengler, S., Zillner, E., Dittrich, Th. (2013), Density of surface states at CdSe quantum dots by fitting temperature-dependent surface photovoltage transients with random walk simulations, *J. Phys. Chem. C* 117, pp. 6462–6468.

Fossum, J. G., Mertens, R. P., Lee, D. S., Nus, J. F. (1983), Carrier recombination and lifetime in highly doped silicon, *Solid-State Electronics* 26, pp. 569–576.

Frame, A. A., Townsend, T. K., Chamousis, R. L., Sabio, E. M., Dittrich, Th., Browning, N. D., Osterloh, F. E. (2011), Photocatalytic water oxidation with nonsensitized IrO_2 nanocrystals und visible and UV light, *J. Am. Chem. Soc.* 133, pp. 7264–7267.

Franz, W. (1958), Einfluß eines elektrischen Feldes auf eine optische Absorptionskante, *Zeitschr. Naturforsch. A* 13, pp. 484–489.

Frenkel, J. (1933), Conduction in poor electronic conductors, *Nature* 132, pp. 312–313.

Fujishima, A., Honda, K. (1972), Electrochemical photolysis of water at a semiconductor surface, *Nature*, 238, pp. 37–38.

Funes, M., Zabel, P., Dittrich, Th., Durantini, E. N., Otero, L. (2010), Interaction induced transition in the nanoporous TiO_2/Pd-porphyrin system, *Phys, Stat. Sol. C* 7, pp. 280–283.

Füssel, W., Schmidt, M., Angermann, H., Mende, G., Flietner, H. (1996), Defects at the Si/SiO_2 interface: their nature and behavior in technological processes and stress, *Nucl. Instrum. Meth. Phys. Res. A* 377, pp. 177–183.

Garrett, C. G. B., Brattain, W. H. (1955), Physical theory of semiconductor surfaces, *Phys. Rev.* 99, pp. 376–387.

Glatzel, T., Sadewasser, S. (eds.) (2018), *Kelvin Probe Force Microscopy — From Single Charge Detection to Device Characterization*, Springer, Cham.

Goodman, A. M. (1961), A method for the measurement of short minority carriers diffusion lengths in semiconductors, *J. Appl. Phys.* 32, pp. 2550–2552.

Gonzáles-Pedro, V., Sima, C., Marzari, G., Boix, P. P., Giménez, S., Shen, Q., Dittrich, Th., Mora-Seró, I. (2013), High performance PbS quantum dot sensitized solar cells exceeding 4% efficiency: the role of metal precursors in the electron injection and charge separation, *Phys. Chem. Chem. Phys.* 15, pp. 13835–13843.

Griffith, D. J. (2004), *Introduction into Quantum Mechanics*, 2nd ed., Prentice Hall, Singapore.

Gross, D., Mora-Seró, I., Dittrich, Th., Belaidi, A., Mauser, C., Houtepen, A. J., Da Como, E., Rogach, A. L., Feldmann, J. (2010), Charge separation in type II tunneling multilayered structures of CdTe and CdSe nanocrystals directly proven by surface photovoltage spectroscopy, *J. Am. Chem. Soc.* 132, pp. 5981–5983.

Gur, I., Fromer, N. A., Geier, M. L., Alivisatos, A. P. (2005), Air-stable all-inorganic nanocrystals solar cells processed from solution, *Science* 310, pp. 462–465.

Guyot-Sionnest, P., Wang, C. (2003), Fast voltammetric and electrochromic response of semiconductor nanocrystal thin films, *J. Phys. Chem. B* 107, pp. 7355–7359.

Gysin, U., Glatzel, T., Schmölzer, T., Schöner, A., Rehsanov, S., Bartolf, H., Meyer, E. (2015), Large area scanning probe microscope in ultra-high vacuum demonstrated for electrostatic force measurements on high-voltage devices, *Beilstein J. Nanotechnol.*, 6, pp. 2485–2497.

Hagfeldt, A., Grätzel, M. (1995), Light-induced redox reactions in nanocrystalline systems, *Chem. Rev.* 95, pp. 49–68.

Hall, R. N. (1952), Electron-hole recombination in germanium, *Phys. Rev.* 87, pp. 387.

Hamam, K. J., Alomari, M. I. (2017), A study of the optical band gap of zinc phthalocyanine nanoparticles using UV-Vis spectroscopy and DFT function, *Appl. Nanosci.* 7, pp. 261–268.

Hamers, R. J., Cahill, D. G. (1991), Ultrafast time resolution in scanned probe microscopies: surface photovoltage on Si(111)-7x7, *J. Vac. Sci. Technol. B* 9, pp. 514–518.

Hartig, P., Dittrich, Th., Rappich, J., (2002a), Surface dipole formation and non-radiative recombination at p-Si(111) surface during electrochemical deposition of organic layers, *J. Eletcroanal. Chem.* 524–525, pp. 120–126.

Hartig, P., Rappich, J., Dittrich, Th. (2002b), Engineering of Si surfaces by electrochemical grafting of p-nitrobenzene molecules, *Appl. Phys. Lett.* 80, pp. 67–69.

Heilig, K. (1968), Messungen der Oberflächen-Photospannung am Si und Ge, *Experimentelle Technik der Physik* 16, pp. 135–148.

Heilig, K. (1974), Determination of surface properties by means of large signal photovoltage pulses and the influence of trapping, *Surf. Sci.* 44, pp. 421–437.

Heilig, K., Flietner, H., Reineke, J. (1979), Investigation of energetic surface state distributions at real surfaces of silicon after treatment with HF and H_2O using large-signal photovoltage pulses, *J. Phys. D: Appl. Phys.* 12, pp. 927–940.

Henzler, M., Göpel, W. (1991), *Oberflächenphysik des Festkörpers*, Teubner Studienbücher Physik, B. G. Teubner, Stuttgart.

Henzler, M., Clabes, J. (1974), Structural and electronic properties of stepped semiconductor surfaces, *Japan. J. Appl. Phys. Suppl.* 2, pp. 389–396.

Hermann-Geppert, I., Bogdanoff, P., Radnik, J., Fengler, S., Dittrich, Th., Fiechter, S. (2013), Surface aspects of sol–gel derived hematite films for the photoelectrochemical oxidation of water, *Phys. Chem. Chem. Phys.* 15, pp. 1389–1398.

Higashi, G. S., Becker, R. S., Chabal, Y. J., Becker, A. J. (1991), Comparison of Si(111) surfaces prepared using aqueous solutions of NH4F versus HF, *Appl. Phys. Lett.* 58, pp. 1656–1658.

Hoex, B., Gielis, J. J. H., van de Sanden, M. C. M., Kessels, W. M. M.. (2008), On the c-Si surface passivation mechanism by the negative-charge-dielectric Al_2O_3, *J. Appl. Phys.* 104, pp. 113703-1–113703-7.

Horiuchi, N. (2010), Milestone 15: Lasers for the masses, *Nature Milestones/ Photons*, May 2010, p. 1.

Hou, X., Wang, X., Liu, B., Wang, Q., Wang, Z., Chen, D., Shen, G. (2014), SnO_2@TiO_2 heterojunction nanostructures for lithium-ion batteries and self-powered UV photodetectors with improved performances, *ChemElectroChem* 1, pp. 108–115.

Jacoboni, C., Canali, C., Ottaviani, G., Alberigi Quarante, A. (1977), A review of some charge transport properties of silicon, *Solid-State Electron.* 20, pp. 77–89.

Jaegermann, W. (1986), Adsorption of Br_2 on n-$MoSe_2$: Modelling photoelectrochemistry in the UHV, *Chem. Phys. Lett.* 126, pp. 301–305.

Johnson, E. O. (1957), Measurement of minority carrier lifetimes with the surface photovoltage studies with germanium, *J. Appl. Phys.* 28, pp. 1349–1353.

Johnson, E. O. (1958), Large-signal surface photovoltage studies with germanium, *Phys. Rev.* 111, pp. 153–166.

Jones, M., Nedeljkovic, J., Ellingson, R. J., Nozik, A. J., Rumbles, G. (2003), Photoenhancement of luminescence in colloidal CdSe quantum dot solutions, *J. Phys. Chem. B* 107, pp. 11346–11352.

Juma, A. O., Azarpira, A., Steigert, A., Pomaska, M., Fischer, C.-H., Lauermann, I., Dittrich, Th. (2013), Role of chlorine in In_2S_3 for band alignment at nanoporous-TiO_2/In_2S_3 interfaces, *J. Appl. Phys.* 114, pp. 053711-1–053711-5.

Kalita, J. M., Wary, G. (2016), Estimation of band gap of muscovite mineral using thermoluminescence (TL) analysis, *Physica B* 485, pp. 53–59.

Kasupke, N., Henzler, M. (1976), Surface state properties of clean cleaved silicon as derived from the temperature dependence of the surface photovoltage, *Surf. Sci.* 54, pp. 111–120.

Kato, Y., Mori, Y., Morizane, K. (1979), Chromium-doped GaAs vapor phase epitaxy, *J. Cryst. Growth* 47, pp. 12–20.

Kegelmann, L., Wolff, C. M., Awino, C., Lang, F., Unger, E. L., Korte, L., Dittrich, Th., Neher, D., Rech, B., Albrecht, S. (2017), It takes two to tango – double-layer selective contacts in perovskite solar cells for improved device performance and reduced hysteresis, *ACS Appl. Mater. Interfaces* 9, pp. 17245–17255.

Keldysh, L. V. (1958), The effect of a strong electric field on the optical properties of insulating crystals, *Soviet Physics JETP* 34, pp. 788–790.

Keldysh, L. V. (1979), Coulomb interaction in thin semiconductor and semimetal films, *Pisma Zh. Eksp., Teor. Fiz.* 29, pp. 716–719.

Kiskinova, M., Pirug, G., Bonzel, H. P. (1983), Coadsorption of potassium and CO on Pt(111), *Surf. Sci.* 133, pp. 321–343.

Kobayashi, E., Boccard, M., Jeangros, Q., Rodkey, N., Vresilovic, D., Hessler-Wyser, A., Döbeli, M., Franta, D., De Wolf, S., Morales-Masis, M., Ballif, C. (2018), Amorphous gallium oxide grown by low-temperature PECVD, *J. Vac. Sci. Technol. A* 36, pp. 021518-1–021518-7.

Kronik, L., Shapira, Y. (1999), Surface photovoltage phenomena: theory, experiment and applications, *Surf. Sci. Rep.* 37, pp. 1–206.

Lagowski, J., Baltov, I., Gatos, H. C. (1973), Surface photovoltage spectroscopy and surface piezoelectric effect in GaAs, *Surf. Sci.* 40, pp. 216–226.

Lam, Y. W. (1971), Surface state density and surface potential in MIS capacitors by surface photovoltage measurements. I, *J. Phys. C: Appl. Phys.* 4, pp. 130–135.

Lauinger, T., Schmidt, J., Aberle, A. G., Hezel, R. (1996), Record low surface recombination velocities on 1 Ωcm p-silicon using remote plasma silicon nitride passivation, *Appl. Phys. Lett.*, 68, pp. 1232–1234.

Lebedev, E., Dittrich, T., Petrova-Koch, V., Karg, S., Brütting, W. (1997), Charge carrier mobility in poly-(p-phenylenevinylene) studied by time-of-flight technique, *Appl. Phys. Lett.* 71, pp. 2686–2688.

Lennard-Jones, J. E. (1931), Cohesion, *Proc. Phys. Soc.* 43, pp. 461–482.

Li, R., Zhang, F., Han, H., Zhang, F., Wang, D., Li, C. (2014), Highly efficient photocatalysts constructed by rational assembly of dual-cocatalysts separately on different facets of $BiVO_4$, *Energy Environ. Sci.* 7, pp. 1369–1376.

Li, R., Zhang, F., Wang, D., Yang, J., Li, M., Zhu, J., Zhou, X., Han, H., Li, C. (2013), Spatial separation of photogenerated electrons and holes among {010} and {110} facets of $BiVO_4$, *Nature Commun.* 4, pp. 1432-1–1432-7.

Lin, X., Ennaoui, A., Levcenko, S., Dittrich, Th., Kavalakkatt, J., Kretschmar, S., Unold, T., Lux-Steiner, M. Ch. (2015), Defect study of $Cu_2ZnSn(S_xSe_{1-x})_4$ thin film absorbers using photoluminescence and modulated surface photovoltage spectroscopy, *Appl. Phys. Lett.* 106, pp. 013903-1–013903-3.

Lin, X. Z., Dittrich, Th., Fengler, S., Lux-Steiner, M. Ch., Eannaoui, A. (2013), Correlation between processing conditions of $Cu_2ZnSn(S_xSe_{1-x})_4$ and modulated surface photovoltage, *Appl. Phys. Lett.* 102, pp. 143903-1–143903-3.

Lord Kelvin, G. C. V. O. D. C. L. LL. D. F. R. S. M. R. I. (1898), V. Contact electricity of metals, *Phil. Mag. Ser.* 5, pp. 82–120.

Lukeš, F. (1972), Oxidation of Si and GaS in air at room temperature, *Surf. Sci.* 30, pp. 91–100.

Macor, L., Gervaldo, Fungo, F., Otero, L., Dittrich, Th., Lin, C.-Y., Chi, C. C., Fang, F.-C., Lii, S.-W., Wong, K.-T., Tsai, C.-H., Wu, C.-C. (2012), Photoinduced charge separation in donor-acceptor spiro compounds at metal and metal oxide surfaces: application in dye-sensitized solar cell, *RSC Adv.* 2, pp. 4869–4878.

Many, A., Goldstein, Y., Grover, N. B. (1965), *Semiconductor Surfaces*, North-Holland Publishing Company, Amsterdam.

Margaritondo, G., Brillson, L. J., Stoffel, N. G. (1980), Secondary illumination effects in the photoemission spectra of GaAs and CdS, *Solid state Commun.* 35, 277–280.

Marshall, J. M., Street, R. A., Thompson, M. J. (1986), Electron drift mobility in amorphous Si:H, *Phil. Mag. B* 54, 51–60.

McClung, F. J., Hellwarth, R. W. (1961), Giant optical pulsations from Ruby, *J. Appl. Phys.* 33, pp. 828–830.

Melitz, W., Shen, J., Kummel, A. C., Lee, S. (2011), Kelvin probe force microscopy and its application, *Surf. Sci. Rep. C* 66, pp. 1–27.

Michels, W. C., Curtis, N. L. (1941), A pentode lock-in amplifier of high frequency selectivity, *Rev. Sci. Instrum.* 12, pp. 444–447.

Meyer, B. K., Politi, A., Peppin, D., Becker, M., Hering, P., Klar, P. J., Sander, Th., Reindl, C., Benz, J., Eickhoff, M., Heiliger, C., Heinemann, M., Bläsing, J., Krost, A., Shokovets, S., Müller, C., Ronning, C. (2012), Binary copper oxide semiconductors: From materials to devices, *Phys. Stat. Sol. B* 249, pp. 1487–1509.

Miller, A., Abrahams, E. (1960), Impurity conduction at low concentrations, *Phys. Rev.* 120, pp. 745–755.

Mora-Seró, I., Anta, J. A., Dittrich, Th., Garcia-Belmonte, G., Bisquert, J. (2006a), Continuous time random walk simulation of short-range electron transport in TiO_2 layers compared with transient surface photovoltage measurements, *J. Photochem. Photobiol.* 182, pp. 280–287.

Mora-Seró, I., Bisquert, J., Dittrich, Th., Belaidi, A., Susha, A. S., Rogach, A. L. (2007), Photosensitization of TiO_2 layers with CdSe quantum dots: correlation between light absorption and photoinjection, *J. Phys. Chem. C* 111, pp. 14881–14892.

Mora-Seró, I., Dittrich, Th., Belaidi, A., Garcia-Belmonte, G., Bisquert, J. (2005), Observation of diffusion and tunneling recombination of dye-photoinjected electrons in ultrathin TiO_2 layers by surface photovoltage transient, *J. Phys. Chem. B* 109, pp. 14932–14938.

Mora-Seró, I., Dittrich, Th., Garcia-Belmonte, G., Bisquert, J. (2006b), Determination of spatial charge separation of diffusing electrons by transient photovoltage measurements, *J. Appl. Phys.* 100, pp. 103705-1–103705-6.

Mora-Seró, I., Dittrich, Th., Susha, A. S., Rogach, A. L., Bisquert, J. (2008), Large improvement of electron extraction from CdSe quantum dots into a TiO_2 thin layer by N3 dye adsorption, *Thin Solid Films* 516, pp. 6994–6998.

Mora-Seró, I., Gross, D., Mittereder, T., Lutich, A. A., Susha, A. S., Dittrich, Th., Belaidi, Caballero, R., Langa, F., Bisquert, J., Rogach, A. L. (2010), Nanoscale interaction between CdSe or CdTe nanocrystals and molecular dyes fostering or hindering directional charge separation, *Small* 6, pp. 221–225.

Murray, C. B., Norris, D. J., Bawendi, M. G. (1993), Synthesis and characterization of nearly monodispersive CdE (E = S, Se, Te) semiconductor nanocrystallites, *J. Am. Chem. Soc.* 115, pp. 8706–8715.

Mwabora, J. M., Ellmer, K., Belaidi, A., Rappich, J., Bohne, W., Röhrich, J., Dittrich, Th. (2008), Reactively sputtered TiO_2 layers on SnO_2:F substrates: a Raman and surface photovoltage study, *Thin Solid Films* 516, pp. 3841–3846.

Naikaew, A., Prajongtat, P., Lux-Steiner, M. Ch, Arunchaiya, M., Dittrich, Th. (2015), Role of phase composition for electronic states in CH3NH3PbI3 prepared from $CH_3NH_3PbI_3/PbCl_2$ solution, *Appl. Phys. Lett.* 106, pp. 232104-1–232104-3.

Nazeeruddin, M. K., Kay, A., Rodicio, I., Humphrey-Baker, R., Müller, E., Liska, P., Vlachopoulos, N., Grätzel, M. (1993), Conversion of light to electricity by cis-X_2Bis(2,2'-bipyridyl-4,4'-dicarboxylate)ruthenium(II) charge-transfer sensitizers ($X = Cl^-$, Br^-, I^-, CN^-, and SCN^-) on nanocrystalline TiO_2 electrodes, *J. Am. Chem. Soc.*, 115, pp. 6382–6390.

Nelson, J. (1999), Continuous-time random-walk model of electron transport in nanocrystalline TiO_2 electrodes, *Phys. Rev. B* 59, pp. 15374–15380.

Neppl, S., Gessner, O. (2015), Time-resolved X-ray photoelectron spectroscopy techniques for the study of interfacial charge dynamics, *J. Electron Spectr. Rel. Phen.* 200, pp. 64–77.

Nekrashevich, S. S., Gritsenko, V. A. (2014), Electronic structure of silicon dioxide (a review), *Phys. Sol. State.* 56, pp. 207–222.

Neppl, S., Shavorskiy, A., Zegkinoglu, I., Fraund, M., Slaughter, D. S., Troy, T., Ziemkiewicz, M. P., Ahmed, M., Gul, S., Rude, B., Zhang, J. Z., Tremsin, A. S., Glans, P.-A., Liu, Y.-S., Wu, C. H., Guo, J., Salmeron, M., Bluhm, H., Gessner, O. (2014), Capturing interfacial photoelectrochemical dynamics with picosecond time-resolved x-ray photoelectron spectroscopy, *Faraday Discussions* 171, pp. 219–242.

Nimoiya, S., Adachi, S. (1995), Optical properties of cubic and hexagonal CdSe, *J. Appl. Phys.* 78, pp. 4681–4689.

Nonnenmacher, M., O'Boyle, M. P., Wickramasinghe, H. K. (1991), Kelvin probe force microscopy, *Appl. Phys. Lett.* 58, pp. 2921–2923.

Nowotny, M. K., Bak, T., Nowotny, J. (2006a), Electrical properties and defect chemistry of TiO_2 single crystal. I. Electrical conductivity, *J. Phys. Chem. B* 110, pp. 16270–16282.

Nowotny, M. K., Bak, T., Nowotny, J. (2006b), Electrical properties and defect chemistry of TiO_2 single crystal. II. Thermoelectric power, *J. Phys. Chem. B* 110, pp. 16283–16291.

Nowotny, J., Bak, T., Nowotny, J. (2006c), Electrical properties and defect chemistry of TiO_2 single crystal. IV. Prolonged oxidation kinetics and chemical diffusion, *J. Phys. Chem. B* 110, pp. 16302–16308.

Nowotny, M. K., Bak, T., Nowotny, M. K., Sheppard, L. R. (2006d), TiO_2 surface active sites for water splitting, *J. Phys. Chem. B* 110, pp. 18492–18495.

Nowotny, M. K., Bogdanoff, P., Dittrich, Th., Fiechter, S., Fujishima, A., Tributsch, H. (2010), Observations of p-type semiconductivity in titanium dioxide at room temperature, *Mater. Lett.* 64, pp. 928–930.

Nowotny, M. K., Dittrich, Th. (2009), unpublished.

O'Regan, B., Grätzel, M. (1991), A low-cost, high-efficiency solar cell based on dye-sensitized colloidal TiO_2 films, *Nature* 353, pp. 737–740.

Orita, M, Ohta, H., Hirano, M., Hosono, H. (2000), Deep-ultraviolet transparent conductive β-Ga2O3, *Appl. Phys. Lett.* 77, pp. 4166–4168.

Osterloh, F. (2013), Inorganic nanostructures for photoelectrochemical and photocatalytic water splitting, *Chem. Soc. Rev.* 42, pp. 2294–2320.

Otero, M., Dittrich, Th. Rappich, J., Heredia, D. A., Fungo, F., Durantini, E., Otero, L. (2015), Photoinduced charge separation in organic-inorganic hybrid system: C_{60}-containing electropolymer / CdSe-quantum dots, *Electrochim. Acta* 173, pp. 316–322.

Overstraeten, Van, R. J. Mertens, R. P. (1987), Heavy doping effects in silicon, *Solid-State Electron.* 30, pp. 1077–1087.

Pankove, J. I. (1971), *Optical Processes in Semiconductors*, Dover Publications, Inc., New York.

Parvan, V., Mizrak, A., Majumdar, I., Ümsür, B., Calvet, W., Greiner, D., Kaufmann, C. A., Dittrich, Th., Avancini, E., Lauermann, I. (2018), $Cu(In,Ga)Se_2$ surface treatment with Na and NaF: a combined photoelectron spectroscopy and surface photovoltage study in ultra-high vacuum, *Appl. Surf. Sci.* 444, pp. 436–441.

Pelton, M., Andrews, J. J., Fedin, I., Talapin, D. V., Leng, H., O'Leary, S. K. (2017), Nonmonotonic dependence of Auger recombination rate on shell thickness for CdSe/CdS core/shell nanoplatelets, *Nano Lett.* 17, pp. 6900–6906.

Poindexter, E. H. (1995), Physical chemistry of hydrogenous species in the $Si-SiO_2$ system, *Z. Naturforsch.* 50a, pp. 653–665.

Prajongtat, P., Dittrich, Th. (2015), Precipitation of $CH_3NH_3PbCl_3$ in $CH_3NH_3PbI_3$ and its impact on modulated charge separation, *J. Phys. Chem. C* 119, pp. 9926–9933.

Prajongtat, P., Dittrich, Th., Hinrichs, K., Rappich, J. (2018), Thickness of AVA^+ controls the direction of charge transfer at TiO_2/PbI_2 interfaces, *J. Phys. Chem. C* 122, pp. 5020–5025.

Prajongtat, P., Wargulski, D. R., Unold, T., Dittrich, Th. (2016), Photochemically driven modulated charge transfer at local contacts between $CH_3NH_3PbI_3$ and carboxylated multiwalled carbon nanotubes, *J. Phys. Chem. C* 120, pp. 3876–3881.

Rappich, J., Dittrich, Th. (2002), Electrochemical passivation of Si and SiGe surfaces, *Thin Films: Non-Crystalline Films for Device Structures* (ed. Francombe, M. H.) 29, pp. 135–259.

Rauscher, S., Dittrich, Th., Aggour, M., Rappich, J., Flietner, H., Lewerenz, H. J. (1995), Reduced interface state density after photocurrent oscillations and electrochemical hydrogenation of n-Si(111): a surface photovoltage investigation, *Appl. Phys. Lett.* 66, pp. 3018–3020.

Ren, H., Dittrich, Th., Ma, H., Hart, J. N., Fengler, S., Chen, S., Li, Y., Wang, Y., Cao, F., Schieda, M., Ng, Y. H., Xie, Z., Bo, X., Koshy, P., Sheppard, L., Zhao, C., Sorrell, C. C. (2019), Manipulation of charge transport by metallic $V_{13}O_{16}$ decorated on bismuth vanadate photoelectrochemical catalyst, *Adv. Mater.* 44, pp. 1807204-1–1807204-9.

Rocchi, M. (1989), Research on monolithic GaAs MESFET circuits at LEP, *Philips Tech. Rev.* 44, pp. 302–309.

Rogach, A. L., Franzl, T., Klar, T. A., Feldmann, J., Gaponik, N., Lesnyak, V., Shavel, A., Eychmüller, A., Rakovich, Y. P., Donegan, J. F. (2007), Aqueous Synthesis of thiol-capped CdTe nanocrystals: State-of-the-art, *J. Phys. Chem. C.* 111, pp. 14628–14637.

Rogach, A. L., Katsikas, L., Kornowski, A., Su, D., Eychmüller, A., Weller, H. (1996), Synthesis and characterization of thiol-stabilized CdTe nanocrystals, *Ber. Bunsenges. Phys. Chem.* 100, pp. 1772–1778.

Rogach, A. L., Kotov, N. A., Koktysh, D. S., Susha, A. S., Caruso, F. (2002), II-VI semiconductor nanocrystals in thin films and colloidal crystals, *Coll. Surf. A: Physicochem. Engin. Asp.* 202, pp. 135–144.

Rogach, A. L., Kornowski, A., Gao, M., Eychmüller, A., Weller, H. (1999), Synthesis and characterization of a size series of extremely small thiol-stabilized CdSe nanocrystals, *J. Phys. Chem. B* 103, pp. 3065–3069.

Rosa, E. B. (1908), The self and mutual inductances of linear conductors, *Bull. Bureau Standards* 4, pp. 301–344.

Rothschild, A., Komem, Y., Levakov, A., Ashkenasy, N., Shapira, Y. (2006), Electronic and transport properties of reduced and oxidized nanocrystalline TiO_2 films, *Appl. Phys. Lett.* 82, pp. 574–576.

Rühle, S., Cahen, D. (2004), Contact-free photovoltage measurements of photoabsorbers using a Kelvin probe, *J. Appl. Phys.* 96, pp. 1556–1562.

Ruske, F. (2019), measurement by ellipsometry, unpublished.

Scanlon, W. W. (1958), Intrinsic optical absorption and the radiative recombination lifetime in PbS, *Phys. Rev.* 109, pp. 47–50.

Schmidt, W. (1902), Bestimmung der Dielektricitätsconstanten von Krystallen mit elektrischen Wellen, *Ann. Phys.* 314, pp. 919–937.

Scholes, G. D., Rumbles, G. (2006), Excitons in nanoscale systems, *Nature Materials* 5, pp. 683–696.

Seah, M. P., Dench, W. A. (1979), Quantitative electron spectroscopy on surfaces: A standard data base for electron inelastic mean free paths in solids, *Surf. Interf. Anal.* 1, pp. 2–11.

Shargaieva, O., Lang, F., Rappich, J., Dittrich, Th., Klaus, M., Meixner, M., Genzel, C., Nickel, N. H. (2017), Influence of the grain size on the properties of $CH_3NH_3PbI_3$ thin films, *ACS Appl. Mater. Interfaces* 9, pp. 38428–38435.

Shay, J. L., Tell, B., Kasper, H. M., Schiavone, L. M. (1973), Electronic structure of $AgInSe_2$ and $CuInSe_2$, *Phys. Rev. B* 7, pp. 4485–4490.

Shavorskiy, A., Neppl, S., Slaughter, D. S., Cryan, J. P., Siefermann, K. R., Weise, F., Lin, M.-F., Bacellar, C., Ziemkiewicz, M. P., Zegkinoglu, I., Fraund, M., Khurmi, C., Hertlein, M. P., Wright, T. W., Huse, N., Schoenlein, R. W., Tyliszcziak, T., Coslovich, G., Robinson, J., Kaindl, R. A., Rude, B. S., Ölsner, A., Mähl, S., Bluhm, H., Gessner, O. (2014), Sub-nanosecond time-resolved ambient-pressure x-ray photoelectron spectroscopy setup for pulsed and constant wave X-ray light source, *Rev. Sci. Instrum.* 85, pp. 093102-1–093102-8.

Shockley, W., Read, W. T. (1952), Statistics of the recombination of holes and electrons, *Phys. Rev.* 87, pp. 835–842.

Steinrisser, F., Hetrick, R. E. (1971), Electron beam technique for measuring microvolt changes in contact potential, *Rev. Sci. Instrum.* 42, pp. 3014–308.

Stoughton, S., Showak, M., Mao, Q., Koirala, P., Hillsberry, D. A., Sallis, S., Kourkoutis, L. F., Nguyen, K., Piper, L. F. J., Tenne, D. A., Podraza, N. J., Muller, D. A., Adamo, C., Schlom, D. G. (2013), Adsorption-controlled growth of $BiVO_4$ by molecular-beam epitaxy, *APL Materials* 1, pp. 042112-1–042112-8.

Strauss, U., Rühle, W. W., Köhler, K. (1993), Auger recombination in intrinsic GaAs, *Appl. Phys. Lett.* 62, pp. 55–57.

Streicher, F., Sadewasser, S., Lux-Steiner, M. C. (2009), Surface photovoltage spectroscopy in a Kelvin prove force microscope under ultrahigh vacuum, *Rev. Sci. Instrum.* 80, pp. 013907-1–013907-6.

Supasai, T., Rujisamphan, N., Ullrich, K., Chemseddine, A., Dittrich, Th. (2013), Formation of a passivating $CH_3NH_3PbI_3/PbI_2$ interface during moderate heating of $CH_3NH_3PbI_3$ layers, *Appl. Phys. Lett.* 103, pp. 183906-1–183908.

Sze, S. M. (1981), *Physics of Semiconductor Devices*, 2nd ed., John Wiley & Sons, New York.

Tang, H., Prasad, K., Sanjies, R., Schmid, P. E., Lévy, F. (1994), Electrical and optical properties of TiO_2 anatase thin films, *J. Appl. Phys.* 75, pp. 2042–2047.

Tauc, J. (1957), Generation of an emf in semiconductors with non-equilibrium current carrier concentrations, *Rev. Mod. Phys.* 29, pp. 308–324.

Timoshenko, V. Yu., Duzhko, V., Dittrich, Th. (2000), Diffusion photovoltage in porous semiconductors and dielectrics, *Phys. Stat. Sol. (a)* 182, pp. 227–232.

Torabi, S., Jahani, F., Van Severen, I., Kanimozhi, C., Patil, S., Havenith, R. W. A., Chiechi, R. C., Lutsen, L., Vanderzande, D. J. M., Cleij, T. J., Hummelen, J. C., Koster, L. J. A. (2015), Strategy for enhancing the dielectric constant of organic semiconductors without scarifying charge carrier mobility and solubility, *Adv. Funct. Mater.* 182, pp. 150–157.

Ulman, A. (1996), Formation and structure of self-assembled monolayers, *Chem. Rev.* 96, pp. 1533–1554.

Vandewal, K. (2016), Interfacial charge transfer states in condensed phase systems, *Annu. Rev. Phys. Chem.* 67, pp. 113–133.

Varshni, Y. P. (1967), Band-to-band radiative recombination in groups IV, VI and III-V semiconductors (II), *Phys. Stat. Sol.*, 20, pp. 9–36.

von Hauff, E., Dyakonov, V., Parisi, J. (2005), Study of field effect mobility in PCBM films and P3HT:PCBM blends, *Sol. Energ. Mats. Sol. Cells* 87, pp. 149–156.

Wang, X. (1990), Sensitive digital lock-in amplifier using a personal computer, *Rev. Sci. Instrum.*, 61, pp. 1999–2001.

Widdra, W., Bröcker, D., Gießel, T., Hertel, I. V., Krüger, W., Liero, A., Noack, F., Petrov, V., Pop, D., Schmidt, P. M., Weber, R., Will, I., Winter, B. (2003), Time-resolved core level photoemission: Surface photovoltage dynamics of the $SiO_2/Si(100)$ interface, *Surf. Sci.* 543, pp. 87–94.

Yablonovich, E., Allara, D. L., Chang, C. C., Gmitter, T, Bright, T. B. (1986), Unusually low surface recombination velocity on silicon and germanium surfaces, *Phys. Rev. Lett.* 57, pp. 249–252.

Yamada, S. (1960), On the electrical and optical properties of p-type cadmium telluride crystals, *J. Phys. Soc. Jpn.* 15, pp. 1940–1944.

Yamamoto, S., Matsuda, I. (2013), Time-resolved photoelectron spectroscopies using synchrotron radiation: Past, present, and future, *J. Phys. Soc. Jpn.* 82, pp. 021003-1–021003-18.

Yu, G., Gao, J., Hummelen, J. C., Wudl, F., Heeger, A. J. (1995), Polymer photovoltaic cells: enhanced efficiency via a network of internal donor-acceptor heterojunctions, *Science* 270, pp. 1789–1791.

Zabel, P., Dittrich, Th., Funes, M., Durantini, E. N., Otero, L. (2009), Charge separation at Pd-porphyrin/TiO₂ interfaces, *J. Phys. Chem. C* 113, pp. 21090–21096.

Zhu, J., Fan., F., Chen., R., An., H., Feng, Z., Li, C. (2015), Direct imaging of highly anisotropic photogenerated charge separations on different facets of a single BiVO₄ photocatalyst, *Angew. Chem. Int. Ed.* 54, pp. 9111–9114.

Zhu, J., Pang, S., Dittrich, Th., Gao, Y., Nie, W., Cui, J., Chen., R., An., H., Fan., F., Li, C. (2017), Visualizing the nano cocatalyst aligned electric fields on single photocatalyst particles, *Nano Lett.* 17, pp. 6735–6741.

Zidon, Y., Shapira, Y., Dittrich, Th. (2007a), Modulated charge separation at tetraphenyl-porphyrin/Au interfaces, *Appl. Phys. Lett.* 90, pp. 142103-1–142103-3.

Zidon, Y., Shapira, Y., Dittrich, Th. (2007b), Illumination induced charge separation at tetraphenyl-porphyrin/metal oxide interfaces, *J. Appl. Phys.* 102, pp. 053705-1–053705-5.

Zidon, Y., Shapira, Y., Dittrich, Th., Otero, L. (2007c), Light-induced charge separation in thin tetraphenyl-porphyrin layers deposited on Au, *Phys. Rev. B* 75, pp. 195327-1–195327-6.

Zidon, Y., Shapira, Y., Shaim, H, Dittrich, Th. (2008), Interactions at tetraphenyl-porphyrin/InP interfaces observed by surface photovoltage spectroscopy, *Appl. Surf. Sci.* 254, pp. 3255–3261.

Zillner, E., Dittrich, Th. (2011), Surface photovoltage within a monolayer of CdSe quantum dots, *Phys. Stat. Sol. RRL* 5, pp. 256–258.

Zillner, E.. (2013), Charge separation in contact systems with CdSe quantum dot layers, Dissertation, Freie Universität Berlin, pp. 1–137.

Zillner, E., Fengler, S., Niyamakom, P., Rauscher, F., Köhler, K., Dittrich, Th. (2012), Role of ligand exchange at CdSe quantum dot layers for charge separation, *J. Phys. Chem. C* 116, pp. 16747–16754.

Zisman, W. A. (1932), A new method of measuring contact potential differences in metals, *Rev. Sci. Instrum.* 3, pp. 367–370. [Instead of measuring directly the current, an audio signal was balanced to silence.]

Index

9 781786 347657